Petroleum Science and Technology

Chang Samuel Hsu · Paul R. Robinson

Petroleum Science and Technology

 Springer

Chang Samuel Hsu
Petro Bio Oil Consulting
Tallahassee, USA

Department of Chemical and Biomedical
Engineering
Florida A&M University/Florida State
University
Tallahassee, FL, USA

China University of Petroleum—Beijing
Changping, Beijing, China

Paul R. Robinson
Katy Institute for Sustainable Energy
Katy, TX, USA

ISBN 978-3-030-16277-1   ISBN 978-3-030-16275-7   (eBook)
https://doi.org/10.1007/978-3-030-16275-7

This Springer imprint is published by the registered company Springer Nature Switzerland AG.
The registered company address is: Gewerbestrasse 11, 6330 Cham, Switzerland

Preface

Driven largely by economic growth in the developing world, the use of petroleum continues to expand. Use of renewable fuels and chemicals is expanding even more rapidly, but petroleum will remain important for decades to come. We wrote this book with the hope that it will help to educate students and professionals who are striving to exploit this valuable resource efficiently, cleanly, and safely.

Our first book together is *Practical Advances in Petroleum Processing*, which appeared in 2013. *Practical Advances* was followed 4 years later by a more comprehensive work: *Springer Handbook of Petroleum Technology*. The Handbook expanded the content to include upstream exploration (discovery) and production (recovery). Its success is demonstrated by the fact that it was downloaded over 80,000 times within 16 months after publication, according to Bookmeatrix.[1] Chapters were contributed by leading authorities from integrated oil companies, catalyst suppliers, licensors, consulting ventures, and academic researchers around the globe.

The present book has a different focus. It is designed to serve as a textbook and reference. It includes areas not discussed in the previous books: midstream operations and petroleum-derived chemicals.

The contents are based on materials used for the course Petroleum Science and Technology, which has been taught at Florida A&M University/Florida State University since 2012. It is divided into 18 chapters to cover fundamentals, upstream exploration and production, downstream refining for fuels and lubricants, petrochemicals and their derivatives, and mid-stream operations. In the last chapter, regulations for safe operations and environmental protection are described along with examples of incidents from throughout the energy and chemical industries.

---

[1] http://www.bookmetrix.com/detail_full/book/9c19f1e2-b01d-4a26-b277-2a487a16570b#citations.

We wish to thank all of the chapter authors of our previous two books. Most of all, we wish to thank our devoted, magnificent wives, Grace Miao-Miao Chen 陳妙妙 and Carrie Robinson, for putting up with our absences—mental if not physical—during so many nights and lost weekends throughout the preparation of this book.

Tallahassee, USA/Beijing, China   Chang Samuel Hsu
Katy, USA   Paul R. Robinson
October 2018

Contents

List of Figures

List of Tables

Part I
Petroleum—Properties, Composition and Classification

Chapter 1
Characteristics and Historical Events

1.1 What Is Petroleum?

There are several ways to answer this question:

In appearance, petroleum is a naturally occurring liquid found beneath the Earth's surface. Petroleum liquid usually is associated with reservoir gas. Together, they are also known as crude oil and natural gas, or simply oil and gas. Some crude oils are as clear as vegetable oils. Others are green, brown, or black. Some flow freely like water. Others are viscous and don't flow at all unless they are heated. Some are solid and are recovered by mining. Tar sand, for example, is a solid combination of clay, sand, water, and bitumen (heavy black viscous oil). These characteristics vary with location, depth, and age of the field.

The English word "petroleum" means "rock oil" or "oil from stone," which is derived from the Greek *pétra*, meaning "rock", and *oleum*, meaning "oil." In Chinese, the characters for petroleum are 石油, pronounced *shi-yóu* which also means "stone (rock) oil".

In composition, petroleum is a complex mixture of countless organic molecules derived from ancient living organisms. The molecules are mostly hydrocarbons, including lesser amounts of heteroatom-containing hydrocarbons, contaminated with various amounts of inorganic matter.

Geologically, petroleum is a fossil hydrocarbon associated with certain geological formations, related more or less to natural gas, bitumen from oil sands or tar sands, kerogen from shale, and coal. The word "petroleum" refers to the liquid form, or "crude oil", but often includes natural gas and bitumen.

Economically and politically, petroleum liquid (crude oil) is an important energy (and material) source. In 2017, it accounted for 34.2% of world energy consumption, as shown in Fig. 1.1 [1]. Its close cousin, petroleum gas (natural gas), accounted for 23.4%. For energy production, the third most important fossil hydrocarbon is coal (27.6%). In additional to serving as our primary source of liquid fuels, petroleum is a raw material from which we produce: lubricants, petrochemicals, construction materials, and thousands of consumer products.

© Springer Nature Switzerland AG 2019
C. S. Hsu and P. R. Robinson, *Petroleum Science and Technology*,
https://doi.org/10.1007/978-3-030-16275-7_1

Fig. 1.1 World Energy
Consumption by
Source—2017 [1]

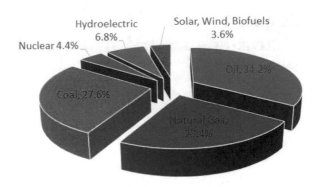

1.2   Fossil Hydrocarbons (Fossil Fuels)

When discussing petroleum, it is useful to introduce other fossil hydrocarbons, also known as fossil fuels. These valuable resources appear in many forms which are described below.

1.2.1   Natural Gas

Natural gas in reservoirs contains mostly methane (CH_4). Like crude oil, the origin of natural gas is biological, and oil and gas are formed by the same natural forces. Hence, most (but not all) petroleum reservoirs contain both oil and gas. We commonly talk about "oil and gas" together. As ancient biomass is transformed into fossil hydrocarbons, different oil- and gas-generation windows correspond to different residence time at different depths. Methane is formed when liquid petroleum is "over-matured" due to excess thermal stress in deep formations.

Methane can also be formed from the anaerobic decomposition of natural wetlands, rice paddies, emissions from livestock, organic wastes in landfills and biomass burning (forest fires, charcoal combustion, etc.), as biogenic methane. The radioactive carbon isotope, ^{14}C, is present in biogenic methane, but absent in fossil methane in natural gas.

Natural gas that contains only traces of other compounds is *dry gas*. If natural gas contains significant amounts of ethane, propane, butanes, and higher hydrocarbons, it is called *wet gas*. The heavier components can be recovered individually or as condensate or natural gas liquid (NGL) in natural gas processing plants. *Natural gasoline* is a condensate fraction comprised mostly of pentane. *Sour gas* contains hydrogen sulfide, and *acid gas* contains carbon dioxide and/or hydrogen sulfide. Sour-gas processing plants coproduce elemental sulfur, which is used to make sulfuric acid and fertilizers. Some natural gas contains commercial quantities of inert gases—helium (the product of α-decay in radioactive minerals underground), neon and/or argon. Almost all commercial helium comes from natural gas plants as a byproduct [2].

1.2.1.1 Gas Hydrates (Clathrates)

Methane is also present in another form as methane hydrate (clathrates) [3], where methane is incarcerated in a cluster of water molecules as an ice-like material. Vast amounts of methane hydrate deposits occur on the deep ocean floor on continental margins and in places north of the arctic circle. It is estimated that methane hydrate deposits contain around 6.4 trillion (6.4 × 10^{12}) tonnes of methane [4]—twice as much carbon as all other fossil fuels on earth. However, the necessary technology for industrial production of the hydrates is not yet available. Hydrocarbon clathrates cause problems for the petroleum industry, because they can form inside gas pipelines, often resulting in obstructions.

The production of methane hydrate is fundamentally different than the extraction of oil and natural gas. The conventional recovery is based on the hydrocarbons flowing naturally through the pores of reservoirs to the production well. Hydrates, on the other hand, are solid. They must be dissociated first before methane can be extracted. Several methods have been tested, such as hot water injection to break down the hydrate and release methane, depressurization by drilling into the deposits to release pressure for releasing methane, and carbon dioxide injection to replace methane in the clathrate (molecular cage). Carbon dioxide hydrates are more stable than methane hydrates. Carbon dioxide captured from natural gas and coal power plants and injected into hydrates for storage is a strategy of carbon dioxde sequestration to reduce its emission into atmosphere. Each technique has its challenges and limitations. Hence, which of these methods will be best suited for production at industrial scales is still uncertain.

1.2.2 Liquids (Crude Oils)

Crude oil is the common name for liquid petroleum. Crude oils, also called crudes, are complex mixtures. There are hundreds of different crudes with significantly different compositions. Crudes typically are named for their source country, reservoir, and/or some distinguishing physical or chemical property. Table 1.1 presents selected physical and chemical properties for ten crude oils.

The lightest liquid is *condensate*, which is essentially natural gas liquids, with boiling points in the gasoline range. *Light crude oils* have low boiling points, low densities (specific gravities), low viscosities, low sulfur, and low or negligible amounts of nitrogen and other hetero-atom compounds. In sweet crudes such as Tapis (a Malaysian crude), the sulfur content is low (0.028%). Sour crudes have more sulfur, which gives them a tart taste. Synthetic crude oil is produced from coal, kerogen, or natural bitumen. Processing costs are higher for conventional or synthetic crudes with high density and large amounts of sulfur, nitrogen, and trace contaminants. Shengli and many other Chinese crudes are very high in nitrogen, which can present special challenges during processing.

Table 1.1 Selected properties of ten crude oils [5]

Crude oil	API gravity[a]	Residue[b] (vol.%)	Sulphur (wt%)	Nitrogen (wt%)
Alaska North Slope	27.1	53.7	1.2	0.2
Arabian Light	33.8	54.2	1.8	0.07
Arabian Heavy	28.0	46.6	2.8	0.15
Athabasca	8	50.8[c]	4.8	0.4
Brent (North Sea)	39	38.9	0.3	0.10
Boscan (Venezuela)	10.2	82.8	5.5	0.65
Kuwait	31.4	49.5	2.3	0.14
Shengli (China)	24.7	72.5	0.8	0.41
Tapis Blend (Malaysia)	45.9	26.3	0.028	0.018
West Texas	40.2	36.4	0.3	0.08

[a]**API Gravity** is related to specific gravity by the formula
°API = (141.5/(specific gravity @ 60 °F))—131.5
[b]Unless otherwise stated, cut point = 343 °C-plus (650°F-plus) for atmospheric resids
[c]Cutpoint = 525 °C (913 °F) for vacuum resids of extra-heavy crude oils and bitumen

Heavy crude oils possess high density—close to that of water. *Extra Heavy oils* have densities greater than water. Both oils have very high boiling points and high viscosities (>1000 centipoise (cP)). They also contain high concentrations of heteroatom-atom containing hydrocarbons.

Distillation yields are an exceptionally important property of petroleum, because distillation is the key step in separating crude oil into useful fractions, which determine the value of crude oil. Crudes containing larger amounts of light, low-boiling fractions—naphtha, kerosene, and gas oil (diesel)—are more valuable. Table 1.2 shows distillation data for four common crudes. The naphtha content of Brent is twice as high as Ratawi, and its vacuum residue content is 60% lower. Bonny Light crude yields the most middle distillate and the least amount of vacuum residue.

1.2.3   Bitumen, Asphalt, Tar

Colloquially, the terms "bitumen," "asphalt," and "tar" are used interchangeably to describe certain black, semi-solid mixtures of hydrocarbons. "Pitch" is an archaic term for the same kind of substance. Geologists say "bitumen" when referring to natural deposits, such as the famous La Brea Tar Pits. In the United States, bitumen produced by crude oil refining is called "asphalt." Outside the United States, bitumen from refineries is called "refined bitumen" or simply "bitumen". The severely degraded oil on oil sand, or tar sand, is also called "bitumen".

Like extra heavy oils, natural bitumens have specific gravities greater than 1.0 (API gravity <10), but the viscosities are higher (>10,000 cP). Under ambient conditions,

Table 1.2   Distillation yields for four selected crude oils [5]

Source field	Brent	Bonny light	Green canyon	Ratawi
Country	**Norway**	**Nigeria**	**USA**	**Mid East**
API gravity	38.3	35.4	30.1	24.6
Specific gravity	0.8333	0.8478	0.8752	0.9065
Sulfur, wt%	0.37	0.14	2.00	3.90
Yields, wt% feed				
Light ends	2.3	1.5	1.5	1.1
Light naphtha	6.3	3.9	2.8	2.8
Medium naphtha	14.4	14.4	8.5	8.0
Heavy naphtha	9.4	9.4	5.6	5.0
Kerosene	9.9	12.5	8.5	7.4
Atmospheric gas oil	15.1	21.6	14.1	10.6
Light VGO	17.6	20.7	18.3	17.2
Heavy VGO	12.7	10.5	14.6	15.0
Vacuum residue	12.3	5.5	26.1	32.9
Total naphtha	30.1	27.7	16.9	15.8
Total middle distillate	25.0	34.1	22.6	18.0
Naphtha plus distillate	**55.1**	**61.8**	**39.5**	**33.8**

natural bitumen is a soft and/or sticky solid, but when heated it flows. In practical terms, it is recovered as a solid but transported and processed as a liquid by adding diluents to lower viscosity. It is important to distinguish between natural bitumen and refined bitumens. The latter are specialty products with rather tight specifications. Refined bitumen is used primarily for paving and construction. Tar sands (also known as oil sands) contain much of the world's recoverable oil.

The largest bitumen/extra heavy oil deposits are in Venezuela with a total of 298 billion barrels. In 2nd place is Alberta, Canada, where the proven reserves are 173 billion barrels; in addition, the province holds 1.4 billion barrels of conventional crude [6]. In comparison, for that same year, proven reserves of conventional crude oil in Middle East were 804 billion barrels [7]. In tar sands, bitumen is associated with sand and clay, from which it can be recovered with hot water or steam. Venezuela's oil sands are technically "extra heavy oil" deposits since they don't contain bitumen. The viscosities of Canadian tar sands vary widely, ranging from 10,000 to 600,000 cP, while those of Venezuelan tar sands are more uniform, typically ranging from 4000 to 5000 cP [7]. In the United States, tar sands are found primarily in Eastern Utah, mostly on public lands. These deposits contain 12 to 19 billion barrels of recoverable oil.

1.2.4   Solids

Compared to liquids and gases, solids are harder to recover, transport and refine. Liquids and gases can be pumped through pipelines and into refineries with relative ease. Slurries of coal and water can be transported as fluids, but the water must be removed and eventually purified at considerable expense before the coal can be burned or gasified. Solid coal is consumed on a large scale to produce heat, steam and electricity. These days, coal-powered transportation vehicles are rare. Coal-burning steam ships and railway locomotives are less efficient than their oil-powered counterparts. Typically, the specific energy of petroleum is 90% greater than a ton of bituminous coal and 40% greater than a ton of anthracite [8]. Even if for some reason a railroad or shipping company wanted to burn coal, doing so wouldn't be practical due to the present lack of coaling stations. Coal is the most widely used fuel in China and many other countries for power generation. Sulfur, ash, and trace metals in the coal cause severe contamination of air with smog, acid rain, mercury, and particulates. Upon combustion, the sulfur becomes sulfur oxides (SOx), primarily SO_2. High-temperature combustion generates nitrogen oxides (NOx) from the nitrogen and oxygen in air. Smog (photochemical smog) is generated by sunlight-induced reactions between nitrogen oxides (NOx) and volatile organic hydrocarbons (VOC); reaction products include ground-level ozone, an especially noxious pollutant. Sulfate particulates, ranging in size from 1 to 20 microns, can be carried by winds hundreds of miles, eventually returning to the earth as dry or wet "acid deposition." Wet deposits are commonly called "acid rain," which also can contain NOx. The combination of smog, particulates, and acid rain can be deadly, especially in large cities.

Kerogen is the solid organic matter in sedimentary rocks. Unlike bitumen, it doesn't flow even when heated. But at high-enough temperatures—e.g., 900 °F (480 °C)—it decomposes into gases, liquids, bitumen, and refractory coke. Huge amounts of kerogen are trapped in oil shale deposits. Fenton et al. [9] estimated that 1.3 trillion barrels of shale oil could be recovered from the world's oil shale reserves. Table 1.3 presents composition information on Green River oil shale from the western United States. About 91% of the kerogen is hydrogen and carbon, but only 15% of the shale is kerogen. Shale oil—synthetic crude from oil shale—tends to contain high amounts of arsenic, a severe poison for refinery catalysts, and mercury. Usually, the arsenic is removed in existing hydrotreating units with special high-nickel chemisorption catalysts, which trap the arsenic by forming nickel arsenides.

Coal is another non-petroleum hydrocarbon resource. It is a black or brown combustible rock composed mostly of carbon, hydrocarbons and ash. Generally, it is classified into four ranks—anthracite, bituminous, sub-bituminous, and lignite. Anthracite is relatively rare, containing 86–97% carbon and has a high heating value. Bituminous coal is far more common. It contains 45–86% carbon and is burned to generate electricity. It is also used extensively in the steel and iron industries. Sub-bituminous coal contains 35–45% carbon, and lignite contains 25–35% carbon. Lignite is crumbly, has high moisture content and relatively low heating value. Over

Table 1.3 Typical composition of Green River oil shale

Kerogen content: 15 wt%[a]	
Kerogen Composition, wt% of kerogen	
Carbon	80.5
Hydrogen	10.3
Nitrogen	2.4
Sulfur	1
Oxygen	5.8
Total	100
Minerals, wt% of mineral content	
Carbonates	48
Feldspars	21
Quartz	15
Clays	13
Analcite & pyrite	3
Total	100

[a]Equivalent to 25 gallons oil per ton of rock

the years, special circumstances have driven the large-scale conversion of coal into liquids, both directly and indirectly. Direct processes convert coal into various combinations of coal tar, oil, water vapor, gases, and char. The coal tar and oil can be refined into high-quality liquid fuels [10].

Developed in 1925, the Fischer-Tropsch (F-T) process is the main indirect route for converting coal into liquids. The coal is first gasified to make synthesis gas (syn gas)—a balanced mixture of CO and hydrogen. Over F-T catalysts, synthesis gas is converted into a full range of hydrocarbon products, including paraffins, alcohols, naphtha, gas oils, and synthetic crude oil. The F-T process was used extensively in Germany between 1934 and 1945. In South Africa, an improved version of the F-T process is used on a large scale to manufacture chemicals and fuels.

Synthesis gas is also derived from natural gas via steam-methane reforming. It can be converted into hydrogen and petrochemicals such as methanol. Worldwide, vast amounts of hydrogen are used to produce ammonia via the Haber-Bosch process.

1.3   Use of Petroleum: A History

Examination of artifacts shows that humans were using petroleum long before writing emerged as a means of conveying knowledge and recording events from one generation to the next. According to archaeologists, bitumen was used for hafting spears as early as 70,000 BC near Umm el Tlel, in present-day Syria [11]. Neanderthals used bitumen, too. A paper by Cârciumaru et al. provides evidence that Neanderthals in

Romania also hafted spears with bitumen between 28,000 and 33,000 BC; the dates are based on uncalibrated ^{14}C dating [12, 13].

Jane McIntosh's excellent book [14] about the ancient Indus valley shows that baskets were water-proofed with bitumen before 5500 BC in Mehrgarh, an ancient site located in present-day Pakistan between the cities of Quetta, Kalat and Sibi.

Bitumen is mentioned in some of the earliest records, specifically those written on tablets in about 3200 BC and discovered in the ancient city of Sumer. Bitumen use is also mentioned in Egyptian pictographs that were written at roughly the same time. Sumer was the leading city of the Sumerian civilization, which arose in about 3500 BC in Mesopotamia—"the land between two rivers"—and lasted until about 1900 BC. The two rivers are the Tigris and Euphrates, located in present-day Iraq.

Sumerian writings describe the use of bitumen for mortar, to cement eyes into carvings, for building roads, for caulking ships, and in other waterproofing applications. The asphalt came from nearby oil pits, and great quantities of it were found on the banks of the river Issus, one of the tributaries of the Euphrates.

The Greek historian Herodotus mentioned the use of bitumen in Babylon (1900 to 1600 BC), including for construction of the famous Tower [15, 16].

From about the same time (3200 BC), Egyptian writings describe the use of pitch to grease chariot wheels and asphalt in mummification, primarily to water-proof the strips of cloth in which the mummies were wrapped. The Egyptians' primary source of bitumen was the Dead Sea, which the Romans called Palus Asphaltites (Asphalt Lake).

As far back as 1500 BC, while drilling for brine, Chinese miners discovered natural gas, which was used as a communal source for lighting and heating. The Chinese were also drilling pioneers. Confucius wrote in 600 BC about using bamboo poles to build pipelines and drill 100-foot natural gas wells.

In 347 AD, oil was being produced from bamboo-drilled wells in China. Bitumen was slowly boiled to get rid of lighter fractions, leaving behind a thermoplastic material with which scabbards and other items were covered. Statuettes of household deities were cast with this type of material in Japan, and probably also in China.

Ancient Persian tablets tell about using bitumen and its fractions for lighting, topical ointments, and flaming projectiles. It is likely that the light fractions were recovered with simple batch distillation apparatus similar to those described by Zosimus, an alchemist who lived at the end of the 3rd and beginning of the 4th century AD [17]. By 500 BC, it was known that light fractions, such as naphtha could be used not just for illumination, but also as a supplement to asphalt, making the latter easier to handle.

Greek fire was invented during the reign of Constantine IV Pogonatus (668–685) by Callinicus of Heliopolis, a Jewish refugee from Syria. This formidable incendiary weapon was hurled onto enemy ships from siphons and burst into flame on contact with air. It could not be extinguished with water. In 673 AD, Greek ships used the weapon to defend Constantinople (today's Istanbul, Turkey), crippling the Arab fleet that was attacking the city [18]. The composition of Greek fire was kept as a top secret and remains a matter of speculation and debate. However, Greek fire is believed to be a viscous liquid composed of naphtha, liquid petroleum, bitumen and quicklime.

In more recent times, the French extracted oil from oil sands in the 1700s. In the United States and Canada, oil appeared in brine wells and was recovered by skimming.

The modern petroleum era began in the 1840s. In 1847, James Oakes built a "rock oil" refinery in Jacksdale, England, to recover "paraffin oil" for lamps. Benjamin Silliman Jr., a chemist hired by the Pennsylvania Oil Rock Company, determined that 50% of a petroleum sample could be distilled into burning oils and 40% could be employed for lubrication and gas lighting. The first modern oil well was drilled in 1848 by F. N. Semyenov in Azerbaijan. Eleven years later, an actual oil refinery was constructed near the well to convert the raw materials into desired products. A large milestone in petroleum products was reached when Canadian geologist Abraham Gesner distilled kerosene from crude oil in 1848. The operation replaced the need for whale oil for lamps and heating. The first true oil well in North America was drilled in Petrolia, Ontario in 1858.

In the 1840–1850s, most home-based lamps burned whale oil or other animal fats. Historically, whale-oil prices had always fluctuated wildly, but they peaked in the mid-1850s. By some estimates, due to the over-hunting of whales, in 1860 several species were almost extinct. Whale oil sold for an average price of US$1.77 per gallon between 1845 and 1855. In contrast, lard oil sold for about US$0.90 per gallon. Lard oil was more abundant, but it burned with a smoky, smelly flame. Michael Dietz invented a flat-wick kerosene lamp in 1857. The Dietz lamp was arguably the most successful of several devices designed to burn oils other than animal fats [19]. Ignacy Łukasiewicz independently developed practical kerosene lamp. By 1858–59, Łukasiewicz lamps were replacing other forms of illumination in Austrian railway stations.

The availability of kerosene got a sudden boost on August 27, 1859, when Edwin L. Drake struck oil with the well he was drilling near Titusville, Pennsylvania. By today's standards, the well was shallow—about 69 feet (21 meters) deep and it produced only 35 barrels per day. Drake was able to sell the oil for US$20 per barrel, a little less than the price of lard oil and 70% less than the price of whale oil.

Drake's oil well was not the first—according to one source, the Chinese beat Drake by thousands of years—but it may have been the first for which the goal was oil production, and it may have been the first well of any kind drilled through rock with a steam powered rotary engine. In any event, the Drake well certainly triggered the Pennsylvania oil rush. Figure 1.2 shows some of the closely spaced wells that sprang up in 1859 in the Pioneer Run oil field a few miles from Titusville.

According to a report issued in 1860 by David Dale Owens, the state geologist of Arkansas: "On Oil Creek in the vicinity of Titusville, Pennsylvania, oil flows out from some wells at the rate of 75–100 gallons in 24 hours already fit for the market. At least 2000 wells are now in progress and 200 of these are already pumping oil or have found it."

According to *The Prize*, a prize-winning book by Daniel Yergin: "When oil first started flowing out of the wells in western Pennsylvania in the 1860s, desperate oil men ransacked farmhouses, barns, cellars, stores, and trash yards for any kind of barrel—molasses, beer, whiskey, cider, turpentine, sale, fish, and whatever else was

Fig. 1.2 Pioneer Run oil field in 1859. Photo used with permission from the Pennsylvania Historical Collection and Museum Commission, Drake Well Museum Collection, Titusville, PA

handy. But as coopers began to make barrels especially for the oil trade, one standard size emerged, and that size continues to be the norm to the present. It is 42 gallons."

The United States has produced about 3 billion barrels since the Drake Well. In 1870, America was the world's leading oil producer, and oil was America's 2nd biggest export. The first cargo of oil was exported from American in 1861, and by 1870 Russia and the United State were the two leading countries for petroleum development. In the 1930s, the United States was the leading producer of oil, but was also a major consumer and therefore not a major exporter. Oil companies began to expand their exploration interests into countries in the Middle East, Africa, Europe, and to Canada.

In 1879, Thomas Edison invented the electric light bulb, which slowly but surely began to replace kerosene as an illuminant. In the late 19th Century and the early 20th Century, the world's navies began to switch from coal to fuel oil. The Anglo-Persian Oil Company (later part of BP) was established to provide fuel for the British navy. In 1889, Gottlieb Daimler, William Mayback and (separately) Karl Benz built the first gasoline-powered automobiles, creating a niche market for naphtha. Internal combustion engines grew in popularity, but were too expensive for most people. Henry Ford changed that. In 1908, the Ford Motor Company began selling Model T automobiles for $950 each. By 1910, 50,000 cars filled the roads of America and the growing demand for fuels pushed refining to its limits. Refining at this moment was still primitive up until the introduction of cracking. Cracking was discovered when engineers realized heavier fractions of crude oil could be cooked until they cracked into lighter components. In the 1920s, the creation of Prohibition Act allowed for experts in alcohol distillation to find jobs elsewhere, namely, the petroleum industry. Bringing knowledge previous learned in the spirits industry, these technologists revolutionized the petroleum field with a multitude of developments and enthusiasm for research. Petroleum refining was well established at this point.

John D. Rockefeller and his partners started concentrating on oil refining, instead of drilling for oil, in 1867. His company grew by taking over other local refineries in Cleveland, Ohio and established the Standard Oil Company in 1870. Through horizontal integration his wealth soared as kerosene and gasoline grew in importance, and he became the richest person in the country. At the peak, Standard Oil controlled 90% of all oil in the United States. In 1911, citing violations of federal anti-trust laws by Supreme Court, the U.S. Justice Department ordered the breakup of Rockefeller's Standard Oil into 34 entities, including five regional "majors", shown in Fig. 1.3.

Among these, Standard Oil of New York (SOCONY then Mobil) and Standard Oil of New Jersey (Esso then Exxon) were merged into ExxonMobil in 1999. Standard Oil of Ohio (Sohio) and Standard Oil of Indiana (Amoco) were acquired by BP in the 1990s. Standard Oil of California (SOCAL) changed the name to Chevron, which merged with Gulf Oil in 1985 and acquired Texaco in 2001. In the 1920s, the seven leading global companies (also known as seven sisters) shown in Fig. 1.4 dominated the oil industry worldwide. Three of the 5 largest oil companies are based in the U.S. and are spinoffs from Standard Oil; the remaining two are in Europe. Table 1.4 lists the historical events of discovery, recovery, refining and usage of crude oil from prehistorical era to 1911.

Organization of Petroleum Exporting Countries (OPEC) was founded in Bagdad, Iraq by Iran, Iraq, Kuwait, Saudi Arabia and Venezuela, with headquarters in Vienna, Austria. In addition to the five founding members, there are 10 other members currently: Algeria, Angola, Ecuador, Equatorial Guinea, Gabon, Libya, Nigeria, Qatar, the Republic of Congo and United Arab Emirates. Indonesia suspended OPEC membership in 2016 for not agreeing with oil production cut. As of 2018, the 15 countries accounted for 44% of global production and 81.5% of the world's "proven" reserves. Two-third of OPEC's oil production and reserves are in the six Middle-Eastern coun-

Fig. 1.3 Five major spinoff companies from Standard Oil in 1911

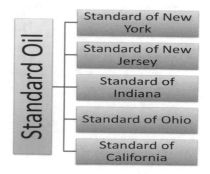

Table 1.4 Historical events of the discovery, recovery, refining and usage of crude oil

Date	Description
≈ 70,000 BC ≈ 5500 BC 3200 BC	According to archeologists, bitumen was used for hafting spears in Umm el Tlel, in present-day Syria Baskets were water-proofed with bitumen before 5500 BC in Mehrgarh, in present-day Pakistan Written descriptions of bitumen use correspond to the advent of writing. Sumerian writing describe the use of asphalt as an adhesive for making mosaics, for lining water canals, sealing boats, and build roads. Egyptian writings describe the use of pitch to grease chariot wheels, and asphalt to embalm mummies
1500 BC 600BC	In China, natural gas is used as a light and heating source Confucius writes about using bamboo poles to build pipelines and drill 100-foot natural gas wells
347 AD c. 672 AD	Oil was produced from bamboo-drilled wells in China Byzantine Greeks invented formidable "Greek fire" by mixing petroleum products with quicklime, which was used to defend Constantinople from the Arab attack
1200–1300 AD	The Persians mine seep oil near Baku (now in Azerbaijan)
1500–1600 AD	Seep oil from the Carpathian Mountains is used in Polish street lamps. The Chinese dig oil wells more than 2000 feet (600 meters) deep
1735 AD	Oil is extracted from oil sands in Alsace, France
Early 1800s	Oil is produced in United States from brine wells in Pennsylvania
1847	James Oakes builds a "rock oil" refinery in Jacksdale, England. The unit processes 300 gallons per day to make "paraffin oil" for lamps. James Young builds a coal-oil refinery in Whitburn, Scotland
1848	F.N. Semyenov drills the first "modern" oil well near Baku
1849	Canadian geologist Abraham Gesner distills kerosene from crude oil
1854	Ignacy Lukasiewicz drills oil wells up to 150 feet (50 meters) deep at Bóbrka, Poland
1857	Michael Dietz invents a flat-wick kerosene lamp (Patent issued in 1859)
1858	Ignacy Lukasiewicz builds a crude oil distillery in Ulaszowice, Poland. The first oil well in North America is drilled near Petrolia, Ontario, Canada

(continued)

Table 1.4 (continued)

Date	Description
1859	Colonel Edwin L. Drake triggers the Pennsylvania oil boom by drilling a well near Titusville, Pennsylvania that was 69-feet deep and produced 35 barrels-per-day
1859	An oil refinery is built in Baku (now in Azerbaijan)
1860–61	Oil refineries are built near Oil Creek, Pennsylvania; Petrolia, Ontario, Canada; and Union County, Arkansas. The most desired product is kerosene for illumination
1863	John. D. Rockefeller, M.B. Clark and Samuel Andrews finance an oil refinery in Cleveland, Ohio
1870	Rockefeller and Andrews charter the Standard Oil Company
1879	Thomas Edison patents the electric light bulb. Over time, the light bulb decreased the market for illumination kerosene
1901	The Spindle Top gusher marks the discovery of the giant East Texas oil field
1908	Oil is discovered in Iran. The Anglo-Iranian Oil Company (AOC) is formed. AOC later became BP
1908	Advent of the Model T Ford brings gasoline-powered automobile travel to the average American
1911	The U.S. Justice Department orders the breakup of Standard Oil in 34 companies, including 5 "regionals"

tries that surround the oil-rich Persian Gulf. The OPEC cartel offset the dominance of "seven sisters." It especially influences crude oil prices.

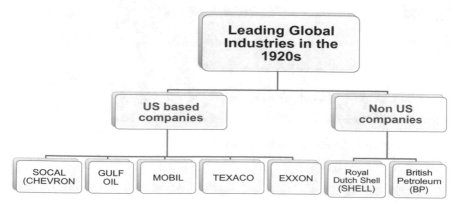

Fig. 1.4 Seven major global companies of the 1920s

Today, about 90% of transportation fuel needs are met by oil. While petroleum accounted for 33% of worldwide energy consumption in 2015, it was responsible for only 3.9% of worldwide electricity generation [1]. Figure 1.5 shows the world's energy consumption since 1820 [20].

Petroleum's worth as a portable, energy-dense fuel powering the vast majority of vehicles and as the base of many industrial chemicals makes it one of the world's most important commodities. The condition of the oil industry depends on several key parameters, such as overall supply and demand, the number of vehicles in the world and the kinds of fuel they burn, net energy gain (economically useful energy provided minus energy consumed), the political stability of oil exporting nations, and the ability to defend oil supply lines.

The top three oil producing countries are the United States, Russia (Eurasia), and Saudi Arabia. Figure 1.6 shows the crude oil productions of these three countries in 2010–2014. The U.S. surpassed Russia and Saudi Arabia as world's largest producing country in 2014, after technology advances in hydraulic fracturing to produce large quantities of gas and oil from shale.

About 48% of the world's readily accessible reserves are located in the Middle East, with 46% coming from the Arab Five: Saudi Arabia, UAE, Iraq, Iran and Kuwait. As mention above, a large portion of the world's total oil exists as bitumen in Canada and extra heavy oil in Venezuela.

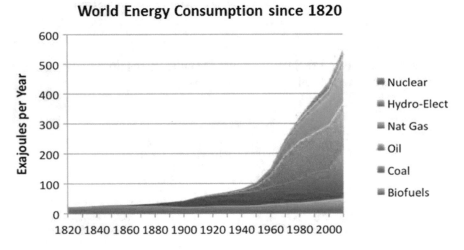

Fig. 1.5 World Energy Consumption since 1820 [20]

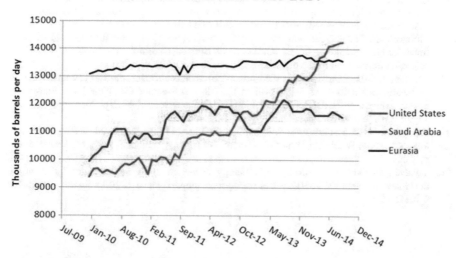

Fig. 1.6 Crude Oil Production of United States, Saudi Arabia and Eurasia (Russia and the Former Soviet Union) in 2010–2014

References

1. BP Statistical Review of World Energy 2018 Full Report. https://www.bp.com/en/global/corporate/energy-economics/statistical-review-of-world-energy/downloads.html. Retrieved 8 Oct 2018
2. How products are made: Helium. https://en.wikipedia.org/wiki/Clathrate_hydrate
3. http://www.madehow.com/Volume-4/Helium.html. Retrieved 15 Sept 2016
4. Buffett B, Archer D (2004) Global inventory of methane clathrate: sensitivity to changes in the deep ocean. Earth Planet Sci Lett 227(3–4):185–199
5. Gray JH, Handwerk GE, Kaiser MJ (2007) Petroleum refining—technology and economics, 5th Edition, CRC Press
6. (a) United States Energy Information Administration, http://www.eia.doe.gov/international/oilreserves.html. Retrieved 1 Mar 2015; (b) http://gulfbusiness.com/top-10-countries-with-the-worlds-biggest-oil-reserves/ Retrieved 22 Sept 2018
7. http://www.oilsandsmagazine.com/news/2016/2/15/why-venezuela-is-albertas-biggest-competitor. Retrieved 22 Sept 2018
8. Wikipedia: Energy density: http://en.wikipedia.org/wiki/Energy_density. Retrieved 20 Nov 2011
9. Fenton DM, Hennig H, Richardson RL (1980) The Chemistry of shale oil and its refined products, presented at symposium on oil shale, tar sands and related materials, American Chemical Society Annual Meeting, San Francisco
10. Speight JG (ed) (2008) Synthetic fuels handbook: properties, process, and performance, McGraw-Hill Professional
11. Wikipedia: History of writing: https://en.wikipedia.org/wiki/History_of_writing. Retrieved 23 Sept 2015
12. Boëda E, Bonilauri S, Connan J, Jarvie D, Mercier N, Tobey M, Valladas H, al Sakhel H, Muhesen S (2008) Middle palaeolithic bitumen use at Umm el Tlel around 70 000 BP. Antiquity 82: 853–86

13. Cârciumaru M, Ion R-M, Niţu E-C, Ştefănescu R (2012) New evidence of adhesive as hafting material on middle and upper palaeolithic artefacts from Gura Cheii-Râşnov Cave (Romania). J Archaeol Sci. https://doi.org/10.1016/j.jas.2012.02.016
14. McIntosh JR. The ancient indus valley. New Perspectives (Understanding Ancient Civilizations) 1st Edition, 2008, ABC-CLIO, Inc. Santa Barbara, California, 57
15. Asphalt: https://en.wikipedia.org/wiki/Bitumen, Retrieved 23 Sept 2015
16. Abraham H (2015) Asphalts and allied substances: their occurrence, modes of production, uses in the arts, and methods of testing (4th ed.). 1938. Van Nostrand Co., New York. Viewed via https://archive.org/details/asphaltsandallie031010mbp. Retrieved 23 Sept 2015
17. Taylor FS. The origins of greek alchemy, 1937, Ambix 1, 40
18. Wikipedia: Greek fire: https://en.wikipedia.org/wiki/Greek_fire. Retrieved 31 Aug 2015
19. Pees ST. (2004) Whale Oil Versus the Others. Oil History, Samuel T. Pees, Meadville, Pennsylvania
20. Tverberg G. Our finite world: world energy consumption by source since 1820. http://ourfiniteworld.com/2012/03/12/world-energy-consumption-since-1820-in-charts/. Retrieved 4 Jan 2015

Chapter 2
Crude Assay and Physical Properties

Around the Earth, different living organisms perished and drifted to the bottom of soil or bodies of water, along with other inorganic materials. Over hundreds of millions of years, under the influence of high pressure and temperature, some carbon-containing portions of the dead organisms escaped full oxidation and ended up as organic matter in sedimentary rocks, as kerogen, oil, or gas. Due to differing climates, organisms, and depositional environments, the compositions of oil depend drastically with region and can vary very significantly. Hence, it is important to determine physical and chemical characteristics of crude oil through a crude oil assay (crude assay).

2.1 Crude Assay

There are hundreds of different crude oils produced in the world with large variations in properties and composition. Crudes typically are named for their source country and reservoir. Both physical and chemical properties of crude oils vary with location, depth, and the age of the oil field.

Characteristics of crude oil that can indicate these differences include: boiling point, calorific value (heating value), API gravity (density), pour point, cloud point, freeze point, aniline point, smoke point, flash point, viscosity, color and fluorescence, Reid vapor pressure, refractive index, sulfur content, nitrogen content, metal content, salt content, micro carbon residue, optical activities, and total acid number [1]. These crude oil bulk properties measured by testing laboratories, along with distillation and product fractionation data at pilot plants, are compiled into a crude oil assay (or crude assay) to characterize a specific crude oil.

A crude assay can be an inspection assay or a comprehensive assay (full assay). Testing can include characterization of the whole crude oil and/or its various boiling fractions produced from fractional distillation. High quality assay data are important to refinery planners and oil traders, who select crudes by comparing their properties with refinery specifications and constraints as well as meeting environmental and other standards. Sulfur content and total acid number (TAN) are common constraints,

© Springer Nature Switzerland AG 2019
C. S. Hsu and P. R. Robinson, *Petroleum Science and Technology*,
https://doi.org/10.1007/978-3-030-16275-7_2

because they relate to corrosion in piping and process equipment. The refiners make changes in plant operations based on the crude assay data, to meet process and product requirements. Hence, crude assays provide the basis upon which companies and traders negotiate contracts, which include pricing and possible penalties due to impurities and other undesired properties. However, traders might purchase off-spec crudes for subsequent trading.

The assays determine selected chemical and physical properties of whole crudes and several distilled fractions. The fractions correspond to boiling ranges for common fuels. Full assays are extensive, and especially important for new crudes. For inspection assays, just a few tests are conducted. In a whole crude assay, analyses are done by combining atmospheric and vacuum distillation runs to provide a true boiling-point (TBP) distillation data. The extent of an assay depends on customer need and affordability. At minimum, the assay should contain a distillation curve, typically a TBP curve, and a specific gravity curve. The common assay inspections are discussed below.

2.2   Distillation

Distillation is the primary method of separating crude oil into useful products. It utilizes differences in boiling points (volatility) of fractions. Boiling point distribution of the components, determined by fractional distillation, is the most important characteristic of petroleum. Large scale and continuous distillation is also the most basic process in any refinery to separate crude oil into fractions of different boiling ranges, shown in Fig. 2.1. Table 2.1 lists the major distillation products (fractions) with their boiling ranges.

Naphtha of high octane-number is a main component of gasoline. It is composed of hydrocarbons mainly from C_5 to C_{10}, with some C_{11} and C_{12}. The carbon numbers of heavy naphtha extend to mid-teens, overlapping with those of kerosene. Kerosene is a middle distillate fraction of crude oil with C_{10} to C_{16}. It is mainly used for heating oil, jet fuel, and diesel. It is the first fraction to show an appreciable increase in the cyclic hydrocarbons that dominate the heavier fractions. Aromatics in kerosene range from 10 to 40%. Light gas oils, C_{14} to C_{18}, are used in both jet fuels and diesel fuels. The C_{20} to C_{50} range contains diesel fuels, heating oils, lubricating oils, paraffin waxes and some asphalts. Residuum (or resid) includes resins, asphaltenes and waxes, and is the most complex and least understood fraction of petroleum. The wax fraction in most residua is about half that in the lube oil fraction. Asphaltenes are dark brown to black amorphous solids. The resins may be light to dark colored, thick, viscous substances to amorphous solids. The resins and the asphaltenes contain about half of the total nitrogen and sulfur in crude oil. Heavy crude oils invariably have more nitrogen and sulfur. Residua frequently contain over 5% oxygen. Resins are the highest of the fractions in oxygenated compounds, while asphaltenes are highest in sulfur compounds.

Fig. 2.1 Relationship between boiling points under atmospheric pressure and under 40 mmHg vacuum [3]

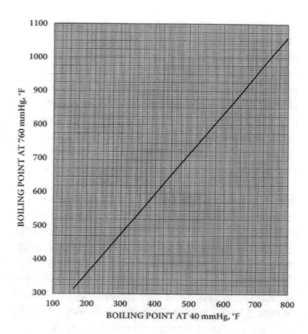

Table 2.1 Boiling range and its corresponding distillation product

Boiling range		Distillation product
°F	°C	
<85	<29	Petroleum Gas (butane and lighter)
90–200	30–90	Light Naphtha (Gasoline, Petrochemicals)
200–400	90–200	Heavy Naphtha (Gasoline)
300–525	150–275	Kerosene (Jet fuel, Heating oil)
350–650	180–345	Distillate (Diesel, Heating oil), Straight-run gas oil
650–1050	345–565	Vacuum gas oil (Lubricant oil)
>1050	>566	Vacuum residuum (Residual fuel oil, Asphalt, Coke)

The amount of valuable products, or the yields of fuels and lubricant oils that can be produced from a crude oil simply by distillation is a major factor in determining the value of the oil.

Smaller scale or laboratory scale distillation methods are described in the *Annual Book of ASTM Standards* [2]. ASTM stands for the American Society for Testing Materials, now called ASTM International. The methods include ASTM D2892 for atmospheric distillation and ASTM D5236 for vacuum distillation. The ASTM D2892 (15/5 distillation) method uses a 15 theoretical plate fractionation column with a 5:1 reflux ratio. It determines accurate boiling points and yields of distillation fractions. The highest temperature is limited to ~350 °C (650 °F) to avoid thermal decomposition. In order to extend the boiling range, distillation is operated under ~40 mmHg vacuum for higher boiling components. Hence, the ASTM D5236 (Hivac distillation) method is done in a vacuum potstill to produce vacuum gas oil, which boils in the traditional lubricant oil range, as well as vacuum resid. A true boiling point (TBP) curve for the distribution of boiling point versus accumulated weight percent distilled is created by combining atmospheric (ASTM D2892) and vacuum (ASTM D5236) distillation runs. Since the components boil at lower temperatures under vacuum, their boiling points need to be adjusted for the difference in pressure to give atmospheric equivalent boiling points (AEBP). Figure 2.1 shows the linear relationship between the AEBP and the boiling points under 40 mmHg vacuum. The correction enables the extension of TBP curves from 350 °C (~650 °F) to above 565 °C (1050 °F). The determination of TBP curves is very time-consuming (>8 h) and requires a large volume of sample, for example, 10 L.

Figure 2.2 shows the TBP curve obtained from distillation. Atmospheric distillation ends at 350 °C and vacuum distillation ends at atmospheric equivalent temperature at 565 °C. The curve is extrapolated beyond 565 °C to estimate final boiling point (FBP) of the whole crude; the extrapolation can be rather arbitrary, sometimes using molecular modeling based on hypothetical structures postulated from constituents measured by nuclear resonance spectroscopy (NMR) and vibrational spectroscopy (infrared and Raman) of the whole oil or vacuum residuum.

More accurate extended TBP curves could be obtained by molecular distillation, or short-path vacuum distillation. The sample, such as vacuum residue (boiling point

Fig. 2.2 TBP curve obtained from atmospheric and vacuum distillation [5]

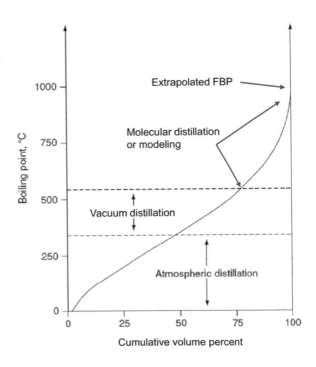

>565 °C), is exposed at increasingly higher temperatures in a distillation column operated in high vacuum (around 10^{-4} mmHg) and a short distance, about 2 cm, between evaporator and collector. The hot plate covered by a thin film of sample is typically suspending next to a cold plate with a line of sight in between. In molecular distillation, fluids are in molecular flow regime, i.e., the free path of the molecules is comparable to the size of the equipment. The final temperature for heavy oils, however, cannot be reached before the sample turns into char that is completely involatile. An example of extending TBP curve from 565 to 700 °C (1290 °F) has been demonstrated by Lopes et al [4].

No viable refining technology for vacuum residues based on distillation has been developed. Extraction would become a preferable method if additional fractionation is desired.

In many plant operations, material balances, physical property correlations, and computer process simulations are based on TBP of the streams of interest.

Single plate distillation or analytical distillation was also developed by ASTM—Method D86 for atmospheric distillation and Method D1160 for vacuum distillation at 40 mmHg [6]. The data were then converted to TBP by mathematical correlation. Since the distillations are carried out using Hempel columns, the combination is also referred to as Hempel distillation. A rather large volume of sample is also required.

An alternative quick and robust method with a minute amount of sample was developed. Whole crude simulated distillation (SimDist) is described by method ASTM D2887 (also D7900 and D7169). It uses high temperature gas chromatography (HTGC) with a thermally stable aluminum-clad nonpolar GC column and flame ionization detector (FID) to determine true distillation curve and predict distillate yields. The GC oven temperatures can be as high as 538 °C, thus making possible the analysis of up to C_{120} hydrocarbons present in petroleum and related materials.

Figure 2.3 illustrates a SimDist result. The GC oven is temperature programmed from room temperature to a maximum temperature under the flow of a carrier gas, typically helium. The vapor of components diffuses in and out of the stationary phase coated inside the GC column. A nonpolar GC column is essentially a boiling point column to separate components by their boiling points. The series of normal paraffins (alkanes) serve as boiling point reference based on their individual boiling points and retention times. Since FID measures the weight percent of the component, the area under a specific boiling range represents the yield of that range. The data are compared with a crude assay True Boiling Point (TBP) curve, shown in Fig. 2.4. It should be noted that boiling points of multiple-ring aromatic components as determined by simulated distillation can differ substantially (20–100 °F or 10–50 °C) from the true boiling points of the pure components. This is partially due to the difference of the vapor-pressure-temperature equilibrium of aromatic compounds compared to other hydrocarbon types. Hence, the deviation of SimDist data is greater in heavy fractions where high concentrations of aromatics are present. In practice, the yields of distillation fractions determined by SimDist need to be correlated or calibrated with the yields from atmospheric and/or vacuum distillation.

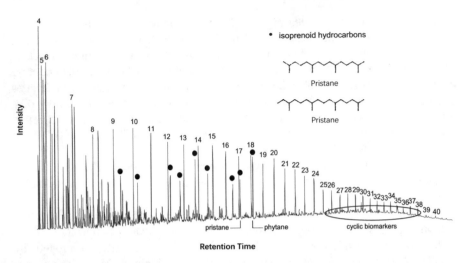

Fig. 2.3 Simulated distillation by gas chromatography (the numbers on top of the peaks are carbon numbers of normal paraffins)

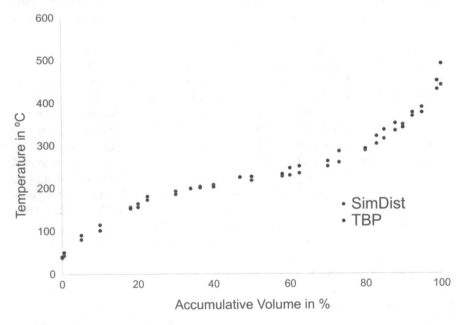

Fig. 2.4 Comparison of SimDist data with TBP curve

2.3 Physical Testing

The following lists some of the most important tests for physical properties of crude oil or its fractions.

- **Specific gravity and API gravity**

Specific gravity, or relative density, is the ratio of the mass of a given volume of the oil to the mass of an equal volume of water at a specific temperature, which is often 60 °F. A commonly used tool for measuring specific gravity includes a temperature bath heated to the desired temperature, a metal tube to fill with the test sample, and a hydrometer which indicates the specific gravity. This test determines specific gravity which is then converted to API gravity, a measurement that has become a standard in the petroleum industry.

API gravity is another notation of the density of the oil, which was established by the American Petroleum Institute (API) as a measure of how heavy or light a petroleum liquid is compared to water. It is defined by ASTM D287/1298 as

$$°API = (141.5/\text{specific gravity at } 60°F) - 131.5$$

The specific gravity of water at 60 °F is 1.0; hence, its °API is 10. The °API of crude oils ranges from <10 for asphaltic crude to >50°API for condensate. Most crudes are in the 20–45°API range. Condensates range between 50 and 70°API. The API gravity of heavy oils falls in the range of 10–15° or <20°API, and for bitumen it falls in the range of 5–10°. In 2010, the World Energy Council defined "extra heavy oil" as crude oil having a gravity less than 10° API and reservoir viscosity no more than 10,000 cP, with a lower API limit of 4°, so it sinks in water rather than floating on it. API gravity is designed so that its value is more or less proportional to the commercial value of the crude oil. A denser oil has a lower °API, and hence, the commercial value of the oil is lower.

- **Viscosity**

Viscosity is a measurement of fluid resistance to shearing flow using a small capillary tube (viscometer) for the sample to flow through. This measures the kinematic viscosity, that is, the ratio of dynamic (shear) viscosity to the density of the fluid, in centistoke (cSt) at a given temperature. It is a very important oil property.

In oil production, viscosity determines the flow of oil and gas through the reservoir, thus, the amount and rate at which oil and gas can be produced.

In engines, viscosity determines how easily the oil is pumped to the working components, how easily it passes through filters, and how quickly it drains back to the engine. The lower the viscosity, the easier all this will happen. That is why cold starts are so critical to an engine: because the oil is cold and so relatively thick that it loses its lubricity. A fluid's viscosity is directly related to its load-carrying capabilities. The greater the viscosity, the greater the loads it can withstand. The viscosity must be adequate to separate moving parts under normal operating conditions (temperature

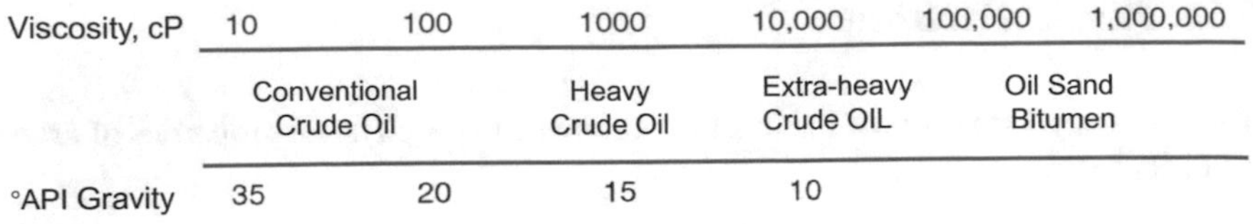

Viscosity, cP	10	100	1000	10,000	100,000	1,000,000
	Conventional Crude Oil		Heavy Crude Oil	Extra-heavy Crude OIL	Oil Sand Bitumen	

°API Gravity	35	20	15	10

Fig. 2.5 Viscosity and density of crude oils. cp: centipoise [4]

and speed). Knowing that a fluid's viscosity is directly related to its ability to carry a load, one would think that the more viscous a fluid, the better it is. The fact is, the use of a high-viscosity fluid can be just as detrimental as using too light an oil. Viscosity index (VI), that is, the change of viscosity with temperature, is a critical property of lubricant base oil. The higher the VI, the lower is the viscosity change with temperature. High performance and synthetic lube oils have high VI's.

For the same homologous series of hydrocarbons, the greater the molecular weight of the compound, the greater the viscosity. When the molecular weight is similar, the viscosity of a cyclic molecule is larger than that of the chain-like molecule. The greater the number of rings is, the greater the viscosity is. Therefore, the "ring structure is the viscosity carrier".

Viscosity is a parameter of oil mobility, also an indispensable physical property for crude oil recovery and refining processes. The viscosity of the oil decreases as its temperature increases.

The viscosity and density of crude oils are closed related, as shown in Fig. 2.5. Oil sand (tar sand) bitumen is the most viscous and has the highest density (lowest API gravity) . On the other hand, conventional crude oils have less viscosity with lower density (or API gravity) . Oil sand was defined by the United States Congress in 1976 as rock types that contain extremely viscous hydrocarbons which cannot be recovered by conventional oil production methods, including enhanced recovery. Oil sand bitumen and extra heavy crude oils are therefore referred to as unconventional oils.

Viscosity for conventional crude oils ranges from 10–100 mPa (884–934 kg/m^3), and for heavy crude oils from 1000 to 10,000 mPa (966–1000 kg/m^3). Tar sand bitumen have viscosities that ranges from 100,000 to 1 million mPa and density over 1000 kg/m^3.

● **Total Sulfur (atom) content**

The sulfur (atom) content of crude oils is in the range of 0.1–5.0 wt%, measured by x-ray fluorescence (ASTM D4294 or D5291). The terms "sweet" and "sour" are historical terms which refer to the taste of crude oil as a function its sulfur content. Indeed, early prospectors would taste oil to determine its quality. Low sulfur oil actually tasted sweet. Crude oil is currently considered sweet if it contains less than 0.5% sulfur.

Sulfur compounds can affect the activity of catalysts. As mentioned earlier, combustion converts organic sulfur compounds into sulfur oxides (SO$_x$), which can react with moisture (H_2O) to form fine sulfuric acid and sulfate particulates (aerosols) in air. The particulates are transported hundreds of miles, eventually returning to the earth as wet or dry acid deposition. The deposition, also known as acid rain, harms buildings, trees and other plants, fish, and land animals. Similarly, organic nitrogen compounds are converted by combustion into nitrogen oxides (NO_x), which reacts with volatile organic compounds (VOC) by photochemical reactions to form smog. Both SO_x and NO_x form particulate matter (PM) in air. PM with diameters smaller than 2.5 μm (PM 2.5) are inhalable particles that can reach bronchial tubes to damage lungs, causing respiratory ailments.

Sweet crude is easier to refine and safer to extract and transport than sour crude. Because sulfur is corrosive, light crude also causes less damage to refineries and thus results in lower maintenance costs over time.

Major locations where sweet crude is found include the Appalachian Basin in Eastern North America, Western Texas, the Bakken Formation of North Dakota and Saskatchewan, the North Sea of Europe, North Africa, Australia, and the Far East, including Malaysia and Indonesia. African crudes tend to be relatively sweet: Bonny Light, the main Nigerian crude, contains about 0.16 wt% sulfur. Brega, the main Libya crude, contains about 0.2 wt% sulfur.

Sour crude oils have more than 0.5% total organic sulfur not including dissolved hydrogen sulfide. Sour crude also contains more carbon dioxide. Most sulfur in crude oil is actually bonded to carbon atoms as sulfur-containing hydrocarbons. Nevertheless, high quantities of hydrogen sulfide in sour crude can pose serious health problems or even be fatal.

- **Hydrogen sulfide**

Hydrogen sulfide is famous for its "rotten egg" smell, which is only noticed at low concentrations. At moderate concentrations, hydrogen sulfide can cause respiratory and nerve damage. At high concentrations, it is instantly fatal. Hydrogen sulfide is so much of a risk that sour crude has to be stabilized via removal of hydrogen sulfide before it can be transported by pipelines and oil tankers. Sour crude is more common in the Gulf of Mexico, Mexico, South America, and Canada. Middle Eastern crudes tend to be relatively sour.

- **Mercaptans**

Mercaptans (thiols) smell like as garlic or rotten eggs and can be toxic. A trace amount can be used as an odorant of natural gas for detection; natural gas is odorless in pure form. Mercaptan sulfur is measured by potentiometric titration of an isopropanol solution of a hydrocarbon sample containing a small amount of NH_4OH. The solution is then titrated with a silver nitrate solution. Mercaptans are removed from sour crudes and oxidized into sulfides through the Merox process or into elemental sulfur by LOCAT® and other methods, discussed in Sect. 7 of Chapter 9, to reduce odor prior to transportation.

- **Nitrogen content**

  Nitrogen-containing compounds can cause poisoning of acidic catalysts due to their basicity. Nitrogen content is normally determined by oxidative combustion and chemiluminescence detection (ASTM D3228 or D4629).

- **Total Acid Number (TAN)**

  Total Acid Number (TAN) is the amount of KOH in mg that is needed to neutralize the acid in a gram of oil dissolved in toluene/isopropanol/water (ASTM D664). Typically, values are 0.05–6.0 mg KOH per gram of the sample. High TAN crudes are purchased and processed carefully due to possible corrosion problems.

- **Metal Content**

  Metal Content ranges from a few to several hundred ppm. It is measured by inductively coupled plasma atomic emission spectroscopy (ICP-AES) (ASTM D5708) or x-ray fluorescence (ASTM D5863). Nickel and vanadium are common in crude oils; they can severely affect catalyst activity.

- **Salt Content**

  Salt Content is measured by conductivity of a crude oil sample dissolved in water compared to reference salt solutions (ASTM D3230) to determine crude oil corrosivity that can lead to shorter life times of pipes and pumps in the refinery. Desalting is needed when the salt content is greater than 30 ppm to bring it down to 2 ppm.

- **Micro Carbon Residue (MCR)**

  Micro Carbon Residue (MCR) is measured by the Conradson carbon (CCR or ConCarbon) method ASTM D189.

  MCR is a measurement of hydrocarbon mixtures' tendency to leave carbon deposits (coke) when burned as fuel or subjected to intense heat in a processing unit such as a catalytic cracker. The ConCarbon test involves destructive distillation, i.e., subjection to high temperature, which causes cracking, coking, and drives off any volatile hydrocarbons produced, and weighing the residue which remains. A somewhat similar test, Ramsbottom carbon, also measures mixtures tendency to form coke.

- **Calorific Value (Heating Value)**

  In the fuel fractions of petroleum, the heating value is of importance. Paraffins have higher calorific values than aromatics. The average value for a crude oil is 2100–2230 kcal/kg, compared with 1170–1670 kcal/kg of bituminous coal.

- **Pour Point**

  The pour point of a liquid is the temperature at which it becomes semi solid and loses its fluidity. It is the lowest temperature at which the oil no longer moves. The pour point is determined as $3°$ above the point at which a sample no longer moves when inverted (ASTM D97). It is important for pipeline transportation from source to loading ports. In crude oil a high pour point is generally associated with a high paraffin content, typically found in crude derived from a larger proportion of plant material as in Type III kerogen.

- **Cloud Point**

In the petroleum industry, cloud point refers to the temperature below which wax in a sample, particularly jet fuel and diesel, forms a cloudy appearance. The presence of solidified waxes thickens the oil and clogs fuel filters and injectors in engines. The wax also accumulates on cold surfaces (e.g. causing pipeline or heat exchanger fouling) and forms an emulsion with water. Therefore, cloud point indicates the tendency of the oil to plug filters or small orifices at cold operating temperatures.

In crude or heavy oils, cloud point is synonymous with wax appearance temperature (WAT) and wax precipitation temperature (WPT). It is determined by ASTM D2500 or D5773 method as the temperature at which a haze appears in a sample due to formation of wax crystals by cooling.

- **Freeze Point**

Freeze Point is the temperature at which crystals start to form in hydrocarbon liquid and then disappear when heated (ASTM D2386).

- **Aniline Point**

Aniline Point is the lowest temperature at which aniline and an oil are completely miscible. It is measured by ASTM D611. The mixture is heated and stirred until homogeneous, then it is cooled with stirring until the two liquids separate. The temperature at which such separation occurs is the aniline point. For clear samples, the aniline point is that at which the mixture suddenly becomes cloudy. In darker samples, the apparatus becomes more complicated, as it is less obvious when the mixture is cloudy. It is used in some specifications as an indication of aromatic content. The lower the aniline point, the greater is the aromatic content.

- **Smoke Point**

Smoke Point is performed on jet fuels and kerosene cuts to determine clean burning by measuring flame height with a standard wick, expressed in mm, in a lamp without smoke forming (ASTM D1322). To test for smoke point, a sample is burned in a lamp that is precisely calibrated using hydrocarbons with known smoke points. The maximum height of the flame without production of smoke is recorded to the nearest 0.5 mm. This is important in determining the composition of crude oil. Samples with lower smoke points are more aromatic. Smoke point is usually most important in jet fuel. Smoke can shorten the lifetime of engine parts, therefore higher smoke points are favorable.

- **Flash Point**

Flash point is the temperature at which combustion occurs when the hydrocarbon mixture is exposed to air or oxygen.

• Color and Fluorecence

The colors of crude oils are as varied as light yellow, green, red, brown and black. Most of the colors come from molecules with aromatic character having large π electron systems in the molecule. The color depends on the degree of conjugation and the size of the conjugated system. Some molecules can absorb light at high frequencies (such as ultraviolet) and re-emit it at a lower frequency (in the visible range), a phenomenon known as fluorescence. In general, the large the aromatic molecules in the heavier fractions, the stronger is the fluorescence at longer wavelengths.

• Refractive Index

Refractive Index (RI) is the ratio of the velocity of light of a specific wavelength in vacuum to the velocity of light in the sample tested (ASTM D1218). RI is dependent upon the density of the oil: the heavier oils have higher reflective indices. It's been used to estimate the content of polynuclear aromatics (PNA, or polynuclear aromatic hydrocarbons, PAH) which generally is higher in heavier oils. PNAs are produced by thermal cracking in refineries and by the burning of fuels. PNAs are a significant component of soot. Some PNAs can interact with DNA and prove to be toxic or carcinogenic. Extended exposure to PNAs can lead to lung, skin, or bladder cancer. It is important to test for PNAs to be aware of the risks associated with burning crude oils with these properties.

• Optical Activities

Some saturated carbon atoms have four different substituents attached to them through single covalent bonds. These asymmetric carbons provide chiral centers to form non-superimposable isomers which as referred as enantiomers. Enantiomers, which result from the inversion of all asymmetric carbon centers, show similar chemical properties, but rotate plane-polarized light in opposite directions. Inversion of less than all of the available asymmetric carbon centers yields diastereomers or epimers. Enzymes in living organisms, which are responsible for biosynthesis of cellular materials, are chiral and generate biomolecules with only one configuration at certain asymmetrical carbon centers. Since petroleum is derived from ancient living organisms, these specific configurations are carried over in saturated biomarker hydrocarbons which are found in immature oils. When an oil reaches maturity, epimers with different optical activity are formed. An example of an epimer is cholestane (C_{27} sterane) which has a chiral center at the C-20 position.

• Reid Vapor Pressure (RVP)

Reid Vapor Pressure (RVP) is a measurement of volatility of a liquid hydrocarbon mixture at 100 °F (ASTM D323). The measurement is taken using an apparatus that includes a liquid chamber that is filled with a chilled sample and a vapor chamber that is heated to 100 °F. The chambers are connected and immersed in a temperature bath at 100 °F until constant pressure is observed. This pressure is the Reid vapor pressure, measured as kilopascals relative to atmospheric pressure (kPa). The vapor pressure of petroleum is important for several reasons. In gasolines, high temperatures and altitudes increase vaporization, which can lead to "vapor lock." While handling crude oil, vapor pressure is of high importance for safety reasons.

Note that RVP differs slightly from true vapor pressure (TVP) of a liquid. TVP is measured in the presence of water vapor and air during sample vaporization in the confined space of the test equipment. Hence, the RVP is the absolute vapor pressure and TVP is the partial vapor pressure.

2.4 Crude Assay Template and Report

Table 2.2 exhibits an example assay report template; there is no standard assay testing grid—each refinery or trading company has its own. Note that not all the tests need to be done for all samples. The tests are requested and performed depending on the needed information for processes and products.

Figure 2.6 shows an example of crude assay report of a benchmark crude, Brent blend, and its fractions [5]. The yields of different fractions (cuts) is expressed by volume %. It can be seen that paraffins are concentrated in the lowest distillation cuts as "paraffinic", with increasing naphthene and aromatic contents in higher distillation cuts as "asphaltic".

2.5 Bulk Properties of Crude Oil

Table 2.3 presents selected bulk physical and chemical properties for 21 crude oils. Athabasca is a heavy oil, with a specific gravity >1.0. In the table, sulfur contents range from 0.14 to 5.3 wt%, and nitrogen contents range from nil to 0.81 wt%. Specific gravities range from 0.798 to 1.014. In sweet crudes such as Tapis, the sulfur content is low. Sour crudes have more sulfur, which gives them a tart taste; in the old days, prospectors did indeed characterize crude oil by tasting it. The crude oils in China (Henan, Liaohe, Shengli, Xinjiang, etc.) have relatively higher nitrogen contents in general compared to the crude oils outside China. The high nitrogen (relative to sulfur) presents operational challenges in refining, where it is converted into ammonia. The usual way to remove ammonia is in the form of ammonium bisulfide (NH_4SH). The NH_4SH dissolves in wash water and is transported to sulfur plants. If there is more NH_3 than H_2S, the NH_3 is not completely removed in this fashion and remains in process gas streams, where it can accelerate corrosion.

Table 2.2 Example crude assay report template

	Whole crude	Light naphtha	Medium naphtha	Heavy naphtha	Kero	AGO	LVGO	HVGO	VR	AR
True Boiling Point, °C	Initial	10	80	150	200	260	340	450	570	340
True Boiling Point, °C	Final	80	150	200	260	340	450	570	End	End
True Boiling Point, °F	Initial	55	175	300	400	500	650	850	1050	650
True Boiling Point, °F	Final	175	300	400	500	650	850	1050		
Yield of Cut (wt% of Crude)		x	x	x	x	x	x	x	x	x
Yield of Cut (vol.% of Crude)		x	x	x	x	x	x	x	x	x
Gravity, °API	x	x	x	x	x	x	x	x	x	x
Specific Gravity	x	x	x	x	x	x	x	x	x	x
Sulfur, wt%	x	x	x	x	x	x	x	x	x	x
Nitrogen, ppm	x		x	x	x	x	x	x	x	x
Viscosity @ 50°C (122°F), cSt	x				x	x	x	x	x	x
Viscosity @ 135°C (275°F), cSt	x					x	x	x	x	x
Freeze Point, °C				x	x	x	x			
Freeze Point, °F				x	x	x	x			
Pour Point, °C	x			x	x	x	x	x	x	x
Pour Point, °F	x			x	x	x	x	x	x	x

(continued)

Table 2.2 (continued)

	Whole crude	Light naphtha	Medium naphtha	Heavy naphtha	Kero	AGO	LVGO	HVGO	VR	AR
Smoke Point, mm				x	x	x				
Aniline Point, °C			x	x	x	x	x	x		
Aniline Point, °F			x	x	x	x	x	x		
Cetane Index, ASTM D976				x	x	x				
Diesel Index			x	x	x	x	x	x		
Characterization Factor (K)	x	x	x	x	x	x	x	x	x	x
Research Octane Number, Clear		x	x	x						
Motor Octane Number, Clear		x	x							
Paraffins, vol.%		x	x	x	x	x	x			
Naphthenes, vol.%		x	x	x	x	x	x	x		
Aromatics, vol.%		x	x	x	x	x	x	x		
Heptane Asphaltenes, wt%	x								x	x
Micro Carbon Residue, wt%	x								x	x
Ramsbottom Carbon, wt%	x								x	x
Vanadium, ppm	x								x	x
Nickel, ppm	x								x	x
Iron, ppm	x								x	x

BRENT16X	Whole crude	Butane and Lighter IBP - 60F	Lt. Naphtha C5 - 165F	Hvy Naphtha 165 - 330F	Kerosene 330 - 480F	Diesel 480 - 650F	Vacuum Gas Oil 650 - 1000F	Vacuum Residue 1000F+
Cut volume, %	100.0	2.9	9.2	21.3	15.6	16.7	24.5	9.7
API Gravity,	40.1	124.6	89.7	56.6	42.8	32.3	22.7	16.1
Specific Gravity (60/60F),	0.825	0.552	0.640	0.752	0.812	0.864	0.918	0.959
Carbon, wt %	86.0	82.4	83.7	86.0	85.9	86.4	86.5	86.4
Hydrogen, wt %	13.6	17.6	16.3	14.0	14.1	13.4	12.9	12.3
Pour point, F	13.1			(132.8)	(71.4)	10.1	82.8	89.7
Neutralization number (TAN), MG/GM	0.064	0.073	0.074	0.074	0.053	0.022	0.075	0.088
Sulfur, wt%	0.347	0.000	0.000	0.001	0.014	0.207	0.578	1.330
Viscosity at 20C/68F, cSt	7.0	0.4	0.5	0.8	1.7	6.9	244.7	3,337,634.5
Viscosity at 40C/104F, cSt	3.5	0.4	0.4	0.6	1.3	4.0	66.1	137,033.9
Viscosity at 50C/122F, cSt	2.6	0.3	0.4	0.6	1.1	3.2	39.5	38,389.0
Mercaptan sulfur, ppm	1.0	0.1	0.4	2.3	3.4	1.2	0.2	0.0
Nitrogen, ppm	898.4	-	-	0.0	0.9	45.1	933.8	5,610.9
CCR, wt%	2.1						0.2	17.8
N-Heptane Insolubles (C7 Asphaltenes), wt%	0.2						-	2.0
Nickel, ppm	1.2							10.9
Vanadium, ppm	6.3							55.9
Calcium, ppm	0.5							
Reid Vapor Pressure (RVP) Whole Crude, psi	10.3							
Hydrogen Sulfide (dissolved), ppm	-							
Salt content, ptb	-							
Paraffins, vol %	39.6	100.0	87.5	46.1	46.7	36.3	24.3	2.8
Naphthenes, vol %	33.2	-	12.5	36.8	37.2	43.0	39.7	20.9
Aromatics (FIA), vol %	27.2	-	-	17.1	16.2	20.7	36.1	76.4
Distillation type, TBP								
IBP, F	(9.1)		60.7	166.0	330.7	480.8	651.3	1,001.1
5 vol%, F	83.7		66.9	174.3	337.2	488.2	663.5	1,011.1
10 vol%, F	144.7		71.9	183.2	344.5	496.4	677.2	1,022.6
20 vol%, F	226.4		81.3	200.0	359.1	512.9	705.3	1,047.5
30 vol%, F	299.4		98.1	215.9	373.7	529.4	734.5	1,075.3
40 vol%, F	391.7		104.3	231.0	388.5	546.1	764.9	1,106.9
50 vol%, F	489.4		110.6	245.7	403.5	562.9	796.8	1,143.6
60 vol%, F	589.6		128.4	260.4	418.6	579.8	830.6	1,187.0
70 vol%, F	698.0		138.5	275.7	433.9	597.0	866.7	1,239.0
80 vol%, F	824.3		147.8	292.3	449.3	614.4	906.1	1,302.8
90 vol%, F	994.3		156.6	310.6	464.6	632.0	949.8	1,382.9
95 vol%, F	1,138.4		160.8	320.2	472.3	641.0	973.9	1,432.4
EP, F	1,431.0		164.6	329.0	479.2	649.1	997.3	1,504.3
Freeze point, F				(121.2)	(59.1)	23.1		
Smoke point, mm				29.1	23.5	16.2		
Naphthalenes (D1840), vol%					3.2	11.5		
Viscosity at 100C/212F, cSt	1.0	0.3	0.3	0.4	0.7	1.4	7.1	527.7
Viscosity at 150C/302F, cSt	0.5	0.2	0.2	0.3	0.5	0.8	2.7	52.6
Cetane Index 1990 (D4737),				30.7	41.9	49.1		
Cloud point, F				(127.7)	(67.9)	15.9		
Aniline pt, F					132.7	146.0	173.5	

Fig. 2.6 Crude assay results of a Louisiana crude oil and its fractions [7]

Among these crude oils, West Texas Intermediate (WTI) and Brent from North Sea are commonly used in the commercial or merchandise communities around the globe as benchmark crudes for the crude oil prices which change constantly depending on supply and demand. Both benchmark crudes have relatively low sulfur and nitrogen compared to other oils. The price of a specific crude oil can be higher or lower than these reference crudes. Value is set primarily by the results of crude assays. In general, the crude oils with low API gravity, high sulfur content and high acid content are sold at deep discounts compared to the reference oils.

Table 2.3 Bulk properties of 21 selected crude oils

Crude oil	API Gravity[b]	Specific gravity	Sulfur (wt%)	Nitrogen (wt%)
Alaska North Slope	26.2	0.8973	1.1	0.2
Arabian Light	33.8	0.8560	1.8	0.07
Arabian Medium	30.4	0.8740	2.6	0.09
Arabian Heavy	28.0	0.8871	2.8	0.15
Athabasca (Canada)	8	1.0143	4.8	0.4
Beta (California)	16.2	0.9580	3.6	0.81
Brent (North Sea)	38.3	0.8333	0.37	0.10
Bonny Light (Nigeria)	35.4	0.8478	0.14	0.10
Boscan (Venezuela)	10.2	0.9986	5.3	0.65
Ekofisk (Norway)	37.7	0.8363	0.25	0.10
Henan (China)	16.4	0.9567	0.32	0.74
Hondo Blend (California)	20.8	0.9291	4.3	0.62
Kern (California)	13.6	0.9752	1.1	0.7
Kuwait Export	31.4	0.8686	2.5	0.21
Liaohi (China)	17.9	0.9471	0.26	0.41
Maya (Mexico)	22.2	0.9206	3.4	0.32
Shengli (China)	24.7	0.9058	0.82	0.72
Tapis Blend (Malaysia)	45.9	0.7976	0.03	nil
West Hackberry Sweet[a]	37.3	0.8383	0.32	0.10
West Texas Intermediate	39.6	0.8270	0.34	0.08
Xinjiang (China)	20.5	0.9309	0.15	0.35

[a]Produced from a storage cavern in the U.S. Strategic Petroleum Reserve
[b]*API Gravity* is related to specific gravity by the formula
$^{\circ}$API = 141.5 ÷ (specific gravity @ 60 °F) − 131.5

The weight percent data of sulfur and nitrogen contents in the crude oils listed in Table 2.3 can also be presented in graphical form as in Fig. 2.7. Both sulfur and nitrogen correlate inversely with API gravity, but for this particular dataset, the correlations are rough due to wide scattering of data points, especially for sulfur.

Table 2.4 exhibits one example of reporting selected assay data for the whole crude and its different residue cuts in a Mexican crude oil. As expected, API gravity decreases with cut point; that is, the deeper cut point, the lower the API gravity. The API of the resid obtained at cut point at 650 °F (lowest cut point) is 17.3 while at 1050 °F (highest cut point) it is 7.1. Sulfur content, nitrogen content, metal content, viscosity, pour point and Conradson carbon (ConCarbon) residue increase with increasing boiling point.

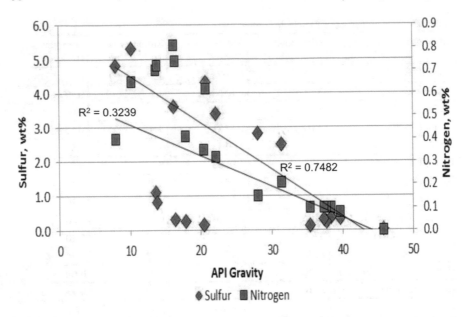

Fig. 2.7 Sulfur and nitrogen versus API gravity for selected crude oils. The top correlation line is for nitrogen and the bottom one is for sulfur, which is more scattered (has a lower R^2 value)

Table 2.4 Properties of a Mexican crude oil and its residua at different cut points [5]

	Whole Crude	Residua		
		650 °F	950 °F	1050 °F
Yield, vol.%	100.0	48.9	23.8	17.9
Sulfur. wt%	1.08	1.78	2.35	2.59
Nitrogen, wt%		0.33	0.52	0.60
API gravity	31.6	17.3	9.9	7.1
Conradson Carbon wt%		9.3	17.2	21.6
Vanadium, ppm		185		450
Nickel, ppm		25		64
Kinematic Viscosity at 100 °F	10.2	890		
Kinematic Viscosity at 210 °F		35	1010	7959
Pour Point, °F	−5	45	95	120

References

1. Hsu CS (1995) Hydrocarbons. In: Encyclopedia of analytical science, premiere edition, Academic Press, London, pp 2028–2034
2. Annual Book of ASTM Standards, section 5: Petroleum products, liquid fuels, and lubricants, Vol. 05.01 and Vol. 05.02. In: American society for testing and materials (ASTM) International, West Conshohocken, PA, 2016
3. Gray JH, Handwerk GE, Kaiser MJ (2007) Petroleum refining—technology and economics, 5th Edition, CRC Press
4. Lopes MS, Savioli Lopes M, Maciel Filho R, Wolf Maciel MR, Median LC (2012) Extension of the TBP curve of petroleum using the correlation DESTMOL. Procidea Eng 42:726–732
5. Speight JG (2006) The chemistry and technology of petroleum, 4th edn. Marcel Dekker, Revised and Expanded
6. Leffler WL (2008) Petroleum refining in nontechnical language. Fourth Edition, PennWell
7. Assays available for download. https://corporate.exxonmobil.com/en/company/worldwide-operations/crude-oils/assays. (Retrieved 20 Sept 2018)

Chapter 3
Chemical Composition

As mentioned, petroleum is not a uniform material. Prior to refining, crude oils contain some amounts of dissolved gas, water, inorganic salts and dirt. After these are removed, most of the remaining compounds are hydrocarbons, some of which contain hetero-atoms—mostly sulfur, oxygen and nitrogen, trace metals (Ni, V, Fe, Cu, etc.) The heteroatom-containing hydrocarbons are often referred to as "non-hydrocarbons" although "hydrocarbons" is commonly used as a general term to include both hydrocarbons and non-hydrocarbons [1, 2]. The hydrocarbon content may be as high as 97% (w/w) in paraffinic crude oil and as low as 50% (w/w) in heavy/asphaltic crude oil and bitumen.

3.1 Elemental Composition

Although petroleum consists basically of compounds of only two elements, carbon and hydrogen, these elements form a large variety of complex molecular structures. Regardless of physical or chemical variations, however, almost all crude oil ranges from 83 to 87% carbon by weight and 10–14% hydrogen. The more viscous bitumens generally vary from 80 to 85% carbon and from 8 to 11% hydrogen.

The elemental composition of petroleum is determined after the removal of water and inorganic mineral materials, such as salt and dirt. Carbon and hydrogen are measured by combustion, which produces CO_2 and H_2O. The products are collected and separated, and their weights are compared to the weight of the test sample. The sulfur and nitrogen contents are measured by the methods mentioned in crude assay physical tests, but other acceptable methods can also be applied to get equivalent results. Oxygen can be present in non-acidic compounds, which are not detected by titration. The overall oxygen content can be reliably measured by neutron activation with instruments not commonly available. Hence, the oxygen content is commonly reported as the difference between the total sample weight and the sum of the weights of carbon, hydrogen, sulfur and nitrogen. However, it is important to understand that some oxygen might come from exposure to air during storage and transportation.

© Springer Nature Switzerland AG 2019
C. S. Hsu and P. R. Robinson, *Petroleum Science and Technology*,
https://doi.org/10.1007/978-3-030-16275-7_3

Table 3.1 Overall elemental composition by weight percent [3]

Element	All petroleum	Tar sand bitumen
Carbon	83–87%	$83.4 \pm 0.5\%$
Hydrogen	10–14%	$10.4 \pm 0.2\%$
Sulfur	0.05–6.0%	$5.0 \pm 0.5\%$
Nitrogen	0.1–2.0%	$0.4 \pm 0.2\%$
Oxygen[a]	0.05–1.5%	$1.0 \pm 0.2\%$
Metals (Ni and V)	<1000 ppm	>1000 ppm

[a]Oxygen is calculated by difference. Caution is needed in sample handling, especially for prolong exposure to air during storage

Table 3.1 shows the average overall elemental composition of all crude oils and tar sand bitumen from Canada. It is noticeable the sulfur, nitrogen, oxygen and metal contents in tar sand bitumen are significantly higher than the average of all petroleum.

3.2   Molecular Composition

More than any other element, carbon binds to itself to form straight chains, branched chains, rings, and complex three-dimensional structures. The most complex molecules are biological—proteins, carbohydrates (including cellulose and hemicellulose), lipids, nucleic acids, and lignin. This is significant, because petroleum was formed from the remains of ancient microorganisms—primarily planktons and algae. However, petroleum is derived from these biomolecules after losing functional groups, leaving behind carbon skeletons that serve as biological markers or biomarkers. The remnants of cellulose, hemicellulose and lignin can be found in another hydrocarbon resource—coal.

Since hydrogen is a much lighter element than the others, oils with higher hydrogen content have lower specific gravities, i.e., higher API gravities. High API gravity oils have a high naphtha and kerosene/middle distillate (light ends) content and low residuum (heavy ends) content whereas low API gravity oils are low in naphtha and kerosene/middle distillate content and high in residuum.

Distillation is the first step in characterizing petroleum. Distillation separates petroleum into different boiling fractions: gas, naphtha, kerosene/jet fuel, light gas oil/diesel, heavy gas oil, lubricating oil, and residuum. The physical and chemical properties of petroleum vary with location, depth, and age of an oil field. The differences in these properties can be attributed to the distributions of the different sizes and types of hydrocarbons.

Petroleum molecules can be classified in several ways. They can be divided into saturated, unsaturated, and polar compounds. They can also be classified as paraffins, olefins, naphthenes, aromatics, polynaphthenes, polyaromatics, naphthenoaromatics, and heteroatom-containing compounds. Saturated hydrocarbons can be acyclic (paraffins or alkanes) or cyclic (cycloparaffins or naphthenes). Olefins are very rare in natural petroleum. They are mainly products from thermal cracking processes in refineries.

Crude oils can also be divided by polarity/solubility as saturates, aromatics, resin and asphaltene (SARA). All of these fractions contain heteroatom-containing hydrocarbons in various amounts.

3.2.1 Alkanes (Paraffins)

Paraffins (alkanes) have a general formula of C_nH_{2n+2}. Figure 3.1 shows molecular structures of some simple alkanes. The simplest paraffin is methane with a single carbon atom. Methane is the major component of natural gas. The next member in the alkane family is ethane with 2 carbon atoms. After that comes propane, with 3 carbon atoms. When the carbon number reaches 4, isomers are possible. Isomers are chemical compounds with the same molecular formula but different structures.

In "normal" paraffins, no carbon atom is connected to more than two other carbon atoms. In simple two-dimensional representations, the carbons are drawn in a straight chain as linear structures. In isoparaffins, at least one carbon atom is connected to three or four other carbon atoms. In two-dimensional diagrams, the structure is branched. Carbon atoms connected to only one other carbon, such as the end-of-chain carbons in n-paraffins, are called primary (1°). Carbon atoms connected to two other carbons are called secondary (2°), those connected to three other carbons are called tertiary (3°), and those connected to four other carbons are called quarternary (4°).

For example, C_4H_{10} includes normal butane (n-C_4), in which all carbon atoms are primary or secondary, and isobutane (methyl propane or i-C_4), in which the central carbon atom is tertiary. C_5H_{12} can have three isomers, normal pentane, isopentane (2-methyl butane) and neopentane (2,2-dimethyl propane), as shown in Fig. 3.1. The central carbon in neopentane is quarternary.

"Isooctane" in the figure is just one of several isooctanes. Its official (IUPAC) name is actually 2,2,4-trimethylpentane. However, this particular molecule is especially important, because it serves as a yardstick of gasoline combustion performance in spark-ignition engines, namely, "Octane Number". For isoparaffins, 2-methyl and 3-methyl alkanes are most abundant. 4-methyl derivatives are present in small amounts, if at all.

Petroleum also contains multi-methyl alkanes. Pristane (2,6,10,14-tetramethylpentadecane) and phytane (2,6,10,14-tetramethylhexadecane) contain isoprene skeleton structures, referred to as isoprenoid hydrocarbons. They are important acyclic biomarkers for indicating the origin and depositional environment of petroleum.

Different isomers have different boiling points and melting points, as illustrated in Tables 3.2 and 3.3. Normal butane (n-butane) boils below the melting point of ice. It is blended into gasoline in cold weather to increase vapor pressure, i.e., Reid vapor pressure, thereby improving the ignition behavior of spark plug internal

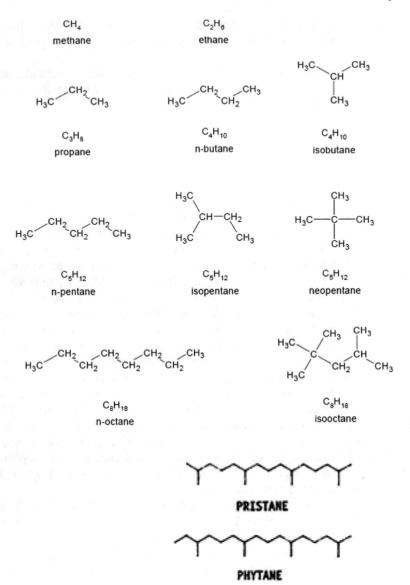

Fig. 3.1 Molecular structures of some simple alkanes

Table 3.2 Boiling points of selected light paraffins

Compound name	Molecular formula	Boiling point	
		°C	°F
Methane	CH_4	−162.2	−259.9
Ethane	C_2H_6	−88.6	−127.4
Propane	C_3H_8	−42.1	−43.7
n-Butane	C_4H_{10}	−0.1	31.7
Isobutane	C_4H_{10}	−11.2	11.9
n-Pentane	C_5H_{12}	36.1	96.9
Isopentane	C_5H_{12}	28.0	82.3
Neopentane	C_5H_{12}	9.5	49.0
n-Octane	C_8H_{18}	125.6	258.0
Isooctane	C_8H_{18}	99.3	210.7

Table 3.3 Melting points of selected C_{16} hydrocarbons

Compound name	Molecular formula	Melting point	
		°C	°F
Hexadecane	$C_{16}H_{34}$	17.9	64.1
5-Methylpentadecane	$C_{16}H_{34}$	−34.2	−29.5
7,8-Dimethyltetradecane	$C_{16}H_{34}$	−86.2	−123.1

Table 3.4 Number of possible isomers of paraffins with a specific carbon number

Carbon number	Boiling point of n-Paraffins		Number of isomers
	°C	°F	
5	36	97	3
10	174	345	75
15	271	519	4347
20	344	651	366,319
25	402	755	36,797,588
30	450	841	4,111,846,763
40	525	977	62,491,178,805,831

combustion engines. However, n-butane has been displaced by ethanol in recently developed reformulated gasolines. Higher carbon number normal paraffins, such as n-hexadecane, have melting points near or above room temperature. They tend to form wax and need to be removed from lubricant oils by dewaxing processes. Hence, the physical properties of components are important to the performance and specifications of refining products.

The number of possible isomers increases exponentially with carbon number, as shown in Table 3.4. Isomeric structures are important for gasoline components for octane number calculations based on molecular composition. They become less important beyond gasoline range (>C_{10}).

3.2.2   Naphthenes (Cycloparaffins)

Naphthenes are cyclic paraffins (cycloparaffins) with ring structures. In petroleum, the rings can have 5-carbon atoms (cyclopentanes) and 6-carbon atoms (cyclohexanes). It should be noted that pure core naphthenes are rare. They generally contain paraffin side chains with either normal or iso-alkyl structures. The molecules without

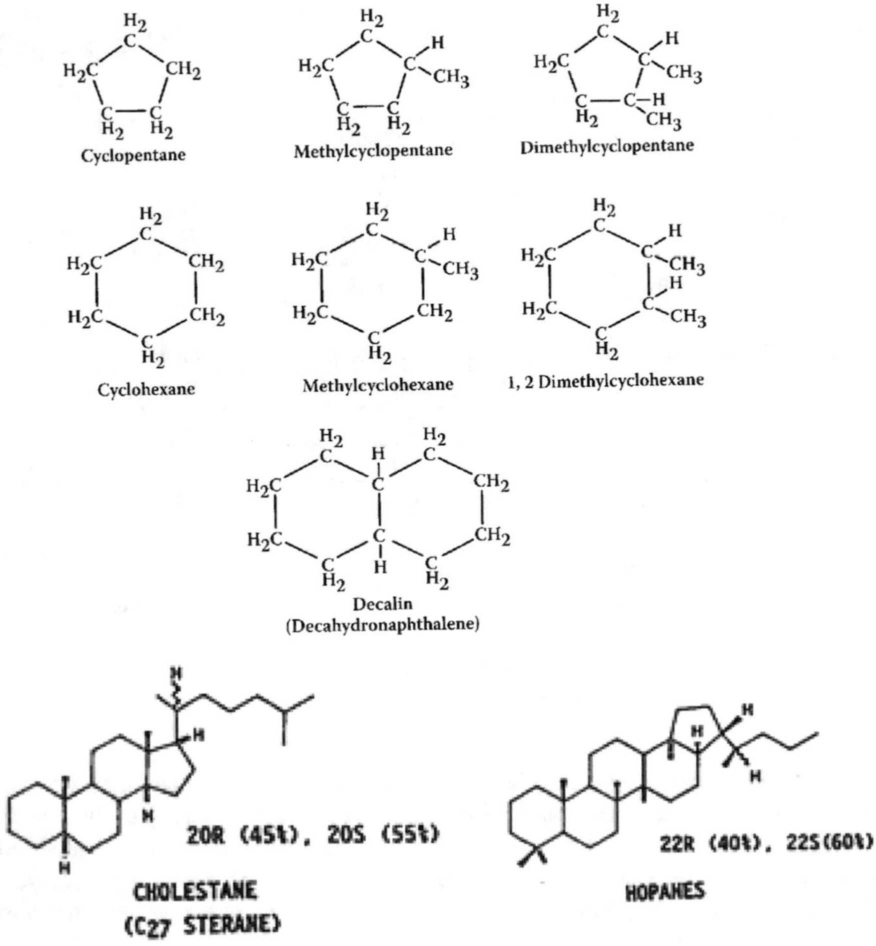

Fig. 3.2 Naphthenes

alkyl substituents are also referred to as "bare" molecules. In refining, naphthenes are formed primarily by the hydrogenation ("saturation") of aromatics.

A homologous series is a group of chemical compounds in which the difference between successive members is a simple structural unit, such as a methylene (–CH$_2$–) group. The homologous series based on cyclopentane and cyclohexane are rather large.

The rings can be connected by a single carbon-carbon bond, as in bicyclohexyls; by sharing two or more carbon atoms, as in decahydronaphthalene (decalin), perhydrophenanthrene, and perhydropyrene; or via heteroatoms, as in dicyclohexyl sulfide. Polyring compounds sharing multiple carbon atoms are called "condensed" molecules. Cholestanes (steranes) and hopanes are two important classes of 4-ring and 5-ring condensed cycloparaffins, used by petroleum scientists as biomarkers that indicate source input, age, maturity, depositional environment, and alteration of crude oils. Figure 3.2 displays some naphthenes for illustration.

3.2.3 Aromatic Hydrocarbons (Aromatics)

Arenes, or aromatics, contain one or more benzene rings. Aromatics are also unsaturated hydrocarbons that will react to add hydrogen or other elements to the ring.

Aromatics tend to be concentrated in heavy high boiling fractions of petroleum such as gas oil, lubricating oil, and residuum. In naphtha fraction, aromatics have the high octane-rating of hydrocarbon types, so they are valuable components in gasoline blends. However, they are undesirable in the lubricating oil range because they have the highest change in viscosity with temperature, that is, lowest viscosity index (VI), of all the hydrocarbons. Asphaltenes are the most aromatic of the fractions. Average oil tends to have more paraffins in the gasoline fraction and more aromatics and asphaltics in the residuum.

The simplest aromatic molecule is benzene, containing 6 carbon atoms and 6 hydrogen atoms with 3 conjugated double bonds around the ring. The simplest 2- and 3-ring aromatics are naphthalene, anthracene, and phenanthrene. Phenanthrenes are commonly found in crude oils while their isomer anthracenes are more common in coal. Aromatics, like naphthenes, can have alkyl (either linear or branched) substituents attached to the core in place of hydrogen at the substitution points. Figure 3.3 presents some aromatics and naphthenes found in crude oils.

Aromatics of greater than 2 rings can have condensed structures as shown in Fig. 3.4. There are two kinds of condensation, *peri-* and *cata*-condensed [4]. These condensed multiring aromatics are generally known as polynuclear aromatics (PNA), heavy polynuclear aromatics (HPNA). or polynuclear aromatic hydrocarbons (PAH). Many of the PAH isomers are carcinogenic, mutagenic or toxic, and are listed as high-priority pollutants by the U.S. Environmental Protection Agency (EPA).

Figure 3.5 shows another set of ring compounds. The –R groups in the figures represent short or long alkyl chains. The long alkyl chains tend to be branched. Some

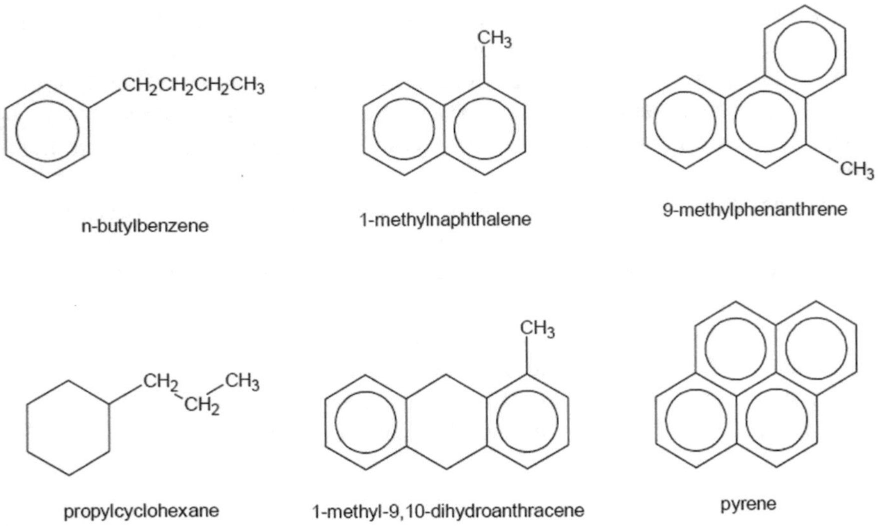

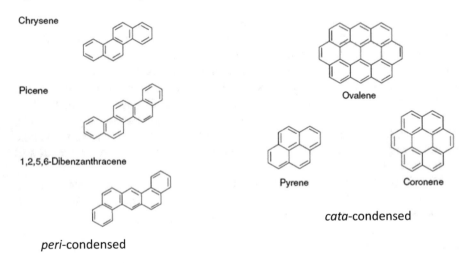

Fig. 3.3   Aromatics and naphthenes found in crude oil

Fig. 3.4   Polynuclear aromatic hydrocarbons

compounds contain both aromatic and naphthenic rings, such as indans, dinaph-thobenzenes and dinaphthonaphthalenes. They are called naphthenoaromatics or hydroaromatics and are formed by partial hydrogenation of aromatic hydrocarbons. In most cases, the hydrogenation of aromatics is reversible; in some refining processes, naphthenoaromatics come from partial dehydrogenation of polynaphthenes.

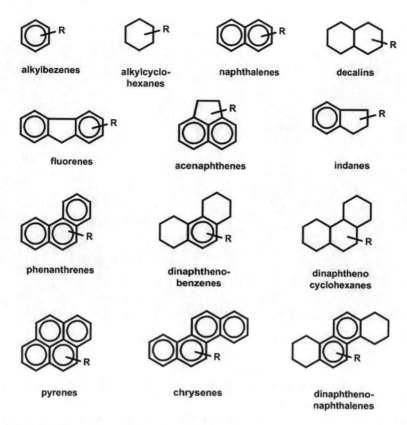

Fig. 3.5 Selected examples of hydrocarbon ring compounds

3.2.4 Heteroatom-Containing Hydrocarbons

Some compounds contain hetero atoms, mainly sulfur, nitrogen and oxygen, as shown in Fig. 3.6. These compounds can be aliphatic and aromatic. Trace amounts of metal chelated compounds, most noticeably nickel and vanadium porphyrins, have also been identified. Heteroatom-containing compounds have been called "non-hydrocarbons" in the literature because they contain elements more than just carbon and hydrogen. They are not straightly hydrocarbons per se.

3.2.5 Sulfur-Containing Hydrocarbons

Sulfur is the third most abundant atomic constituent of crude oils. It is mostly present in the medium and heavy fractions of crude oils. In the low and medium molecular ranges, sulfur is associated only with carbon and hydrogen, while in the heavier

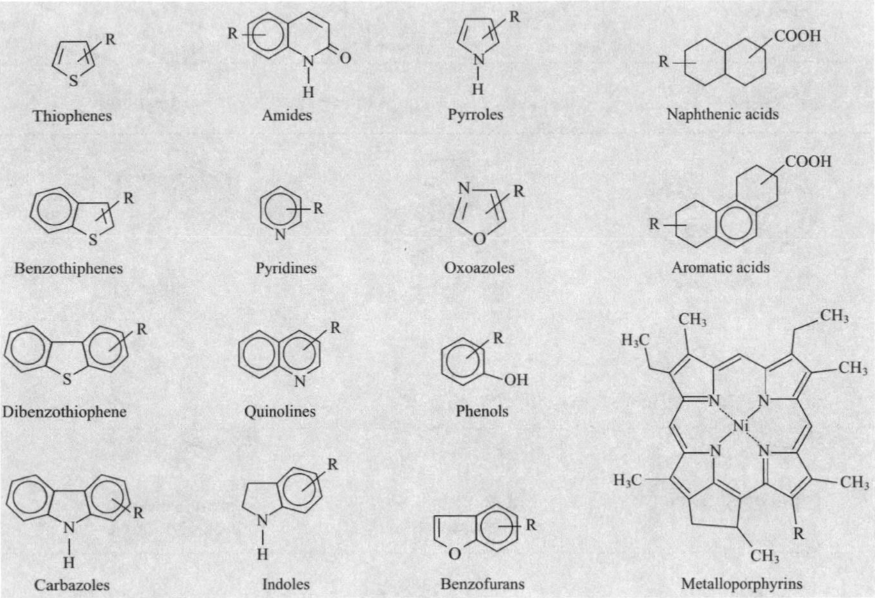

Fig. 3.6 Selected examples of heteroatom-containing compounds

fractions it is frequently incorporated in the large polycyclic molecules that also contain nitrogen and oxygen.

Sulfur is present in petroleum in several forms: elemental sulfur, hydrogen sulfide, mercaptans (thiols), sulfides, disulfides, polysulfides, thiophenes and sulfones/sulfoxides. The first six non-aromatic sulfur components are corrosive to the metal surface of pipes, vessels and equipment. Although the aromatic sulfur components are not as reactive as non-aromatic sulfur components, they will need to be removed during the refining processes for making fuels less harmful to the environment. Without removal, the oxidation of sulfur compounds upon combustion will yield sulfur oxides (SO_2 and SO_3, or SO_X) that will be dissolved in moisture in air forming acid rain.

The simplest sulfur molecule in petroleum is hydrogen sulfide (H_2S), which is present in natural gas or dissolved in the oil. Various forms of elemental sulfur are in crudes delivered to refineries. The simplest organic sulfur molecule is methyl mercaptan, in which a hydrogen atom of H_2S is replaced by a methyl group. Mercaptans having a SH functional group are also called thiols. Other than H_2S, they are the main components responsible for the foul and unpleasant odor of crude oils. They can also be responsible for corrosion of pipelines or transportation vessels. Hence, mercaptan oxidation (for example by the Merox process), $2\,R–SH + [O] \rightarrow R–S–S–R + H_2O$, is a common practice, even at the production site, to improve safety and greatly reduce the foul odor prior to transportation.

Fig. 3.7 Sulfur-containing hydrocarbons commonly present in petroleum

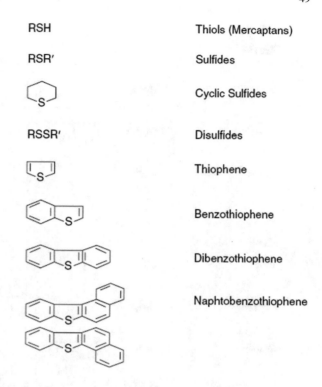

RSH   Thiols (Mercaptans)

RSR′   Sulfides

                             Cyclic Sulfides

RSSR′   Disulfides

                             Thiophene

                             Benzothiophene

                             Dibenzothiophene

                             Naphtobenzothiophene

Thiophenes are more stable than mercaptans, sulfides and disulfides. They are removed by hydrodesulfurization. Figure 3.7 lists some of the representative sulfur organics found in petroleum.

The total sulfur in crude oil varies from below 0.05% (by weight), as in some Pennsylvania oils, to about 2% for average Middle Eastern crudes, and up to 5% or more in heavy Mexican or Mississippi oils. Generally, the higher the specific gravity of the crude oil, the greater is its sulfur content (see Fig. 1.9). The excess sulfur must be removed from crude oil during refining, because otherwise sulfur oxides released into the atmosphere during combustion of the oil would generate sulfur oxides, which would turn to sulfuric or sulfurous acid with moisture and return to the earth as acid rain (a general term for wet deposition), a major pollutant. The sulfates can also attach to the smog particles, particularly PM 2.5), harmful to human health.

3.2.6 Nitrogen-Containing Hydrocarbons

Nitrogen is present in almost all crude oils, usually in quantities of less than 0.1% by weight. As with sulfur compounds, in general the higher the specific gravity of the crude oil, the greater is its nitrogen content (Fig. 1.9). Except for molecular nitrogen (N_2), the simplest stable nitrogen-containing molecule is ammonia (NH_3).

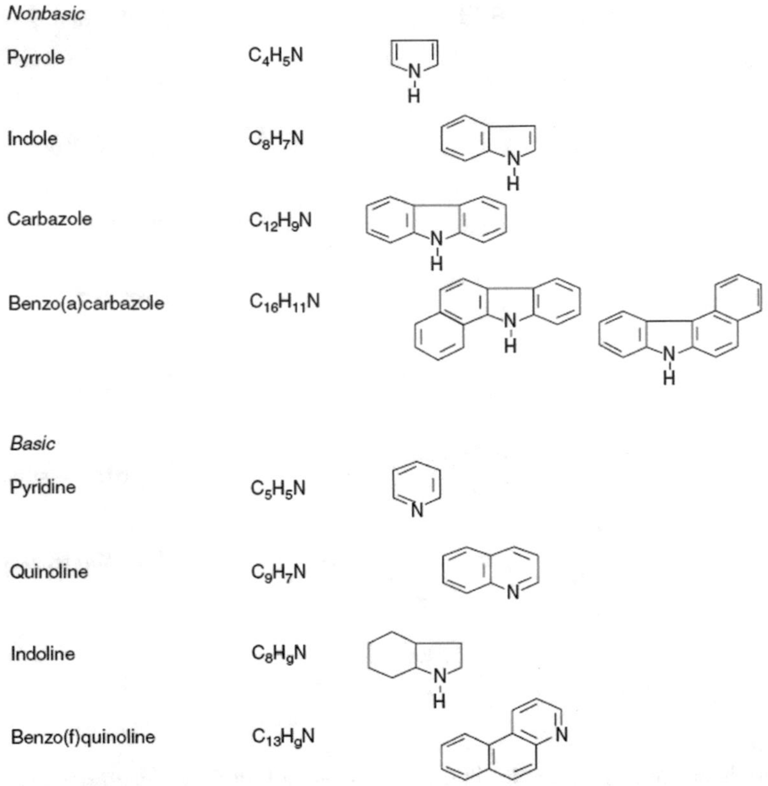

Nonbasic

Pyrrole	C_4H_5N	
Indole	C_8H_7N	
Carbazole	$C_{12}H_9N$	
Benzo(a)carbazole	$C_{16}H_{11}N$	

Basic

Pyridine	C_5H_5N	
Quinoline	C_9H_7N	
Indoline	C_8H_9N	
Benzo(f)quinoline	$C_{13}H_9N$	

Fig. 3.8 Basic and non-basic aromatic nitrogen hydrocarbons

Amines are a class of organic compounds where the nitrogen atoms in ammonia are replaced by alkyl groups. In primary amines, only one hydrogen atom in the molecule is replaced by an alkyl group, linear or branched. Those in which two and three hydrogen atoms are replaced by alkyl groups are secondary and tertiary amines, respectively. There are amines with in which the nitrogen atom is connected to a naphthenic ring, such as cyclohexyl amine.

Aromatic nitrogen compounds are generally divided into basic and non-basic, as determined by titration with perchloric acid in a 50:50 glacial acetic acid and benzene. Basic nitrogen compounds generally contain nitrogen atom with a pair of nonbonding electrons. They are of most concern to the refiners, because many catalysts used in upgrading processes contain acidic sites. Basic nitrogen compounds can react with those sites by coordinating nonbonding electrons with empty d-orbitals on the catalysts (Lewis acid sites) to deactivate the catalysts. Non-basic nitrogen compounds containing nitrogen atoms that are bonded with hydrogen are more benign, but can be converted into basic nitrogen compounds during processing, especially

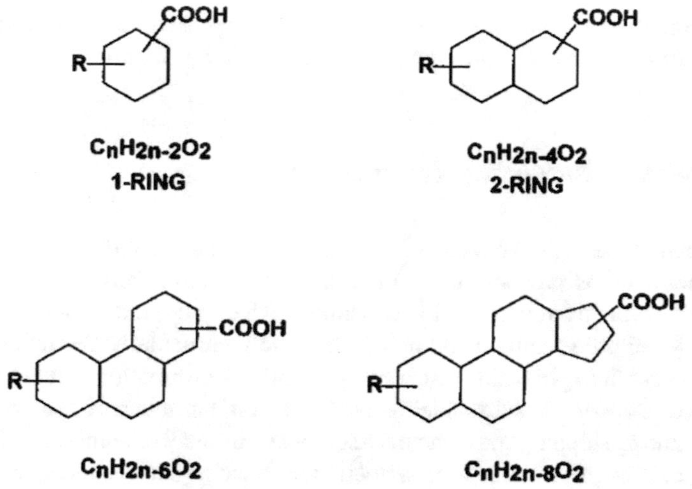

Fig. 3.9 Naphthenic acids

in hydroprocessing (hydrocracking, hydrotreating, etc.). Figure 3.8 displays some of the basic and nonbasic nitrogen compounds.

Burning of nitrogen compounds will produce NO and NO_2, or NOx. They are undesirable compounds in air, where they react with volatile organic compound in the presence of sunlight to produce photochemical smog. Similar to SOx, they also are responsible for acid deposition.

3.2.7 Oxygen-Containing Hydrocarbons

The oxygen content of crude oil is usually less than 2% by weight and is present as part of the heavier hydrocarbon compounds in most cases. For this reason, the heavier oils contain the most oxygen.

Oxygen-containing compounds in crude oils are mainly furans, phenols and naphthenic acids. Other acids, such as fatty acids and aromatic acids, can also be present in some crude oils. The presence of ketones, esters, ethers and anhydrides have been reported in residua, but they can be generated by sample treatment or handling; oxygen compounds can be formed from prolonged exposure to air during or after production of crude oil.

Furans are normally present in crude oils in small amounts and are easy to remove by hydroprocessing. Phenols are key components in coal derived from lignin in ligno-cellulosic biomass, but insignificant in conventional crude oils. Many of the recently discovered and recovered oils from Africa, Asia and offshore contain significant amounts of naphthenic acids, shown in Fig. 3.9, [5] which cause corrosion concerns.

Unlike total oxygen content, which is difficult or expensive to obtain reliably, the acid content can be much more conveniently obtained by titration.

3.2.8 Metal-Containing Hydrocarbons

Many metallic elements are found in crude oils, including most of those present in seawater. This is probably due to the close association between seawater and the organic forms from which oil is generated. Among the most common metallic elements in oil are vanadium and nickel that show particularly deleterious effects on catalysts, such as poisoning, excessive gas and coke formation. Crudes delivered to refineries can contain substantial amounts of iron, but much of that comes from corrosion during shipping and storage. Other metals in small amounts include copper, manganese, lead, mercury, as well as nonmetals, such as arsenic and silicon.

Metals can be present in petroleum as chelates, complexes or other coordination compounds. Although many of these organometallics remain to be characterized, the most recognized are porphyrins, which have a tetrapyrrole structure in which the nitrogen atoms form chelates with the metal atoms (Fig. 3.10). The most common porphyrins contain nickel(II) and vanadium(V) (as vanadyl). Less common are porphyrins containing iron(II) and copper(II). Vanadium is dominant in marine-sourced oils (Ni/V < 1) while nickel is dominant in lacustrine oils (Ni/V > 1) [6]. Metalloporphyrins are present in heavy petroleum fractions: vacuum gas oil and resid. They are serious catalyst poisons.

Arsenic and mercury are present as alkyls. Both are potent catalyst poisons—five to ten times worse than Ni, V, or Fe on a weight basis. Arsenic and mercury alkyls are relatively volatile, so they show up not just in residue, but in lighter fractions as well. Arsenic is common in Green River shale oil and crude oils from Canada, West

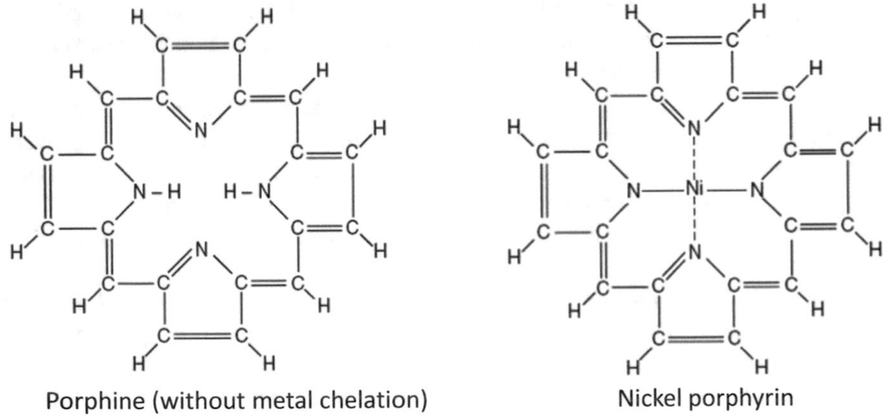

Porphine (without metal chelation)   Nickel porphyrin

Fig. 3.10 Porphine and nickel porphyrin

Africa, and the Ukraine. Mercury is a major problem in natural gas condensate in Thailand.

Silicon is not a metal, strictly speaking, and it is not common in raw crude oil. But organosilicon compounds serve as flow improvers in production facilities, pipelines, and transportation hubs. Silicon degrades the activity and structural integrity of certain refinery catalysts. In one respect, silicon isn't as bad as other metals, because catalysts can tolerate more of it. Ni, V, Cu, and certain forms of iron can be rejected by distillation, but like arsenic and mercury alkyls, silicon compounds are too light for that. The only practical way to remove them is by chemisorption onto special guard materials.

3.2.9 Olefins

Olefins, also known as alkenes, are aliphatic hydrocarbons with one or more C–C double bonds. They are very reactive compared to other hydrocarbon types, including aromatics. They are commonly present in living organisms, but found only trace in a few crude oils because they are usually reduced underground during burial to paraffins with hydrogen or to thiols with hydrogen sulfide. Most of the olefins in refining are generated by thermal processes. Olefins are important feedstocks for high polymers, such as polyolefins and polyesters, also for gasolines through polymerization (poly gasoline) and alkylation (alkylate). They are also important for making a wide variety of chemicals. Figure 3.11 presents a few examples of olefins.

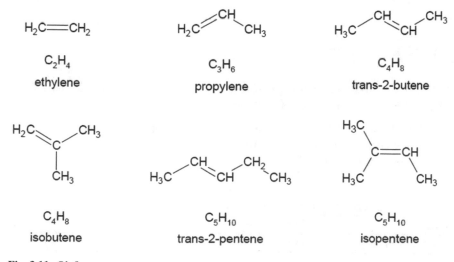

Fig. 3.11 Olefins

3.2.10 Alkynes

Alkynes,which contains triple bonds in the molecule, are not found in petroleum. The simplest alkyne is acetylene which can be produced from partial combustion of methane or natural gas with air or oxygen [7]. It is widely used as a fuel in oxyacetylene welding and cutting metals. Most of alkynes are used as important raw materials for making valuable chemicals and plastics.

3.3 Petroleum Molecule Types: Summary

The molecular composition of crude oil varies widely from formation to formation. Four different types of hydrocarbon molecules appear in crude oil, shown in Table 3.5. The relative percentage of each varies from oil to oil. The properties of the oil are largely determined by the distributions of these compound types and individual molecules.

Petroleum molecules can be lumped into groups, thereby allowing us to create manageable models for physical properties, chemical composition and reaction kinetics. Table 3.6 summarizes all of organic hydrocarbon molecules found in petroleum. They are grouped into different compound classes and hydrocarbon types.

Crude oil also may contain a small amount of decay-resistant organic remains, such as siliceous skeletal fragments, wood, spores, resins, coal, and various other remnants of former life. Sodium chloride also presents in most crudes, and is usually removed with desalting technology.

3.4 Crude Oils in China

The crude oils in China are usually different than the conventional crude oils in other countries, with the following unique properties [8]:

1. High viscosity, high wax content, high freezing point, relatively high specific gravities between 0.85 and 0.95. By definition they are heavy oils.
2. Low atom ratio of H to C.

Table 3.5 Average and range of hydrocarbon types in crude oils [1]

Hydrocarbon	Average (%)	Range
Alkanes (paraffins)	30	15–60%
Naphthenes	49	30–60%
Aromatics	15	3–30%
Asphaltics	6	Remainder

Table 3.6 Classes and types of hydrocarbon molecules found in petroleum

Compound class	Compound type
Saturated hydrocarbons	Normal paraffins Isoparaffins and other branched paraffins Cycloparaffins (naphthenes) Multi-ring cycloparaffins
Unsaturated hydrocarbons	Olefins (rarely found in petroleum. Mainly products from thermal cracking processes)
Aromatic hydrocarbons	Benzenes Condensed aromatics including polynuclear aromatic hydrocarbons (PAHs) Naphthenoaromatics (hydroaromatics)
Saturated heteroatom-containing hydrocarbons	Mercaptans (alkyl and cycloalkyl) Sulfides (alkyl and cycloalkyl) Disulfides (alkyl and cycloalkyl) Amines (primary, secondary and tertiary) Naphthenic acids
Aromatic heteroatom-containing hydrocarbons	Thiophenes (single and multi-ring) Pyrroles (single and multi-ring) Pyridines (single and multi-ring) Furans (single and multi-ring) Phenols (single and multi-ring) Aromatic acids
Metal-containing hydrocarbons	Ni and VO porphyrins

Table 3.7 Comparison of Chinese crude oils with crude oils outside China in sulfur and nitrogen contents [8]

Crude oil in China			Crude oil outside China		
Location	S, wt%	N, wt%	Location	S, wt%	N, wt%
Daqing	0.10	0.16	Canada (Athabasca)	4.8	0.4
Shengli	0.80	0.41	Texas (WTI)	0.34	0.08
Gudao	2.09	0.43	Saudi Arabia	2.55	0.09
Xingjiang	0.05	0.13	Venezuela (Boscan)	5.3	0.65

3. Relatively high nitrogen to sulfur ratio, shown in Table 3.7.
4. High nickel content, low vanadium content, shown in Table 3.8.

Table 3.8 Comparison of Chinese crude oils with crude oils outside China in metal contents [8]

Crude oil	Metal content, μg/g				Ni/V
	Ni	V	Fe	Cu	
Daqing (China)	3.1	0.04	0.7	0.25	78
Shengli (China)	26.0	1.6	13.0	0.1	26
Gudao (China)	21.1	2.0	12.0	<0.2	11
Liaohe (China)	32.5	0.6	9.3	0.3	54
Huabei (China)	15.0	0.7	1.8	<0.3	21
Zhongyuan (China)	3.3	2.4	8.2	0.4	1.4
Iran	8.7	88.8	4.0	0.07	0.10
Saudi Arabia	11.1	31.4	1.9	0.06	0.35
Texas	5.8	20.8	5.8	0.4	0.28
UK	0.59	3.48	0.23	0.26	0.17

References

1. Hsu CS (1995) Hydrocarbons. In: Encyclopedia of analytical science, premiere edition, Academic Press, London, pp 2028–2034
2. Hsu CS (ed) (2003) Analytical advances for hydrocarbon research, Kluwer Academic/Plenum Publishers, New York
3. Wikipedia: Petroleum. https://en.wikipedia.org/wiki/Petroleum. Retrieved 6 Aug 2015
4. Hsu CS, Lobodin V, Rodgers RP, McKenna AM, Marshall AG (2011) Compositional boundaries for fossil hydrocarbons. Energy Fuels 25:2174–2178
5. Hsu CS, Dechert GJ, Robbins WK, Fukuda EK (2000) Naphthenic acids in crude oils characterized by mass spectrometry. Energy Fuel 14(1):217–223
6. Barwise AJG (1990) Role of nickel and vanadium in petroleum classification. Energy Fuels 4:647–652
7. Watt LJ (1951) The production of acetylene from methane by partial oxidation. Univ. of British Columbia. https://open.library.ubc.ca/cIRcle/collections/ubctheses/831/items/1.0059187. Accessed 23 Nov 2018
8. Xu C, Yang C (2009) Petroleum refining engineering. Petroleum Industry Press, Beijing (in Chinese)

Chapter 4
Classification and Characterization

4.1 Petroleum Classification

Petroleum crude oils are broadly classified as paraffinic, asphaltic and mixed crude oils. Paraffinic crude oils are composed of aliphatic hydrocarbons (paraffins), wax (long-chain normal paraffin) and high-grade oils. Paraffinic crude oils can also contain a small amount of asphaltic (bituminous) material. Traditional examples are Pennsylvania grade crude oils. The paraffin wax in this crude oil is the origin of its classification. Relative to other hydrocarbons, at a given temperature and pressure, normal paraffins have a greater tendency to precipitate. This can cause problems during oil production and processing. In the production process, waxy heavy paraffins can build up in the tubing close the surface or they can block perforations. This often happens in depleted reservoirs or reservoirs under gas-cycling conditions. Isoparaffins are great for producing high-grade lubricating oils.

Asphaltic crude oils contain large proportions of asphaltic materials with low or negligible concentrations of paraffins. It has been suggested that a crude oil should be called asphaltic if the distillation residue contains less than 2% wax and paraffinic if it contains more than 5% wax. Asphaltic crude oils are highly viscous, often appearing in a sticky semi-solid or liquid state. They also have larger molecules and high boiling points. Asphaltic oils are cheaper than paraffinic oils because they are heavier and thicker. More effort is needed to convert them into fuels and other designated purposes. Asphaltic crude is also known as pitch, bitumen, petroleum asphalt or asphalt. Some asphaltic crude oils contain a predominance of cycloparaffins, which can be feedstocks for high viscosity lubricating oils, which are more sensitive to temperature changes than paraffinic crudes. Such crude oils are also known as naphthenic crude oils. Such crudes come from the U.S. Midcontinent region and parts of Iran.

Most crude oils are mixed, containing paraffins, naphthenes, and aromatic hydrocarbons. These crudes usually have a lower API gravity and are heavier. Because of their heaviness, naphthenic and aromatic crudes are worth less due to their higher requirements for refinery processing. Naphthenic crudes may also contain metals

C. S. Hsu and P. R. Robinson, *Petroleum Science and Technology*,
https://doi.org/10.1007/978-3-030-16275-7_4

such as vanadium, arsenic, and iron. There are some crude oils which have up to 80% aromatic content, and these are used as aromatic-base oil.

Attempts have been made to define or classify petroleum based on various distillation properties when combined with another property such as density. In 1933, Watson and Nelson [1] of Universal Oil Products (UOP, now a part of Honeywell) introduced a ratio between the mean average boiling point and specific gravity that could be used to indicate the chemical nature of hydrocarbon fractions and, therefore, could be used as a correlative factor. UOP or Watson Characterization Factors (Kw) are calculated with:

$$K_w = (T_B)^{1/3}/G$$

where T_B is mean average boiling point in °K and G is specific gravity at 60 °F. K_w ranges from 10.5 for highly naphthenic crude to 12.9 for highly paraffinic crudes. K_w for highly aromatic hydrocarbons exhibit values of 10.0 or less.

Figure 4.1 depicts a relationship between crude oil API gravity and Watson characterization factor. While not definitive, it can be observed from the trend line that lower gravity crudes tend to be more naphthenic, i.e., have lower K_w values, while higher-gravity crudes tend to be more paraffinic.

Another method of classifying crude oil is the correlation index (CI) developed at the U.S. Bureau of Mines [3], defined as:

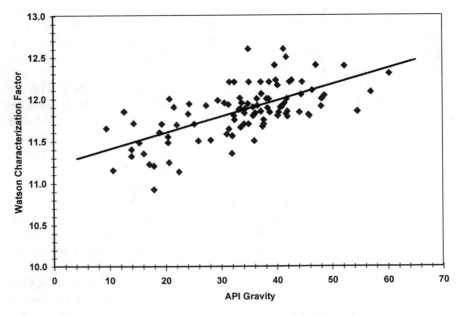

Fig. 4.1 Typical characterization factors versus various crude oil API gravities [2]

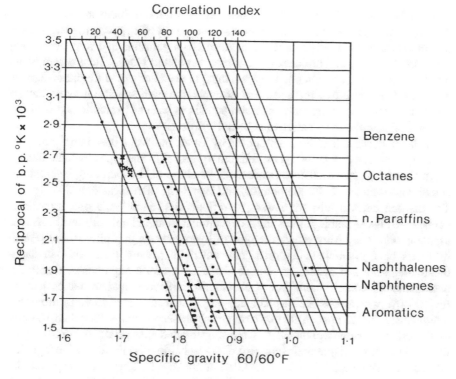

Fig. 4.2 Chemical indexes of pure compounds [4]

$$CI = 48,640/T_B + 473.7\,G - 456.8$$

where T_B is mean average boiling point in °K (°C + 273) and G is specific gravity at 60 °F (15.6 °C).

CI is an indicator of the aromaticity of a crude oil. The CI scale ranges from zero for straight-chain paraffins to 100 for benzene. The CI of a pure compound can be calculated from its boiling point and specific gravity, shown in Fig. 4.2 for a few representative compounds. In the figure, the correlation lines are drawn along with reciprocal of boiling point as y-axis and specific gravity at 60 °F as x-axis. Normal paraffins are along with the CI = 0 line. Aromatic compounds with specific gravity similar to benzene are distributed between CI = 20 and 100 where benzene is located. Naphthenes are in between paraffins and aromatics, with CI between 10 and 50. Cyclohexane has a CI of 51, while polycyclic aromatics have CI higher than 100.

For crude oils, CI values between 0 and 15 indicate a predominance of paraffinic hydrocarbons; values from 15 to 50 indicate a predominance either of naphthenes or of a mixture of paraffins, naphthenes, and aromatics; values above 50 indicate a predominance of aromatics.

Both the Watson Characterization Factor and Correlation Index are solely based on two physical properties: average boiling point and specific gravity (density). They do not define the relative proportions of open chain paraffins and ring naphthenes (both are saturates) and aromatics, the actual compound types. There certainly are overlaps between these compound types, thus, in crude oils of different categories determined by these two methods. For example, the high specific gravity may be due to both naphthenic and aromatic compounds as well as asphaltic and resinous materials. Their chemical nature may be quite different.

A chemical classification was proposed by Tissot and Welte based on the oil contents of various hydrocarbon types in a ternary plot [5]. The types include alkanes (normal and iso), cyclo-alkanes (naphthenes) and aromatics including heteroatom-containing aromatics (NSO compounds), resins and asphaltenes, as shown in Fig. 4.3. The oils are defined into six classes: paraffinic oils, naphthenic oils, paraffinic-naphthenic oils, aromatic-intermediate oils, aromatic- naphthenic oils and aromatic-asphaltic oils. They truly reflect the type of hydrocarbons that oil contains. Paraffinic oils have open-chain hydrocarbon content of greater than 50%. In naphthenic oils, saturated cyclic compounds account for >50% naphthenes. Paraffinic-naphthenic oils contain more than 50% saturated molecules and aromatics account for 25–40% with resins and asphaltenes vary between 5 and 15%. In aromatic-intermediate oils, aromatics are 40–70%, resins and asphaltenes constitute 10–30% of the oil. Aromatic-naphthenic oils are biodegraded oils originated from paraffinic and paraffinic-naphthenic oils. Aromatic-asphaltic oils are mostly biodegraded aromatic-intermediate oils, with a few true aromatic oils. They are heavy and viscous, with resins and asphaltenes content varies between 30 and 60%. The variation trends of hydrocarbon contents due to thermal maturation and biodegradation of the oils are also indicated by the arrows.

There are other ways to classify the crude oil, including:

(1) Classification according to API gravity:

> **Light**—API gravity >32
> **Medium**—API gravity 20–32
> **Heavy**—API gravity 10–20
> **Extraheavy**—API gravity ≤10, viscosity <10,000 cP
> **Bitumen**—API gravity ≤10, viscosity >10,000 cP.

Figure 4.4 shows the classification of heavy oils with API gravity <20. The API gravity for both Extra heavy oils and Bitumen is less than 10. Bitumen's viscosity is greater than 10,000 cP, while Extraheavy oil's viscosity is less than 10,000.

(2) Classification according to sulfur content:

> **Sweet**—<0.5% S,
> **Intermediate**—0.5–1.5% S
> **Sour**—>1.5% S.

Typical sweet crudes include West Texas Intermediate (WTI, which is a so-called benchmark crude traded on the New York Mercantile Exchange) from mostly

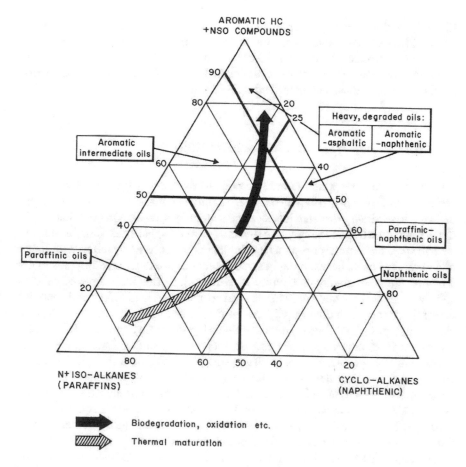

Fig. 4.3 Crude oil chemical classification based on hydrocarbon types [5]

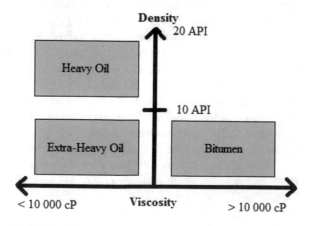

Fig. 4.4 Heavy oil classification

Louisiana and Oklahoma, Nigerian crudes, and Brent crude oils from the North Sea (Brent is a benchmark crude traded on the International Petroleum Exchange). Sour crudes include Alaska North Slope, Venezuelan, and West Texas Sour from fields like Yates and Wasson. Intermediate crudes include California Heavy, such as from the San Joaquin Valley, and many Middle East crudes.

4.2 SARA Analysis for Characterization

The complex mixtures which comprise crude oil are separated by distillation and solubility. Solubility depends largely on polarizability and polarity. A conventional procedure is to separate petroleum into saturates, aromatics, resins and asphaltenes fractions, known as a SARA (Saturates-Aromatics-Resin-Asphalt) analysis. Asphaltenes could be pentane, hexane or heptane insoluble. Lighter solvent is preferred for lighter oils to remove asphaltenes. Saturates and aromatics are determined by adsorption (column) chromatography on alumina or silica gel. Resins are recovered by polar solvents after the collection of Saturates and Aromatics. There are several analytical methods for this, including gravity-driven chromatographic separation, thin layer chromatography (TLC), and high-performance liquid chromatography (HPLC). The results from these analytical techniques do not produce identical results. Hence, they are not interchangeable. For comparison purposes, the same techniques should be applied in the same order every time. The fastest and least expensive method is Iatroscan TIC with a flame ionization detector (FID) to determines the weight percent of each fraction in the oil. The sample is spotted onto a quartz rod coated with sintered silica particles and separated over thin layer surface. The rod is then scanned directly in the FID at rated speed for quantification.

In theory, SARA analysis is applicable to all types of crudes, However, most of the problems in SARA analysis arise from the loss of light ends. Hence, SARA analysis should be confined to heavy oils, or heavier fractions of the oils with lighter fractions determined by other methods, such as gas chromatography (GC). Figure 4.5 shows SARA analytical scheme for crude oil with the fraction boils above 210 °C. The light fraction that boils below 210 °C is distilled off for subsequent GC analysis that can be used to determine the saturates and aromatics contents. Resin and asphaltene are not present or just in negligible amounts in light ends. The fraction that boils above 210 °C is treated with n-hexane in the example shown to precipitate off asphaltene. However, the solvent can also be n-pentane for light crudes or n-heptane for heavy crudes, which may be a better choice because n-hexane may contain azeotropic impurities, isohexanes and cyclohexane that are difficult to remove. The n-hexane insoluble precipitate is Asphaltene, which is subjected to infrared or nuclear magnetic resonance spectroscopy for structure determination. The n-hexane soluble fraction, i.e., deasphaltened oil (DAO) or maltene, is subjected to liquid chromatography separation on alumina or silica gel into saturates, aromatics and resins fractions. Saturates elute off the chromatography column with n-hexane. Benzene or toluene is normally used to elute off aromatics. A polar solvent, such as methanol or pyridine or a mixed

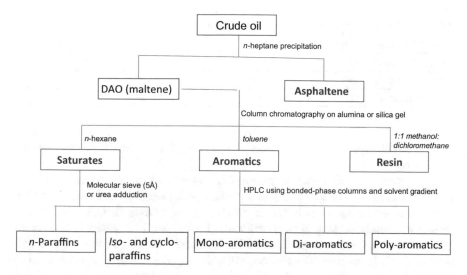

Fig. 4.5 Crude oil SARA analysis scheme

solvent, such as methanol/dichloromethane, is used to elute off the polar materials remaining in the column as resins. The paraffins can be further separated into normal paraffins and isoparaffins/cycloparaffins through molecular sieve (usually with 5Å diameter pores) or urea adduction. Aromatics are further separated by high performance liquid chromatography (HPLC) on a bonded silica get column with solvent gradient into monoaromatics, diaromatics, and polyaromatics fractions. Much better separation between aromatic ring types is achievable by more sophisticated HPLC solvent elution programs using dual columns, such as dinitroaminopropyl (DNAP) and polar amino cyano (PAC) bonded phase, and multiple switches [6].

Asphaltene is insoluble in an alkane solvent, but soluble in benzene or toluene. Structure determinations of asphaltenes vary widely from lab to lab, even for the same oil with the same equipment. Recently developed Fourier-transform ion cyclotron resonance mass spectrometry (FT-ICR MS) has been used to characterize asphaltenes at molecular level [7].

SARA analysis is an industry accepted measurement as a chemical approach to access chemical properties of crude oils by splitting into four fractions. The typical SARA distributions of light, medium and heavy crude oils are shown in Table 4.1. Saturates dominate gas condensates but are very low in biodegraded crudes. Asphaltene, on the other hand, is high in heavy oils and negligible in gas condensates. Asphaltene has an important role in organic deposition during petroleum production and processing. A small change in petroleum composition can cause asphaltene dropout. This can happen in reservoir, wellbore, pipelines and all the way to refinery facilities. The measurement error in SARA analysis can be significant, say. ±20% for saturates and as high as 100% error for asphaltene due to precipitation. The SARA

Table 4.1 Typical SARA distributions of light, medium and heavy crude oils

Class of crude oil	Saturates	Aromatics	Resins	Asphaltenes
Light Crude	92	8	1	0
Medium Crude	78	15	6	1
Heavy Crude	38	29	20	13

content can change during production or over time; hence, remeasurements would become necessary.

The saturates fraction consists of nonpolar compounds including linear (normal paraffins), branched (isoparaffins) and cyclic (cycloparaffins) saturated hydrocarbons. Aromatics are more polarizable, containing one or more aromatic rings. Both resins and asphaltenes have polar constituents. Resins are miscible with light n-alkane (n-C_5, n-C_6 or n-C_7), while asphaltenes are insoluble in even an excess of n-alkane solvent. Asphaltene molecules are colloidally dispersed in crude oil are stabilized by polar molecules of aromatic and resin molecules, preventing any major aggregation of the asphaltene [8].

Asphaltene precipitation is caused by a number of factors including changes in pressure, temperature and liquid-phase composition. Hence, the composition and structure of asphaltene in reservoir and laboratory conditions would be different. The equilibrium between the reservoir crude oil and asphaltene as self-assembled nanocolloidal particles would be destroyed by releasing dissolved gas upon drilling. Drilling, completion, stimulation, and hydraulic fracturing activities can also induce precipitation in the near-wellbore region.

References

1. Watson KM, Nelson EF (1933) Improved methods for approximating critical and thermal properties of petroleum. Ind Eng Chem 25(8):880–887
2. PetroWiki: Crude characterization. http://petrowiki.org/Crude_oil_characterization. Retrieved 9 Oct 2017
3. Smith HM (1940) Correlation index in crude oil analysis, U. S. Bureau of Mines, Tech. Paper 610
4. (a) Smith HM (1960) Correlation index in crude oil analysis, U. S. Bureau of Mines, Tech. Paper 610, 1940; (b) Gruse WC.; Stevens DR. Chemical Technology of Petroleum, McGraw-Hill
5. Tissot BP, Welte DH (1984) Petroleum formation and occurrence, 2nd edn. Springer-Verlag, Berlin-Heidelberg-New York-Tokyo
6. Qian K, Hsu CS (1992) Molecular transformation in hydrotreating processes studied by online liquid chromatography mass spectrometry. Anal Chem 64:2327–2333
7. Hsu CS, Hendrickson CL, Rodgers RP, McKenna AM, Marshall AG (2011) Petroleomics: advanced molecular probe for petroleum heavy ends, special feature: perspective. J Mass Spectrom 46:337–343
8. Aske N, Kallevik H, Johnsen EE, Sjoblom J (2002) Asphaltene aggregation from crude oils and model systems studied by high-pressure NIR spectroscopy. Energy Fuels 16:1287–1295

Part II
Exploration and Production
of Petroleum—Upstream
Science and Technology

Chapter 5
Petroleum System and Occurrence

5.1 Carbon Cycle

Petroleum can be considered as a substance produced from a "leak" in the carbon cycle, as exhibited in Fig. 5.1 by Ourisson et al. [1]. Animals and fish take in oxygen to produce carbon dioxide through respiration while living and by decay after death. Plants and algae, on the other hand, take in carbon dioxide and evolve oxygen through photosynthesis. This carbon cycle can be interrupted or broken if the dead organisms are not completely oxidized back into carbon dioxide. This occurs when oceanic microalgae settle to bottom and accumulate in tremendous quantities, interspersed with layers of inorganic sediment. Incomplete oxidation within the anoxic accumulation is accomplished by saprophytic microorganisms and anaerobic bacteria. Similar situations can occur for land plant and microorganisms, especially in lakes and river deltas.

5.2 Petroleum System

There are several terms one must first understand when studying a petroleum system. The first term is *basin*. A *basin* (*sedimentary basin* or petroleum province) is an area of the Earth where there is a large depression. This depression allows for the space to be filled with sediments and oils. Next is *stratum*. When there is layering of rocks in the ground this is referred to as a *stratum*. A *Reservoir* is an underground collection of oil or gas within porous rock formations. A *play* is a region that when studied shows a specific source, reservoir or trap that has potential for hydrocarbon build up.

A *prospect* is an individual exploration target where a specific trap has been identified and mapped but has not been drilled yet. A *reserve* is an accumulation in which the presence of oil and gas has been verified by drilling. An *outcrop* is bedrock that is visible above ground to show ancient superficial deposits. Being able to see the bedrock on the surface of the Earth simplifies determination of the porosity and

© Springer Nature Switzerland AG 2019
C. S. Hsu and P. R. Robinson, *Petroleum Science and Technology*,
https://doi.org/10.1007/978-3-030-16275-7_5

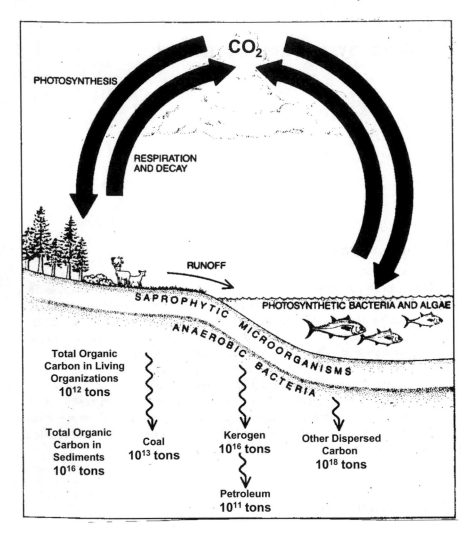

Fig. 5.1 Carbon Cycle [1]

permeability of the rock. These two properties are the most important to the discovery
of oil. Porous rocks such as sandstone, limestone, or dolomite allow petroleum to
build up in the reservoir, but also allow oil and gas to migrate. Impermeable rocks
such as clay or shale trap oil and gas, inhibiting large-scale accumulation and also
preventing further migration.

Figure 5.2 shows the relationship between the geological scales for petroleum
exploration. Once the prospects are identified, the next step is to lease the target
areas for drilling wells for production.

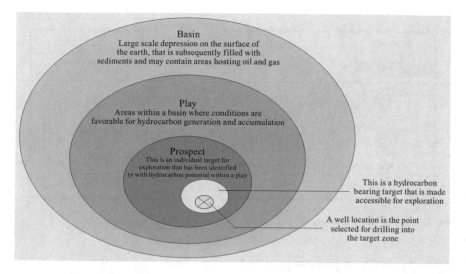

Fig. 5.2 Relative geological scales for petroleum exploration and production [10]

Crude oil and natural gas come from organic-rich sediments, which are formed by the accumulation and subsequent burial of dead organisms deep under the earth's surface in stratified layers. The stratified sediment bed is also referred to as the *formation*. The rock above the organic-rich source rock is called the *overburden*, which is sedimentary rock that encases the source rock, seal rock, and reservoir to provide thermal stress. Under thermal stress by overburden pressures of several hundreds or thousands of atmospheres over several hundred million years, the organic material is converted and compacted into kerogen. Oil and natural gases are released from the kerogen then migrate through porous carrier rock, such as sandstone, or along fault lines, to accumulate under impermeable cap rock or salt domes which functions as seal or trap (reservoir). The raw crude oil in the reservoirs can be mixtures of liquid and gas trapped in rock and sand and other materials.

The petroleum system, as represented in the simplified picture shown in Fig. 5.3, encompasses porous *source rock* containing kerogen, which is the ultimate **source** of petroleum and natural gas, permeable *carrier rock* for oil and gas **migration** from the source to the reservoir, porous *reservoir rock* and impermeable *seal rock* (also known as cap rock) for **accumulations**, and overburden. Source rock and reservoir rock are also permeable. A common transport mechanism is simple diffusion driven by buoyancy and ground-/pore-water flow. All these generation-migration-accumulation processes must be present at the right time and right place for petroleum formation and accumulation.

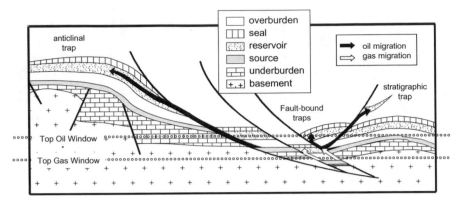

Fig. 5.3 Petroleum System [2]

5.3   Origin and Occurrence

Since the Renaissance, scientists have debated two theories on the formation of coal, petroleum and other fossil hydrocarbons: inorganic through abiogenic processes deep within the Earth versus organic from sediments through paleotransformation during burial [2]. The inorganic-origin hypothesis implied that petroleum appeared on Earth before life.

Scientists had put forward inorganic theories based on the following evidence:

1.   Some inorganic reactions generate a certain amount of hydrocarbons.
2.   The gas and lava which discharge from volcanos contain hydrocarbons.
3.   Many planets contain hydrocarbons. For example, the atmospheres of Jupiter and Uranus contain methane.

In 1953, the historic Miller-Urey experiment showed how simple molecules can be transformed into amino acids, the building blocks of proteins. Miller introduced water, methane, ammonia, and hydrogen into a sterile 5-liter glass flask connected to a smaller flask half filled with liquid water [3]. The water in the smaller flask was heated to generate steam, which was allowed to enter the larger flask. Continuous electrical sparks were fired between two electrodes in the larger flask to simulate lightning. The larger flask was cooled so that the water condensed and trickled into a U-shaped trap at the bottom of the apparatus. After a day, the solution in the trap turned pink. After one week of continuous operation, the flask was cooled, and the liquid in the trap was quickly dosed with mercuric chloride to prevent microbial contamination. Paper chromatography revealed the presence of glycine, α- and β-alanine.

Miller's original experiments yielded 20 different amino acids. Using variations of the Miller experiment, in addition to amino acids, the other major building blocks of life, sugars, nucleic acid bases, and lipids, also have been produced [4]. It is believed that from these building blocks, primitive life evolved.

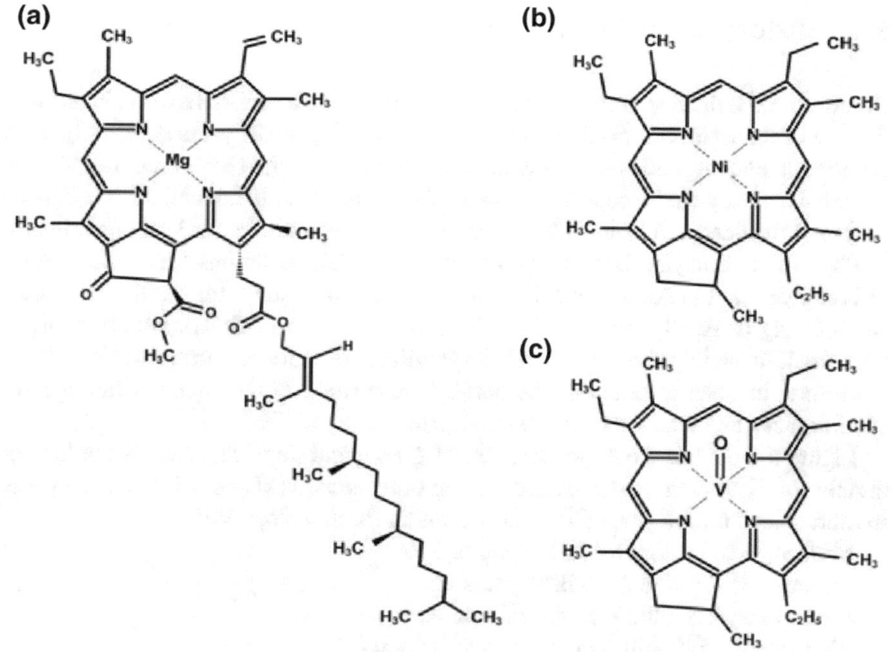

Fig. 5.4 Comparison of Chlorophyll a in plants (a) with porphyrins (b and c) found in petroleum [6]

In the 18th century, fossil evidence led scientists to conclude that coals were derived from plant remains. These conclusions influenced scientists to look into similar explanations for the origin of petroleum. Mikhail Lomonosov gets credit for suggesting that petroleum and bitumen were produced underground from coal at high pressure and temperature [5].

Modern theories, in which petroleum was said to originate from ancient sedimentary, organic-rich rocks, emerged in the 19th century [2]. In 1863, T. S. Hunt of the Canadian Geological Survey concluded that bitumen in some North American Paleozoic rocks resulted from paleo-transformation of ancient marine microorganisms. Further scientific advances in paleontology, geology, and chemistry led to refinement of the model. In 1936, Alfred Treibs linked the chlorophyll in plants to the porphyrins in petroleum, shown in Fig. 5.4 [6]. Additional geochemical evidence was provided by discovery of optical activity in moderately mature oils [7]. Further, there were a host of hydrocarbons that could be traced back to specific biological precursors, known as fossil fuel biomarkers [8].

5.4   History of Earth and Life

Based on the evidence from radiometric age dating, physicists have deduced that the age of the Earth is 4.54 ± 0.05 billion years (4.54 × 10⁹ years ±1%) [9]. This calculation encompasses measuring the concentration of the stable end product of a radioactive decay chain, coupled with knowledge of the half-life and initial concentration of the decaying isotope. In this instance, the end product is lead (specifically ^{206}Pb) and the decaying isotope is uranium (^{238}U). The latter has a half-life of 4.47 billion years and persists in stable minerals, making it ideal for dating the Earth. U/Pb dating is usually performed on the mineral zircon ($ZrSiO_4$), which incorporates uranium and thorium atoms in its crystalline structure but strongly rejects lead. Therefore, one can assume that the entire lead content of the zircon is radiogenic. Similar measurements are made on meteorites,

Figure 5.5 depicts the major division of geological time. The *Eon* is the largest division of the geologic time scale, covering time intervals of about 500 million years to more than a billion years [10]. The 4 eons in Earth history are:

Hadean, ~4.6 billion to 4 billion years ago;

Archean, 4 billion to 2.5 billion years ago;

Proterozoic, 2.5 billion to 545 million years ago; and

Phanerozoic, 545 million years ago to present.

When Earth was newly formed, the surface was molten and constantly bombarded by meteorites in the *Hadean* eon. The moon was formed during this eon. As the planet cooled in the *Archeon* eon, an atmosphere formed, largely from gases spewed from volcanoes. The gases included water vapor, hydrogen sulfide (low temperature)

Fig. 5.5 Major divisions of geological time scale

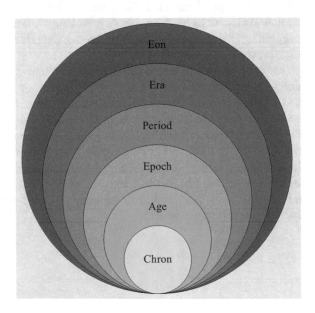

or sulfur dioxide (high temperature), methane, carbon dioxide, carbon monoxide, nitrogen, hydrogen and inert gases. Eventually, the crust cooled further and solidified. Large quantities of liquid water, which is essential for life, appeared on the surface. At this point, the atmosphere lacked free oxygen; the atmosphere was reductive.

From fossil evidence, it appears that life on Earth began as early as 3.6 billion years ago as single-cell organisms that do not possess membrane-bound organelle such as nuclei within the cell (prokaryotes). Cyanobacteria appeared 3.4 billion years ago. These primitive algae captured energy from the Sun and used it to perform photosynthesis, which generates organic matter and oxygen from water and carbon dioxide:

$$\text{Sunlight} + H_2O + CO_2 + \text{ nutrients } \rightarrow \text{``}CH_2O\text{''} + O_2$$

"CH_2O" symbolizes glucose, polysaccharides, and carbohydrates, i.e., organic matter, which capture energy of light quanta ($h\nu$) from sunlight. Oxygen is by-product of this process. Nutrients include nitrates, phosphates, iron, and other elements, which are discharged when the organic matter is digested.

As the Earth's atmosphere became oxidative, more life forms appeared in the *Proterozoic* eon. Multi-cell organisms with cell nuclei (eukaryotes) appeared about 2 billion years ago. Sexual reproduction began around 1.2 billion years ago, and new multicellular life forms continued to evolve.

Entering the *Phanerozoic* eon, simple animals appeared about 545 million years ago. This is when most of the world's petroleum formed and accumulated. Fish and proto-amphibians appeared 500 million years ago, and soon after, plants appeared on land. For the last 400 million years, a variety of insects, seeds, amphibians, reptiles, mammals, dinosaurs, birds, flowers, great apes (Hominidae) and human predecessors (genus Homo) evolved. The evidence of modern humans appeared about 250,000 years ago.

Eons are subdivided into **Eras**, such as Paleozoic, Mesozoic and Cenozoic within the *Phanerozoic* Eon. Eras delineate the major extinctions of living species. For example, the boundary between the end of Paleozoic Era and the beginning of the Mesozoic Era, which occurred 252 million years ago, represents the time when more than 90% of all marine species and about 70% of terrestrial species on Earth became extinct. Similarly, the boundary between the end of the Mesozoic Era and the beginning of the Cenozoic Era delineates the time when about 50% of all species on Earth went extinct, including the famous dinosaurs. The Era is subdivided into smaller segments known as **Periods**. Periods are the most commonly used subdivision of the geologic time scale. For example, the Paleozoic Era is divided into the Cambrian, Ordovician, Silurian, Devonian, Mississippian, Pennsylvanian and Permian periods as shown in Fig. 5.6.

Bacteria and algae undoubtedly are and have been ecological pioneers. Bacteria have enormous versatility in their physiology, with continuous evolution over geological time. Dead bacteria are second only to phytoplankton in contribution of organic matter ultimately buried and preserved in sediments. Due to microscopic or submi-

Fig. 5.6 Eras are subdivided
into Periods

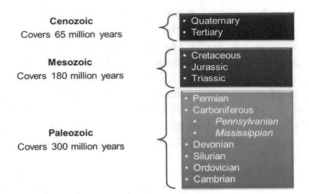

croscopic size and lack of hard part, bacteria are rarely fossilized. Their geological records are often associated with plant tissues and animal and insect remains.

The autotrophic phytoplankton and heterotrophic zooplankton are foundation of the food chain within the pyramid of life. High abundances of fossil phytoplankton groups, including blue green algae, green algae and arcritarchs, appeared in the Paleozoic era, particularly in Ordovician, Sulurian and Devonian periods [11]. More abundant phytoplankton groups, also including dinoflagellates, coccolithophorids, silicoflagellates, euglenids, diatoms, ebridians and discoasters, appeared in the Mesozoic and Cenozoic eras [11].

5.5 Paleo-Transformation

More evidence has been obtained to increase our understanding of how biological molecules were transformed into petroleum molecules. Upon death, some microorganisms, such algae and plankton, which are abundant in ocean, can escape full oxidation (create a leak in carbon cycle) with quick settlement and deposition of debris in layers. These sediments can be pushed more deeply under the surface by tectonic movement and undergo molecular transformation due to thermal stress under anoxic conditions. Most oil found under dry land today was below sea level when the organic deposition occurred. Subsequent tectonic movement lifted and folded the rock into its present morphology [12].

When dead organic matter (diatoms, planktons, pollens and spores) is buried below about 200 m, the organisms decompose in the presence of water and minerals in a process called **diagenesis**. This includes chemical, physical and biological changes to the sediment. The organic matter breaks down into proteins, lipids and carbohydrates (biopolymers) due to increasing temperature and pressure. These are further broken down into amino acids that in turn become fulvic, humic, and humin acids (geopolymers, precursors of kerogen). This transformation in the first few hundred meters of burial occurs at temperatures around 60 °C. Eventually, the decaying mass is consolidated, reached equilibrium under burial conditions, and compacted into the organic-rich sediment known as kerogen. Hydrothermal solutions, meteoric groundwater, porosity, solubility and time are influential factors in the diagenesis process. Hydrocarbon formation begins in this step. The most important hydrocarbon formed is methane. Also produced are carbon dioxide, water and some heavy heteroatomic compounds during later stages of diagenesis.

As the kerogen-laden sediment is buried further, say several kilometers deep, considerable increase in temperature and pressure plays an increasingly important role in the physiochemical paleo-transformation of the organic matter. Tectonics may also contribute to this increase. Temperatures may range from 50–150 °C, and geostatic pressures due to overburden may vary from 300–1500 bars depending on the depth of the sediment; the weight of overburden rock is directly proportional to depth. Under such conditions, the organic matter in kerogen is converted into gaseous and liquid hydrocarbons during **catagenesis**, a process essentially equivalent to thermal cracking. Hydrocarbon release from kerogen first produces liquid petroleum. In the later stage, wet gas and condensate are produced, accompanied by significant amounts of methane, which is the main constituent of natural gas.

If the kerogen is exposed to extreme heat, as it would be in the proximity of magma, sediment goes through a metamorphic path called **metagenesis**. If the temperature exceeds 150 or even 200 °C, then any organic matter which has made it to this point will become methane and carbon residue (pyrobitumen), with no more generation of petroleum.

Figure 5.7 shows the gas and oil products through paleotransformation stages. In general, deposition temperature increases with the depth. In shallow burial, biogenic gas, mainly methane, can be generated by methanogenic archaea in shallow sediments in a temperature range of ~35–45 °C. Catagenesis begins at ~50 °C and ends at ~100 °C when oil generation/release from kerogen is complete. Much of the initial petroleum from oil-prone source rock is liquid that contains associated gas. Above 150 °C, the remaining kerogen with over-mature oil generates mainly thermogenic gas, which is dissolved in condensate initially and becomes drier as the generation near completion. Thermogenic gas contains mostly methane. Thermogenic methane can be differentiated from biogenic methane from carbon isotope distribution that will be discussed later. Underground dry gas can also be generated from subsurface biodegradation by an anaerobic microbial process. Biomarkers, molecular fingerprints of biological origin, are present in immature oil and its precursors.

5.6 Fossil Oil Biological Markers (Biomarkers)

The ubiquitous presence of biological markers (biomarkers) in crude oils and coals provides the evidence of the biological origin of petroleum and coal. Biomarkers are the molecules that retain the basic carbon skeleton of biological precursors. Crude oil compositions vary widely depending on the oil sources, the thermal regime during oil generation, the geological migration and the reservoir conditions. Most of the crude

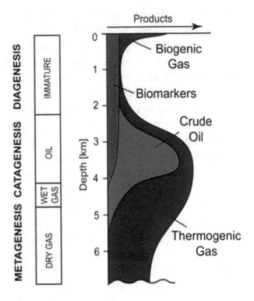

Fig. 5.7 Oil and gas products generated at paleotransformation stages [16]

oil constituents undergo changes in their chemical structure over time as an effect of several factors, among which are biodegradation and weathering. Biomarkers are more degradation-resistant in the environment than other hydrocarbon groups in oil.

Biomarkers are very useful molecules for exploration, exploitation and production in upstream applications. Some representative biomarkers and their biological precursors are shown in Fig. 5.8. With detailed studies, biomarkers can be used for assessment of source input, age, maturity and alteration of petroleum [13, 14]. They can also be used as fingerprinting for oil-oil and oil-source rock correlation during exploration to determine the source of oil, source potential and migration pathways of a source rock to the reservoir. In environmental applications, biomarker analysis is of great importance in forensic investigations. It is used for determination of the source of spilled oil, differentiation and correlation of oils, and monitoring the degradation process and weathering state of oils under a wide variety of conditions [15].

5.7   Kerogen

Upon death and burial, living organisms, such as planktons, algae, spores and pollens, begin to undergo decomposition or degradation without full oxidation. Large biomolecules, such as proteins and carbohydrates, begin to break down. The dismantled components can then condense, accompanied by the formation of mineral components, to form geopolymers. The smallest units are fulvic acids, the medium units are humic, and the largest units are humins. The subsequent sedimentation and

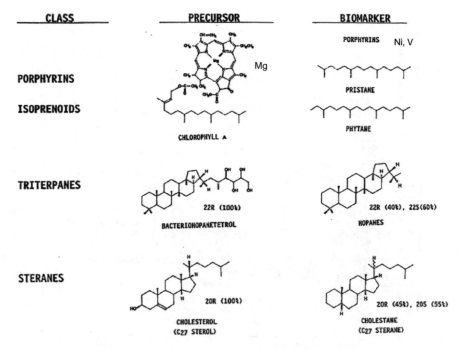

Fig. 5.8 Representative Biomarkers [13]

progressive burial of geopolymers provides ever-increasing pressure and temperature gradients. Upon sufficient geothermal pressure and sufficient geological time,
geopolymers were converted into kerogen.

Kerogen is complex organic matter in sedimentary rock. It releases oil and gas
hydrocarbons upon geothermal heating, with the oil-formation window at 60–160 °C
and the gas formation window at 150–200 °C. Kerogen can not be dissolved in
organic solvents, not even soluble when subsequently treated with non-oxidizing
mineral acids. It has high molecular weight >1000 daltons (Da).

5.7.1 Kerogen Types

There are three main hydrocarbon-producing types of kerogen, expressed by plotting
the hydrogen-to-carbon ratio versus the oxygen-to-carbon ratio in the van Krevelen
diagram shown in Fig. 5.9 [11, 16, 17].

Type I refers to kerogen with high initial H:C ratio >1.25 and a low initial O:C ratio
<0.15. It's also called Sapropelic kerogen. It derives mainly from lacustrine algae
in anoxic lakes and other unusual marine environments, and is formed mainly from
proteins and lipids. Paraffin hydrocarbons are predominant over cyclic structures,

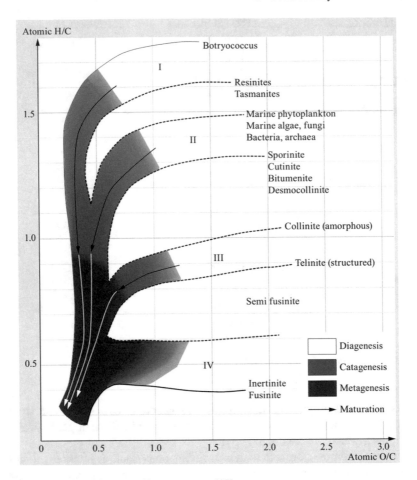

Fig. 5.9 Van Krevelen diagram of kerogen types [17]

naphthenes and aromatics. It has a greater tendency to produce a large yield of volatile and extractable liquid hydrocarbons than any other type of kerogen.

Type II, also called Planktonic kerogen, refers to kerogen a with relatively high H:C ratio <1.25 and low O:C ratio of 0.03–0.18. It is derived from lipids of a mixture of phytoplankton, zooplankton and bacteria deposited in a reducing environment, and tends to produce a mix of gas and oil.

Type IIS kerogen has the same properties as Type II kerogen, but is high in sulfur (8–14%, atomic S/C ≥ 0.04). It generates oil at a lower thermal maturity than typical type II kerogen with less than 6 wt% sulfur.

Type III, or Humic kerogen, has a relatively low initial H:C ratio <1 and a high initial O/C ratio 0.03 to 0.3. Type III kerogen is derived essentially from land plants and contains much identifiable vegetal debris. This type of kerogen is an important source of polyaromatic nuclei and heteroatomic compounds with phenolic and carboxylic acid functional groups. The material is thick, resembling wood and coal. It tends to produce coal and gas.

Type IV (residue) kerogen contains mostly decomposed organic matter in the form of highly condensed polycyclic aromatic hydrocarbons (PAH's). The H:C ratio is <0.5 and it has no potential to produce hydrocarbons.

5.8 Carbon Isotopes

Carbon has three isotopes: 98.89% ^{12}C, 1.11 wt% ^{13}C and trace ($\sim 10^{-12}$) ^{14}C in natural abundance. ^{14}C is radioactive while the former two are stable. ^{14}C is produced by reaction of thermal neutrons from cosmic radiation with ^{14}N in the upper atmosphere. It enters biosphere as CO_2 via photosynthesis. By emitting a β particle (an electron), ^{14}C radioactively decays to form stable ^{14}N.

$$^{14}C(6p + 8n) - \beta \rightarrow ^{14}N \,(7p + 7n)$$

Since dead tissue does not absorb $^{14}CO_2$, ^{14}C can be used for radiometric dating of dead biological materials, such as Egyptian mummies. However, the half-life of ^{14}C is 5570 years, so it essentially disappears after 5 half-lives, or ~30,000 years. The age of natural gas and crude oils is on the order of several hundred million years; hence, ^{14}C cannot be used for the age determination of natural gas and crude oils. The age of petroleum is based on radioactive materials with much long half-lives, such as uranium, in fossils or rocks in close proximity.

As mentioned earlier, biogenic methane from recent decay of biological materials contains radioactive ^{14}C that is absent in the methane in natural gas and reservoirs.

For most stable isotopes, the magnitude of fractionation from kinetic and equilibrium fractionation is very small; for this reason, enrichments are typically reported in "per mil" (‰, parts per thousand). The ratio of the other two stable carbon isotopes, $^{13}C/^{12}C$, is widely used in geochemistry to distinguish between organic and inorganic carbon and to determine sources of gas and oil. It is expressed as the enrichment of ^{13}C ($\delta^{13}C$), which is the $^{13}C/^{12}C$ ratio compared to that of a standard, shown below:

$$\delta^{13}C(\text{per mil}, \%_{00}) = \left[\left(^{13}C/^{12}C\right)_{\text{sample}} / \left(^{13}C/^{12}C\right)_{\text{standard}} - 1 \right] \times 1000$$

The commonly used standard reference material is belemnite from the PeeDee Formation in South Carolina (PDB with $^{13}C/^{12}C = 1.123 \times 10^{-2}$) or Solenhofen limestone (NBS-20 with $^{13}C/^{12}C = 1.1218 \times 10^{-2}$). The Selenhofen limestone is is more commonly used now due to the depletion of PDB.

The $^{13}C/^{12}C$ ratio can be obtained from isotope ratio mass spectrometry, shown in Fig. 5.10, with a precision of 1 part in 10,000. In the analysis, the sample is combusted to CO_2, which is ionized by an electron beam generated by filament at 70 eV energy; results are highly repeatable. The resulting ions are repelled from the

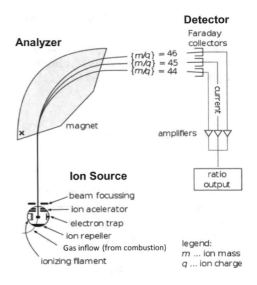

Fig. 5.10 Isotope Ratio Mass Spectrometer [19]

ion source and accelerated into a magnet. The ions of different mass-to-charge (m/q) ratios are deflected at different angles. Three Faraday cage detectors are placed to measure the abundances of $^{12}CO_2$ (m/q = 44), $^{13}CO_2$ (m/q = 45) and $^{14}CO_2$ (m/q = 46). This gives the $^{13}C/^{12}C$ ratio, which is converted into $\delta^{13}C$ and compared to the standard.

The $\delta^{13}C$ value for the bicarbonate in seawater is about the same as that for PDB. The CO_2 in air has 7% less ^{13}C than PDB. The U.S. National Bureau of Standards (NBS) has an oil standard with a $\delta^{13}C$ of -19.4‰ on the PBD scale, which can be used as a secondary standard. The carbonates rocks have a $\delta^{13}C$ range between $+5\text{‰}$ and -5‰. The $\delta^{13}C$ ranges of marine plankton and terrestrial plants are in the ranges of -14 to -28‰ and -14 to 32‰, respectively. Fossil organic matter falls in the range covered by marine plankton and terrestrial plants. Petroleum is typically between -18 and -34‰, another indication of organic derived material.

The $\delta^{13}C$ the biomethane ranges from -55 to -85‰, while for methane formed by thermal cracking at depth (geological or fossil methane) the range is from -25 to -60‰.

5.9 Sulfur Incorporation into Fossil Hydrocarbons

All crude oils and natural gases contain sulfur compounds. Sulfur is essential for life. It is a constituent of many proteins, enzymes and cofactors. The incorporation of sulfur into fossil hydrocarbons can come from both organic and inorganic sources [18]. From dead living organisms, mineralization converts organic sulfur into inorganic forms, such as H_2S, elemental sulfur, polysulfides and sulfide minerals. Oxidation converts H_2S, elemental sulfur and polysulfides to sulfates (SO_4^{2-}). Thermochemical sulfate reduction (TSR) reduces sulfates into sulfides, which can be incorporated into organic compounds.

One form of sulfur in petroleum and natural gas is H_2S, which is highly toxic. It can cause corrosion of metals, including the carbon steels commonly used for pipelines, sea-going tankers and rail cars. For these reasons, H_2S must be removed before oil and gas can be transported. Other forms of sulfur include various forms of elemental sulfur and sulfur incorporated into organics. Most of this sulfur is removed by processing in refineries. Combustion of sulfur-containing petroleum fractions generates sulfur oxides (SOx), which react with moisture in the air and return to the earth as acid rain. Acid rain harms buildings and plants—in the late 20th Century, it devastated the Black Forest in Germany—and it damages animal and human respiratory systems. Hence, sulfur content is an important factor in determining the quality and value of crude oil, due to its impact on the cost of removal. A crude oil is considered "sweet" if it contains less than 0.5% sulfur. In comparison, sour crude oil contains impurity sulfur levels larger than 0.5%.

References

1. Ourisson G, Albrecht P, Rohmer M (1984) The microbial origin of fossil fuels. Sci Am 251:44–51
2. Walters CC (2006) The origin of petroleum. In: Hsu CS, Robinson PR (eds) Practical advances in petroleum processing. Springer, New York
3. Miller SL (1953) Production of amino acids under possible primitive earth conditions. Science 117(3046):528–529
4. Horowitz H (1992) Beginnings of cellular life. Yale University Press, New Haven
5. http://pineislandnews.com/Pine_Island_News_Blog/content/1757-lomonosov-fathered-theory-oil-originates-biological-material. Retrieved on 10 Nov 2016
6. Treibs A (1936) Chlorophyll and hemin derivatives in organic mineral substances. Angew Chem 49:682–686
7. Oakwood TS, Shriver DS, Fall HH, Mcaleer WJ, Wunz PR (1952) Optical activity of petroleum. Ind Eng Chem 44:2568–2570
8. Philp RP. C&E News, 2/10/86, pp 28–43
9. Patterson C (1956) Age of meteorites and the earth. Geochim Cosmochim Acta 10(4):230–237
10. Ali HN (2017) Fundamentals of petroleum geology for exploration. In: Hsu CS, Robinson PR (eds) Springer handbook of petroleum technology. Springer, New York
11. Tissot BP, Welte DH (1984) Petroleum formation and occurrence, 2nd edn. Springer-Verlag, Berlin-Heidelberg-New York-Tokyo
12. Hunt JM (1979) Petroleum geochemistry and geology. W. H. Freeman and Company, San Francisco
13. Hsu CS, Walters CC, Isaksen GH, Schaps ME, Peters KE (2003) Biomarker analysis in petroleum exploration. In: Hsu CS (ed) Analytical advances for hydrocarbon research Kluwer Academic/Plenum Publishers, New York/Boston/ Dordrecht/London/Moscow
14. Peters KE, Moldowan JM (1993) The biomarker guide: interpreting molecular fossils in petroleum and ancient sediment, Prentice Hall
15. Prince RC, Elmendorf DL, Lute JR, Hsu CS, Haith CE, Senius JD, Dechert GJ, Douglas GS, Butler EL (1994) 17*(H),21*(H)-Hopane as a conserved internal marker for estimating the biodegradation of crude oil. Envrion Sci Tech 28:142–145

16. Peters, K. E.; Schenk, O.; Scheirer, A. H.; Wygrala, B.; Hantschel, T (2017) Basin and Petroleum System Modeling. In: Hsu CS, Robinson PR (eds) Springer handbook of petroleum technology, Springer, New York
17. Walters CC (2017) Origin of petroleum. In: Hsu CS, Robinson PR (eds) Springer handbook of petroleum technology. Springer, New York
18. Ivanov MV and Freney JR (1983) The global biogeochemical sulphur cycle. Chichester, John Wiley
19. Wikipedia: Isotope-ratio mass spectrometry. https://en.wikipedia.org/wiki/Isotope-ratio_mass_spectrometry. Retrieved on 10 Nov 2016

Chapter 6
Exploration for Discovery

6.1 Exploration (Looking for Oil and Gas)

In modern times, a required prerequisite to looking for oil is obtaining the rights to do so. Leases must be acquired from land- or mineral-rights owners, and licenses must be obtained, often from multiple legal jurisdictions—local, regional, state or provincial, and/or national governments. Environmental, health and safety standards must be understood and met. To develop large offshore basins can cost tens of billions of dollars, more than even the biggest oil companies can afford on their own. Several platforms might be required, and the cost of a platform in a severe-weather locale, such as the North Sea, can exceed $2 billion; in 1976, the ill-fated Piper Alpha platform cost $1.6 billion—equivalent to about $3 billion in 2015. Therefore, partnerships are formed to share capital costs, risks, and eventual profits.

There are generally two distinct phases in searching for natural resources: prospecting and exploration. Prospecting is the search for unknown deposits. Exploration follows this up with precise investigations and development of the reserves and deposits found. Production can only begin after exploration has demonstrated that sufficient amounts of resources can be extracted.

In ancient times, oils and gas were discovered near natural seeps or leaks to the Earth's surface. Such easy discoveries do not occur in modern days. Various geophysical methods have to be applied in the search for oil and gas reservoirs of commercial value. As discussed in the previous chapter, definitions applicable to petroleum exploration include the following:

- A *Basin* is a large-scale depression on Earth's surface filled with sediments and oils. Basins can change over geological time due to movement of the Earth driven by plate tectonics. The majority of oils in Middle East are marine-sourced. The desert in Saudi Arabia today was under the ocean several hundred million years ago. Similarly, marine fossils are found high in Himalayas. On the other hand, many of today's offshore oils are terrigenous-sourced.

© Springer Nature Switzerland AG 2019
C. S. Hsu and P. R. Robinson, *Petroleum Science and Technology*,
https://doi.org/10.1007/978-3-030-16275-7_6

- A **Play** is a group of oil fields or prospects in the region that are controlled by the same set of geological circumstances with conditions favorable for hydrocarbon accumulation.
- A **Prospect** is an individual exploration target which can be a specific trap that has been identified and mapped but has not been drilled yet.

In the petroleum system described earlier, outcrops provide visible evidence of alternating layers of porous and impermeable rocks. Porous rock (typically sandstone, limestone, or dolomite) provides the reservoir and migration media of petroleum. Impermeable rock (typically clay or shale) acts as a trap to prevent migration.

A book by Charles F Conaway presents a good overview of how we look for and produce petroleum [1]. Commercially significant petroleum systems include the following:

- Organic-rich source rock in which oil and gas were formed.
- Pathways comprised of permeable rock such as sandstone and limestone, or cracks in faults, which allow oil and gas to migrate. Often, the movement of oil and gas is promoted by the flow of underground water.
- Porous, permeable reservoir rock capped by impermeable rock, which prevents fluids from migrating to the surface. Shale and salt domes are common caps.

When petroleum geologists look for oil and gas, they search for geological structures that might serve as traps. In so doing, they rely heavily on reflection seismology. During seismic exploration, explosives or thumper devices send sound waves through the earth. Reflected sound waves are measured with hydrophones (in water) and geophones (on land). Different layers reflect sound in different ways. With the help of sophisticated software, geophysicists transform seismic data into 3-dimensional maps that show the structure of subsurface rock formations.

For petroleum accumulation, there are four major kinds of reservoir traps, three of which (structural traps) are illustrated in Fig. 6.1.

- Anticline traps are, in essence, inverted bowls. They can be symmetrical or asymmetrical. A steep anticline might be called a dome. Anticline traps hold most of the world's conventional crude oil.
- Faults are formed at the boundaries of cracks in the earth's crust. The four major kinds of faults are thrust, lateral, normal and reverse. Thrust and lateral faults are created by horizontal movement. The San Andreas Fault in California is a well-known example of a lateral fault. To be more specific, the San Andreas is a right-lateral, strike-slip fault. Normal and reverse faults are created by vertical movement and are more likely to create fault traps for petroleum. The fault acts as a trap when it pushes an impermeable substance (such as clay or shale) across the fault line through permeable rock, blocking fluid migration.
- A salt-dome trap is created when a mass of underground salt in an evaporite from a former ocean is pushed up by buoyancy and underground pressure into a dome. The salt dome breaks through layers of rock and pushes them aside as it rises. The salt is impermeable, and if it abuts porous rock, it can serve as a reservoir cap.

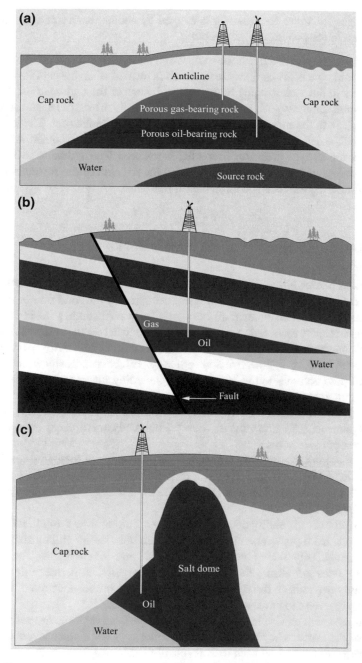

Fig. 6.1 Three types of hydrocarbon traps—anticline (**a**), fault (**b**), and salt-dome (**c**). As with oil and gas, the water layers are not pools, but water-saturated porous rock

• In stratigraphic traps, the reservoir is capped by another reservoir or by layers of rock with lower porosity or permeability.

A fundamental part of a trap is the seal, which prevents hydrocarbons from migrating. The main type is the capillary seal, which is formed when the capillary pressure of the pores in the rock is equal to the buoyant force of the hydrocarbons trying to move upward. Capillary seals are classified based on which type of leak is more likely to form, a leak in the membrane seal or a leak in the hydraulic seal. The membrane seal can leak when the pressure of one side exceeds that of the other. After some oil escapes the trap the pressure differential is restored. The hydraulic seal will leak when the forces of the contained fluid exceed the strength of the surrounding rock, literally fracturing the rock and allowing oil to escape.

6.2 Physical Properties of Reservoir Rocks

Porosity and permeability are the two most important properties of reservoir rocks. Porosity is the fraction of volume that is not occupied by the solid framework of the rock. It may be classified either as absolute or effective. Absolute porosity is the total void volume relative to the bulk volume regardless of whether or not the void volumes are interconnected. Without interconnections, no fluid conductivity (permeability) would occur, leading to low effective porosity. Effective porosity is the ratio of connected pore volumes to the bulk volume, including interconnections. However, not all the interconnected pores may be available for the storage or transmission of oil. Some pores may be filled with water which cannot be replaced by oil, and some oil-filled pores may not allow the oil to move out. Only solution gas drive can force such oil out of these pores.

In oil reservoirs, the porosity represents the percentage of total void space which can be occupied by oil or gas. Hence, it determines the maximum storage capacity of the rock for oil or gas. The porosities of the most frequently found reservoirs are between 10 and 20%.

Permeability is the ability of a rock to allow a fluid to flow through its interconnected pores. Both porosity and permeability are affected by grain size, grain distributions, angularity, the amount of cementation, and consolidation. The presence of porosity does not ensure the rock will have a significant permeability, because permeability depends on the shape, size and degree of interconnection of the pores. However, for a rock to have permeability, it must be porous.

Outcrops which expose bedrock or ancient superficial deposits are visible on the surface of the Earth. They provide visual evidence of alternating layers of porous and impermeable rocks. Porous rock (typically sandstone, limestone, or dolomite) provides the reservoir of petroleum and migration routes for oil and gas while impermeable rock (typically clay or shale) acts as a trap to prevent oil and gas from migration for accumulation.

For oil shale, reservoir rock and source rock are the same because the oil has not migrated. Permeability is practically zero and the exploitation is only possible by mining or hydraulic fracturing.

6.3 Geophysical Methods

Initial attempts to locate petroleum after its initial discovery were limited to above-ground efforts. Areas where petroleum had begun to reach the surface were easy first targets for drilling. In the mid 1800s, it was discovered that certain formations above the ground, such as broad low hills, corresponded to oil reserves. Such features were formed where the Earth had folded, so there was a greater chance of having oil beneath them for the reasons explained in the section on the formation of traps. Still, drilling didn't always find oil, and success rates were very low. Nowadays, above ground techniques are much more sophisticated. Instead of looking at the rock formations from the ground level, aircraft and satellites are used to survey the earth. Aircraft can also measure slight changes in the gravitational force above the earth, allowing geologists to guess at the density of what lies below the surface.

Because drilling to the depths where oil or gas is found is very costly, companies must perform several tests in order to justify drilling in certain prospects. Geophysical methods applied for oil and gas discovery include gravity, seismic, magnetic, electric, electromagnetic and borehole logging (or well logging).

In large scale remote areas, gravity survey methods are often used to determine where a seismic survey should be acquired. The gravity method is a non-invasive, non-destructive remote sensing geophysical technique that measures densities of rocks underground at specific locations for a better understanding of the subsurface geology. The higher the gravity value, the denser is the rock beneath. No energy needs to be put into the ground in order to acquire data.

Seismic exploration is the search for commercially economic subsurface deposits of crude oil, natural gas and minerals by the recording, processing, and interpretation of artificially induced shock waves in the earth. Seismic surveys use the propagation of natural or artificial seismic waves in the subsurface to measure the properties of rocks and fluids and to locate likely rock structures underground in which oil and gas might be found [2]. Seismic waves are generally thought of as having low frequency (in the range of five to eight Hertz) and large amplitude.

Artificial seismic energy is generated on land by vibratory mechanisms mounted on specialized trucks to fire shock waves into the ground. Seismic waves reflect and refract off subsurface rock formations and travel back to acoustic receivers called geophones. It uses the basic relationship between the velocity and travel time (measured in milliseconds) of sound waves to estimate distance (depth) of subsurface structures. A variety of other wave properties is used to characterize rocks and fluids. The travel times of the returned seismic energy, integrated with existing borehole well information, aid geoscientists in estimating the structure (folding and faulting) and stratigraphy (rock type, depositional environment, and fluid content) of subsurface formations, and determine the location of prospective drilling targets by identifying underground structures conducive to oil accumulation.

Although seismic lines are two-dimensional, in other words they represent a cross-section, it is possible these days to collect 3D seismic and 4D seismic data as well, where 3D data represents a volume, and 4D data includes the dimension of time. For

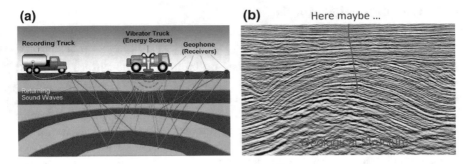

Fig. 6.2 Seismic data acquisition (**a**) and resulting 2D-map for interpretation (**b**)

example, two 3D volumes collected at different points in time would represent 4D data, and it could be used, for example, to look at the effects of steam flooding on a field over a period of time [3]. Fig. 6.2 shows the energy source and collection of seismic data (a) and the resulting 2D map for interpretation (b). The anticline locations can have oil and gas accumulation for targeted exploration. Other geophysical and geochemical measurements are applied to confirm or differentiate other possible accumulations, such as water or inert gas. Magnetic, electric and electromagnetic data, for example, are used for surveys of mineral contents for oil generation potential and depositional environment.

Once an exploratory well is drilled, well logging methods can be performed by lowering a 'logging tool'—a string of one or more instruments - on the end of a wire lowered through the borehole. Logging provides a continuous top-to-bottom record of rock petrophysical properties using a variety of sensors. Logging tools developed over the years measure the resistivity, porosity, natural gamma ray activity, electrical, acoustic, stimulated radioactive responses, electromagnetic behavior, nuclear magnetic resonance, pressure, and other properties of the rocks and their contained fluids for *Formation Evaluation* [4].

A typical well log is shown in Fig. 6.3, where the vertical axis is the depth of the well (the distance from wellhead).

Some of the logs are described below:

- Gravity (density)—a radioactive source of γ-rays (^{40}K, ^{232}Th, ^{238}U) and a Geiger counter as a detector. The degree to which the original radiation is absorbed is a function of rock density. Density values: salt, 2.1–2.2; igneous rocks, 2.5–3.0; sedimentary rock, 1.6–2.8. Gravity methods detect and measure variations in the Earth's gravity field which are associated with the changes of rock density. Some geological structures cause anomalies (disturbances) in normal density distribution. Application of the gravity method consists of calculation of gravity anomalies; that relates to the depth and size of the bodies causing these anomalies.
- Electric Resistivity—the resistivity of a rock layer is a function of its fluid content. Oil and gas filled reservoirs have higher resistivity (lower conductivity) than neighboring formations.

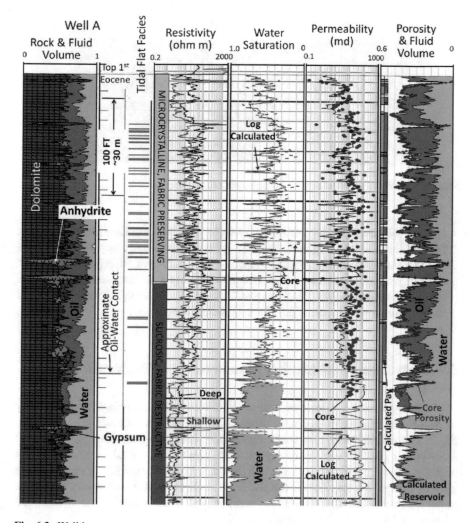

Fig. 6.3 Well log

- Acoustics—similar to surface seismic prospecting. It uses an electric acoustic pulse generator and a receiver separated by an acoustic insulator to define bed and evaluate formation porosity.
- Radioactivity—logging is accomplished by measuring natural γ (gamma) rays and the reflection of neutrons from an artificial source (a mixture of beryllium and radium). Neutron logging determines the relative porosity of the rock formation, and γ-ray measurement is used to define shale.
- Magnetism—identifies specific minerals in a rock formation, especially ferromagnetic minerals.

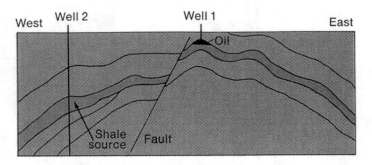

Fig. 6.4 Concept of geochemistry application in exploration [6]

- Nuclear Magnetic Resonance (NMR) Logs—The logs use the Earth's magnetic
  field as the external field and electromagnets for the locally imposed field. It is
  used for the determination of the presence and quantities of different fluids (water,
  oil and gas), pore size distribution, effective porosity and permeability, bound and
  free water saturation, etc. However, NMR logs are less sensitive to matrix lithology
  than conventional (acoustic, density and radioactivity) logs.

6.4 Petroleum Geochemistry for Petroleum Exploration

Petroleum geochemistry deals with the application of chemical principles to the study
of the origin, generation, migration, accumulation, and alteration of petroleum [5].
Fig. 6.4 depicts the basic concept behind geochemistry for oil exploration [6]. If oil
from Well 1 correlates with organic extracts from shale in Well 2, the fault is not
acting as a seal and oil migrates through the fault. If no correlation is found, the fault
may be acting as a trap for oil formed from shale in Well 2, and therefore, it is a
target for further drilling. If no oil is found in Well 2, the extract of the source rock
(shale) can be used for the correlation studies.

   The most important molecules to provide such oil-oil and oil-rock correlations with specificity are petroleum biomarkers, discussed in Sect. 5.6 of the last
chapter. Petroleum biomarkers, although present in trace amounts, are also important
molecules for in-depth geological and geochemical studies to relate the petroleum
with deposition environment and history. Their distribution patterns in crude oils
and source rocks unravel stratigraphic origin, migration pathway, accumulation, and
alteration due to biodegradation or water washing of the existing petroleum deposits
[7]. After all, petroleum is an organic mixture, so the biomarkers can serve as molecular indicators for assessing oil and reservoir quality, especially in collaboration with
geophysical studies and measurements.

6.5 Hydrocarbon Migration [8–10]

There are numerous factors that control the migration processes of hydrocarbons, expansion of kerogens, increase of pressure, and the expulsion out of the source rock. Kerogen generates a hydrocarbon fluids matrix at temperatures of at least 50–70 °C and a minimum depth range of 1000–1500 m. Overpressure creates fracture porosity that allows the oil and gas to escape the impermeable source rock or to be carried by pore water. The expulsion of the oil out of the source rock is also a dynamic process driven by the oil generation itself. A good source rock has a total organic content (TOC) ranging from 3 to 10%. The fluid pressure of the oil within black shales can become high enough to produce microfractures in the rock. Once the micro fractures form, the oil is squeezed out and the source rock collapses. **Primary migration** of oil and gas is movement within the fine-grained portion of the mature source rock. The released petroleum compounds can move through porous carrier rock or faults into a trap. Hence, **secondary migration** is hydrocarbon movement outside the source rock, or movement through fractures within the source rock, shown in Fig. 6.5. Buoyancy of the hydrocarbons occurs because of differences in densities of respective fluids and in response to differential pressures in carrier and reservoir rocks. There is a possibility of **tertiary migration** when petroleum moves from one trap to another or to a seep or even to the surface due to tectonic movement, topographically driven flow, thermal expansion, etc [8]. The migration mostly occurs as one or more separate hydrocarbons phases, liquid or gas, depending on the temperature and pressure conditions. It's also called dismigration [9].

6.6 Petroleum Accumulation

As mentioned, the most essential elements of carrier and reservoir rocks are porosity and permeability [10]. The permeable strata in an oil trap is known as the reservoir rock. The reservoir rock must contain pores or voids to store hydrocarbon fluids, shown in Fig. 6.6. These pores must be interconnected and permeable to fluids and gases. The impermeable cap rocks or salt domes function as traps to stop petroleum migration for accumulation. The majority of petroleum accumulation is found in relatively coarse-grained porous and impermeable clastic rocks, such as sandstones and siltstones, and carbonate rocks. Porosities in reservoir rocks usually range from 5 to 30%. Most of the traps where petroleum is found are formed by tectonic activity, such as faults, folds, salt domes, etc. Fault zones are not always sealed as traps; they might even provide migration avenues. Anticlinal structures are the most common type of traps. Seals are provided by impermeable shales and claystones as cape rocks shaped like inverted bowls, which often appear as hills on the surface. The famous Spindletop well in East Texas was drilled into such a hill. There is always residual water in pores which cannot be displaced by migrating petroleum.

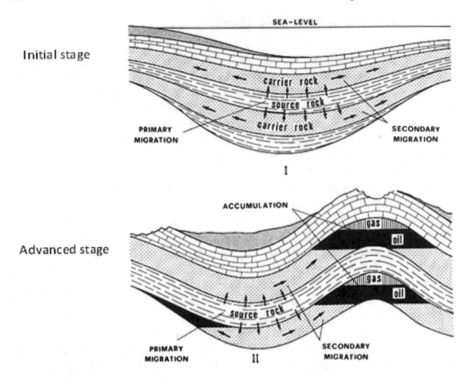

Fig. 6.5 Schematic presentation of *primary* migration and *secondary* migration [10]

Fig. 6.6 Oil and gas in reservoir

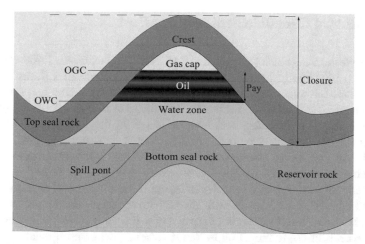

Fig. 6.7   Gas-oil and oil-water contacts in reservoir. OGC:oil-gas contact; OWC: oil-water contact [3]

The process of the filling of reservoir traps and formation of petroleum accumulations often occurs over extended periods of geologic time. At the Earth's surface, the traps are filled from the bottom with oil and gas due to gravity, since the petroleum is lighter in density than the bottom water. The most common traps are the anticline, where petroleum displaces the pore water from the top of the culmination and expands into the flanks. The boundary where the oil-saturated and the water-saturated pore spaces are in contact is referred to as oil/water contact.

Most reservoirs contain both gas and oil. Since the gas has the highest buoyancy, a free gas phase will separate from the oil, accumulating in the apex of the structure as a gas cap. If the source rock is "overmature", more gas is generated than liquid and the reservoir would be predominantly gas. Almost always, there is a large quantity of water in porous rock under the hydrocarbon reservoir, as depicted in Fig. 6.7. The gas above the liquid is under very high pressure. As such pressures, the solubilities of gas-in-oil and oil-in-water are high. During production, safe operation requires controlled degassing of the oil.

To hold the petroleum in the trap, an impermeable cap rock must seal the trap. Some of the best sealing cap rocks are anhydrite or rock salt. The best quality cap rocks hold many of the large petroleum accumulations are in the Middle East. Some petroleum seeps occur where permeable pathways in the form of fractures or faults lead to the surface of the Earth either from mature source rocks or from leaking accumulations.

6.7   Western Canadian Sedimentary Basin

The vast Western Canadian Sedimentary Basin (WCSB) underlies 1,400,000 square kilometres (540,000 square miles) of Western Canada, including southwestern Manitoba, southern Saskatchewan, Alberta, northeastern British Columbia, and the southwest corner of the Northwest Territories (see the map in Fig. 6.8). It consists of a massive wedge of sedimentary rock about 6 km (3.7 miles) thick under the Rocky Mountains in the west, but the layer thins to zero at its eastern margins, shown in Fig. 6.9 [11]. The WCSB contains one of the world's largest reserves of natural gas, estimated at 143 trillion ft^3 with peak production 16 billion ft^3 a day of gas in 2000. It also contains the largest oil sand deposits, mainly in the areas of Athabasca, Cold Lake and Peace River in Alberta, with an estimate of 200 million barrels of recoverable oil. This compares to 298 million barrels of extra heavy oil in Venezuela and 268 million barrels of conventional oil in Saudi Arabia. [12] Conventional oil is being depleted quickly. Oil sands accounted for ~90% of production from the WCSB by 2016. The WCSB also has huge reserves of coal.

The WCSB provides a great opportunity for geochemistry studies. The strata were deposited at different ages, with various degrees of maturity in source rocks. Bitumen on oil sands experienced severe weathering and biodegradation. Thus, the oil, gas and bitumen provide an ideal set of geochemical samples for age, maturity and alteration studies of marine-sourced oils and rocks.

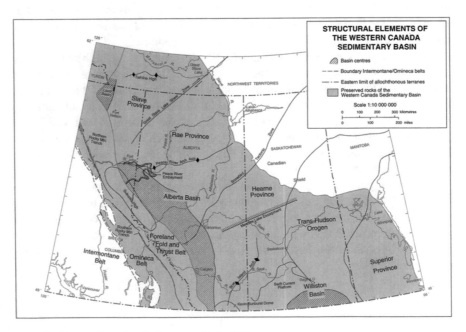

Fig. 6.8 Western Canadian Sedimentary Basin [11]

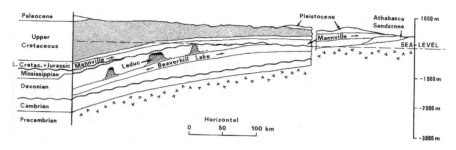

Fig. 6.9 Cross section of Western Canadian Sedimentary Basin from southwest (left) to northeast (right)

References

1. Conaway CR (1999) The petroleum industry: a nontechnical guide, PennWell (Tulsa)
2. Ganssle G (2017) Seismic explorations. In: Hsu CS, Robinson PR (eds) Springer handbook of petroleum technology, Springer, New York
3. Ali HN (2017) Fundamentals of petroleum geology. In: Hsu CS, Robinson PR (eds) Springer handbook of petroleum technology, Springer, New York
4. Hill DH (2017) Formation evaluation, In: Hsu CS, Robinson PR (eds) Springer handbook of petroleum technology, Springer, New York
5. Hunt JM (1979) Petroleum geochemistry and geology. W. H. Freeman and Company, San Francisco
6. Philp RP C&E News, 2/10/86, pp 28–43
7. Hsu CS, Walters CC, Isaksen GH, Schaps ME, Peters KE (2003) Biomarker analysis in petroleum exploration. In: Hsu CS (ed) Analytical advances for hydrocarbon research, Kluwer Academic/Plenum Publishers, New York/Boston/ Dordrecht/London/Moscow
8. PetroWiki: Hydrocarbon Migration: http://petrowiki.org/Hydrocarbon_migration. Retrieved 6 Aug 2016
9. http://wiki.aapg.org/Tertiary_migration. Retrieved 9 Oct 2018
10. Tissot BP, Welte DH (1984) Petroleum formation and occurrence, 2nd edn. Springer-Verlag, Berlin/ Heidelberg/New York/Tokyo
11. Wright GN, McMechan ME, Potter, DEG (2016) Structure and architecture of the western canadian sedimentary basin. http://www.cspg.org/documents/Publications/Atlas/geological/atlas_ 03%20structure_and_architecture_of_the_WCSB.pdf Retrieved 15 Sept 2016
12. http://geab.eu/en/top-10-countries-with-the-worlds-biggest-oil-reserves/. Retrieved 19 Oct 2018

Chapter 7
Production for Recovery

7.1 Drilling for Oil and Gas Recovery

Locating possible oil or gas accumulation is followed by exploratory drilling, during which rock and fluid samples are collected for laboratory testing, and the reservoir structure is further defined. During this process and the subsequent drilling of production wells, measure-while-drilling (MWD) technology analyzes sound waves generated by drill bits to further refine structure maps. Exploratory drilling provides the basis for refined estimates of the volume of reservoirs and their potential economic value. If the refined estimates justify further investment, production wells are drilled into the underground reservoirs.

Each reservoir can have multiple wells to ensure that the extraction rate is economically optimal. Some secondary wells may also be used to pump water, steam, chemicals or gas mixtures into the well to enhance productivity and to raise or maintain reservoir pressure and, consequently, the rate of extraction.

Key components of a drilling rig are shown in Fig. 7.1. To drill a well, a bit is attached to the end of a drill string, which is comprised of sections of steel pipe that are 30-feet (9 m) long. Drilling bits are designed to produce a generally cylindrical hole in the Earth's crust by rotary drilling. Many drill bits are encrusted with natural or man-made diamonds or coated with tungsten carbide steel to increase the hardness and extend the useful life of the bit. The drill string is lengthened as the well gets deeper by attaching additional sections of pipe to the top. Drill collars are thick-walled sections of drill pipe at the bottom of the drill string; collars apply extra weight to the bit. As the bit cuts through rock, a drilling mud is pumped down through the inside of the drill string, out the bit, and up through the space between the drill string and the well bore.

Drilling fluids or drilling muds are designed blend of fluids, solids and chemicals. They aid the drilling of boreholes into the earth in several ways: by providing hydrostatic pressure to prevent formation fluids from entering the well bore, by keeping the drill bit cool and clean during drilling, by controlling downhole pressure, and by carrying out rock cuttings back to the surface. The density of drilling mud is crucial.

© Springer Nature Switzerland AG 2019
C. S. Hsu and P. R. Robinson, *Petroleum Science and Technology*,
https://doi.org/10.1007/978-3-030-16275-7_7

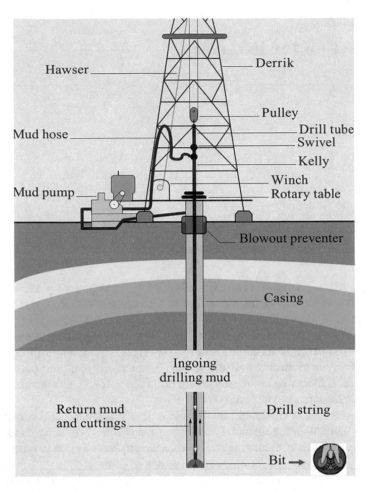

Fig. 7.1 A drilling rig [1]

If the mud density is too low, a well is susceptible to a surface blowout. If the mud density is too high, it can cause an underground blowout—the rupture of the reservoir underground—pushing drilling mud into another formation. The cuttings are removed in shakers, and the fluid goes back to the mud pit, from which it is recycled.

A derrick is used to support drilling apparatus, which must be tall enough to accommodate a 90-foot-long "triple" comprised of three sections of pipe. For offshore drilling, a platform must be built for support of the drilling rig. The drilling system includes the power system (diesel engine and electric generator), the hoisting system, rotating equipment, casing, the circulation system, and the blowout preventers. The derrick, pulleys and hawser must be robust enough to support the lifting and manipulation of the entire drill string, which can weigh hundreds of tons. The kelly is the top joint of a drill string. It has flat sides that fit inside a bushing on the rotary table, which turns the drill string and bit. Note that not all drilling rigs use a kelly system.

As the drill begins tunneling through the ground, chunks of rock and mud are sucked into the pipe by pumps. The chunks are circulated out of the hole and into pits on the surface previously carved by the drilling crew. The derrick is tall enough to allow new sections of pipe to be added as the hole becomes deeper. After a certain depth, depending on the oil reserve, the drill must be pulled out and the well must be cased to prevent the hole from collapsing in on itself. Casing involves placing steel piping into the hole and then pumping cement into the annular space between the outside of the casing and the rock. Centralizers are sections of pipe which keep the casing from resting against the surrounding wellbore wall, as in Fig. 7.2. Once this section of the hole is deemed secure enough, the crew continues to drill deeper with drill pipe of ever narrower diameters. This cycle of drilling, casing and cementing is done until the final pre-determined depth is reached within the reservoir rock.

Throughout the drilling process, the location and direction of the drill bit is determined. Well logs are evaluated and cuttings are analyzed, and if necessary the drilling plan is adjusted accordingly. When drilling is complete and tests have confirmed that

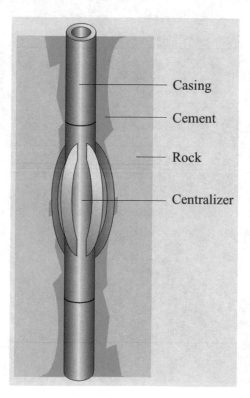

Fig. 7.2 Casing with bow-string centralizers that allow cement to flow through all empty annular space between casing and well bore

the location is correct, preparations to extract the oil commence. A perforating gun with explosive charges is dropped into the well to perforate the casing to allow the flow of oil up the pipe.

7.2   Well Completion

Wells are completed by casing the well bore with steel pipe and cementing the casing into place. Casing prevents the well from collapsing. The outside diameters of casing pipe range from 4.5 to 16 in. (114–406 mm). Cementing is a key step. In a good cement job, the entire annular space between the casing and the well bore is filled with cement. Centralizers keep the casing from resting against the well bore to

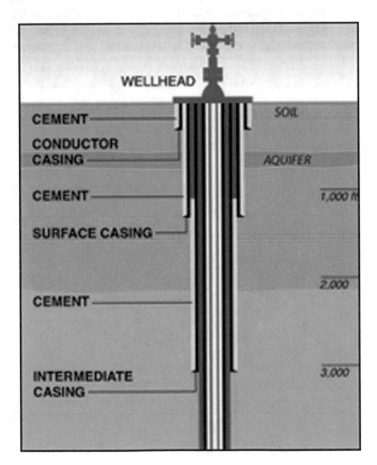

Fig. 7.3 Well completion with cement

block the flow of cement, resulting in a poor cement job and increasing the risk of a blowout. Figure 7.3 shows well completion with cementing.

The integrity of cementing is tested by a positive-pressure test and a negative-pressure test. In the positive-pressure test, the pressure in the steel casing is increased to see if it is intact along with seal assembly. In the negative-pressure test, the pressure is reduced to below atmospheric to test the integrity of the cement at the bottom of the hole. The test is deemed successful if the pressure remains low after the suction is stopped.

A key consideration during completion is the weight of the well casing, which rests on the bottom of the well bore, creating friction that limits the distance/depth (D/D) ratio. Long, shallow wells are especially susceptible to D/D limitations. In extended-reach horizontal drilling (ERHD) technology, [2] a special tool at the bottom of the casing allows the casing to be filled with air instead of mud. This reduces the friction substantially, allowing much higher D/D ratios. The technology is employed in certain wells at Wytch Farm, England. On Platform Irene offshore California, ERHD enabled Unocal to set records, both for horizontal reach (14,671 ft), widest pay zone (5990 ft) and greatest angle of deviation from vertical (76°) [3]. Those records have since been broken, and the technology has advanced, but the invention by Mueller, et al. [2] was a significant first step. After cementing, perforation guns punch small holes through the casing into the reservoir rock, providing a path for the flow of oil and gas into the well. In open-hole completion, the last section of the well is uncased. Instead, the installation of a gravel pack stabilizes the casing and allows fluids to enter the well at the bottom. After perforation, special acid-containing fluids are pumped into the well to increase porosity and stimulate production. Usually, a smaller diameter tube is inserted into the casing above the production zone and packed into place. This provides an additional barrier to hydrocarbon leaks, raises the velocity at which oil flows under a given pressure, and shields the outer casing from corrosive well fluids.

A blowout preventer (BOP), shown in Fig. 7.4, seals the high-pressure drill lines and relieves a sudden increase in well pressure into the atmosphere for uncontrolled gush or gas. The BOP is a collection of safety valves and other devices at the top of a well. It is located on the surface for an onshore well. It can be placed beneath the ocean on the sea floor for an offshore well. When activated, it stops a blowout by sealing off the top of the well.

Underground blowouts are the most common of all well control problems. Many surface blowouts begin as underground blowouts. Prompt, correct reaction to an underground event can prevent a dangerous and costly surface blowout [5]. The severity of the BP Deepwater Horizon oil spill in 2010 was aggravated by the failure of a blowout preventer.

Fig. 7.4   Blowout preventer (BOP) [4]

7.3   Directional Drilling/Horizontal Drilling

Conventional vertical wells are drilled straight down into the earth. Descriptions of such wells in China, drilled with bamboo poles, date back to 1500 BC. Vertically drilled wells are only able to access the targeted gas and/or oil that immediately below the well. Using horizontal or directional drilling techniques, shown in Fig. 7.5, a number of wells can be drilled in different directions from a single well pad, which is much more efficient than having numerous vertical well pads set up to extract oil or gas. This decreases the surface disturbance and reduces the overall cost of well pad setups, replacements and maintenance. This is especially beneficial offshore, because it can decrease the required number of expensive platforms by an order of magnitude. Tremendous savings are realized when offshore oil can be reached from onshore drilling sites, as is the case for many wells in the huge Wytch Farm oil field in the Purbeck District of Dorset, England.

Horizontal drilling technology achieved commercial viability during the late 1980s. Two key components in directional/horizontal drilling plays are mud motors and measurement while drilling (MWD) sensors. Mud motors can rotate the drill

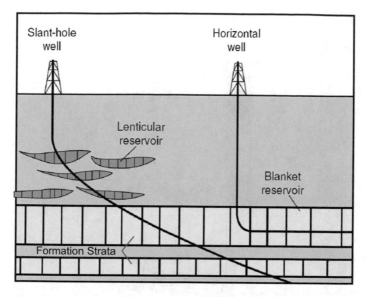

Fig. 7.5 Directional drilling (left) and horizontal drilling (right) wells [6]

bit without rotating the entire length of drill pipe between the bit and the surface. Figure 7.6 shows bent sections of pipe (bent subs) which compel the bit to follow a path that deviates from the previous orientation. MWD sensors are used determine the azimuth and orientation of the bit. A rotary steerable system (RSS) employs specialized downhole equipment to replace conventional directional tools such as mud motors, allowing operators to steer the drill bit in real time.

In the past few years, directional drilling combined with hydraulic fracturing has been applied to tight rock formations, resulting fantastic production of oil and/or natural gas from reservoirs which were otherwise unproductive. Major examples are the Eagle Ford Formation near Three Rivers, Texas, the Bakken Formation of North Dakota, and the Marcellus Shale of the Appalachian Basin. Due to this technology, the United States is now among the top oil producers in the world.

The availability of directional/horizontal drilling also stimulates the development of in situ toe-to-heel air injection (THAI) combustion and steam-assisted gravity drainage (SAGD) production of heavy oil and bitumen to be discussed later.

7.4 Offshore Drilling

The term of "offshore drilling" refers to drilling activities on the continental shelf, although the term can also be applied to drilling in lakes, inshore waters and inland seas. In offshore drilling, a well is drilled below the seabed to explore for and sub-

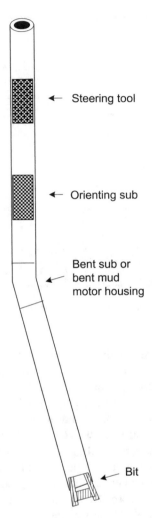

Fig. 7.6 Arrangement of a steering tool, orienting sub, and bent sub for directional drilling

sequently extract petroleum which lies in undersea rock formations. Offshore wells can be drilled from onshore or from platforms.

For offshore wells, two depths are important: the depth of the water, and the distance from the top of the sea bed to the pay zone [7]. Both must be considered when selecting a platform design. Another consideration is metocean, i.e., meteorology and oceanography, involving the quantification of winds, waves, currents and related physical phenomena in the ocean and atmosphere. United Nations Convention on Law of the Sea is used for the determination how far we can drill offshore from the coastline. Figure 7.7 shows the main offshore oil production region in the world as of 2012 [7].

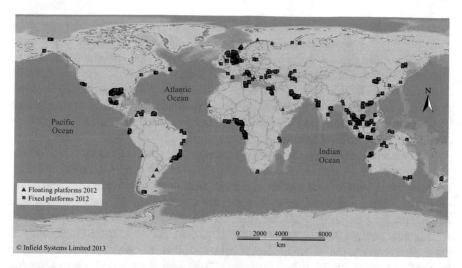

Fig. 7.7 Main offshore oil production regions in the world as of 2012 [7]

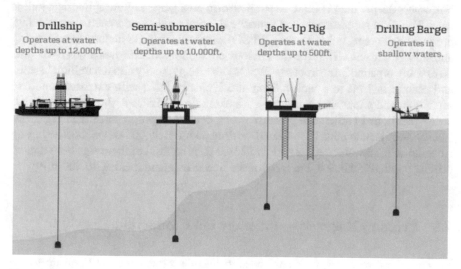

Fig. 7.8 Mobile offshore drilling units

As of 2010, offshore drilling as deep as 3 km below water was possible. The equipment that does this is called a mobile offshore drilling unit (MODU). There are four main types of MODUs in use today for exploration: submersibles, jack ups, drill ships, and semisubmersibles, shown in Fig. 7.8.

In shallow waters, jack-ups and submersible rigs usually are used. Submersible MODUs rest about 30 ft below the water on the sea floor. They are connected through steel posts to drilling platforms on the surface. A jack up sits on top of a floating barge. The rig's legs extend as far as 550 ft to the ocean floor as sort of stilts to raise the rig above the surface and keep it safe from choppy waters; an example is shown in Fig. 7.8. Semisubmersible rigs and drill ships are used in deeper waters. Semisubmersible rigs simply let in enough water to lower the platform to appropriate operating heights. The weight of the lower hull stabilizes the drilling platform, while massive anchors hold it in place. Drill ships that have a drilling rig on the top deck use dynamic positioning equipment to maintain alignment with the drilling site, guided by satellite information and underwater sensors on the subsea drilling template to keep track of the drilling location.

The major production structures and systems are summarized in Fig. 7.9 [4]. Fixed platforms or jackets are used for water less than 1500 ft deep with mild climates. Locations suitable for such platforms are not only constrained by water depth, but also by the wind and wave strength. Compliant towers are designed to sway and move with the stresses of wind and sea. Sea Star platforms are a larger version of submersible designs and are connected to the ocean floor by tension legs. Both of these platforms operate at water depths up to 3500 ft. For deeper waters, floating production systems, tension leg platforms (TLP), and subsea systems transfer the oil and natural gas to production facilities, either by risers or undersea pipelines. For waters deeper than 7000 ft, a spar platform on a giant, hollow cylindrical hull is used. The most sophisticated and versatile floating production systems are floating, production, storage and offloading (FPSO) platforms. FPSOs include an offloading and storage capacity, They are the obvious choice for stranded fields where there is no existing pipeline infrastructure. FPSOs store oil, typically about a million barrels, and offload that oil to a tanker. They are typically fitted with extensive onboard separation and processing equipment. An example of FPSOs is shown in Fig. 7.10.

As mentioned earlier, the greatest *water* depth a jackup can drill in is 550 ft. Many newer jackup units have a rated *drilling* depth of 35,000 ft. On the floating rig side, the deepest water depth so far is 12,000 ft. A handful of these rigs have a rated drilling depth of 50,000 ft, but most of the newer units are rated at 40,000 ft [9].

7.5   Primary Recovery—Primary Oil Production

In primary oil recovery, fluids are pushed into the production well and up to the surface by natural forces: gas drive and water drive. Gas drive is the most efficient. Oil is pushed by natural underground pressure, usually supplied by associated natural gas, liquid expansion and evolution of dissolved gas, shown in Fig. 7.11. Once the wellbore is drilled, the free gas begins to expand. The expansion energy of the gas is what allows it to rise out of the reservoir and travel to the surface. The gas-oil contact plane drops as the oil is depleted. The next most efficient propulsive force is natural water drive, in which the oil is driven upward under hydrostatic pressure and into the well by expansion of water inside the reservoir, also shown in Fig. 7.11. Below the natural resource is an aquifer. The water that drives this kind of recovery can be either located beneath the natural resource or on the edges of the reservoir in which the resource is contained. Once the wellbore is drilled into the reservoir, water

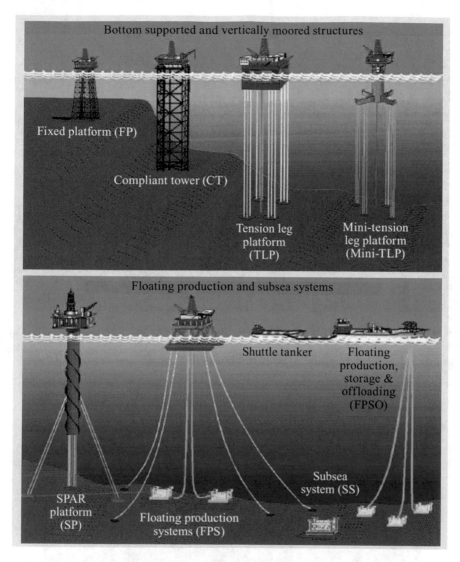

Fig. 7.9 Major offshore production structures and systems [7]

in the aquifer begins to push the hydrocarbons to the surface until they have been completely displaced or until the point at which so much water has accumulated in the well that it is no longer a quality resource.

The efficiency for primary recovery is generally low. Primary recovery typically recovers 5–15% of oil in the reservoir due mostly to microscopic trapping and bypassing of the remaining oil [10].

Fig. 7.10 A colossal offshore platform lights up the night off the coast of Norway [8]

Oil producing well

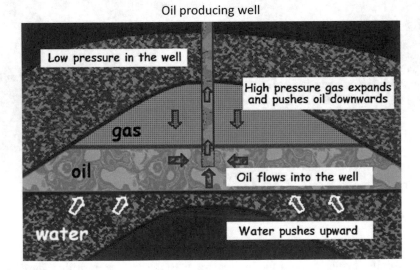

Fig. 7.11 Primary oil recoveries by gas drive and water drive

7.6 Secondary Recovery

In primary recovery, the natural force used to drive the resource to the surface will eventually decrease and become not sufficient. To compensate, an artificial lift system using mechanical energy may be implemented to aid in the recovery of the natural resource in a more economical fashion. This mechanism is secondary recovery. Horsehead pumps or sucker-rod pumps are common surface implements. Submerged pumps also are used. As with primary recovery, a single production well can be used for artificial lift.

Water injection (water flood) or gas injection (gas flood) stimulate production by increasing reservoir pressure or displacement of oil towards the production wells. At least two wells are used: an injection well and one or more production wells. Typically, the gas is reinjected natural gas. Another method is gas lift in which compressed air, water vapor, carbon dioxide or some other gas is injected into the bottom of an active well, reducing the overall density of fluid in the wellbore. Gas injection delivers a necessary amount of compressed gas to a distribution cap, pushing the oil out of the reservoir. Water injection involves pumping water into the reservoir to displace the natural resource to an adjacent production well.

Figure 7.12 generically illustrates the use of a combination of injection and production wells to stimulate production, via both secondary and tertiary recovery. On average, primary and secondary recovery methods combined allow 20–40% of the reservoir oils to be recovered [11].

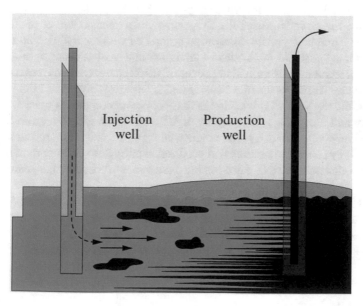

Fig. 7.12 Generic depiction of injection and production wells for secondary and tertiary recovery

7.7 Tertiary Recovery (Enhanced Oil Recovery)

Tertiary recovery, also known as enhanced oil recovery (EOR), is used to extract the remaining oil from reservoirs where primary and secondary recoveries are no longer cost effective. Many sandstone or carbonate reservoirs have low primary and secondary recovery due to poor sweep efficiency for bypassed or unswept oil. EOR may also be used to stimulate production from reservoirs containing very viscous crude oils and from low-permeability carbonate reservoirs. It is designed to reduce viscosity of the crude oil. According to the US Department of Energy (DOE), there are three primary techniques for EOR: *gas injection*, *chemical injection* and *thermal recovery*. Using EOR, 30–60%, or more, of the formation and reservoir's original oil can be extracted, compared with 20–40% using primary and secondary recovery.

7.7.1 Gas Injection

Miscible flooding is considered one of the most effective enhanced oil recovery processes applicable to light-to-medium oil reservoirs. This method can yield as much as 17% of a field's original oil-in-place. It is accomplished with hydrocarbon solvents or gases such as carbon dioxide, natural gas, or liquefied petroleum gas (LPG). The injected fluids are capable of displacing crude oil for recovery from the reservoir rock. Supercritical CO_2 has a viscosity similar to hydrocarbon miscible solvents. Both improve volumetric sweep-out when there is an unfavorable viscosity ratio in the reservoir. However, the CO_2 density is similar to that of oil. Therefore, CO_2 floods minimize gravity segregation compared with the hydrocarbon solvents.

Introducing miscible gases reduces the interfacial tension between oil and water, maintains reservoir pressure, and improves oil displacement. CO_2 is most commonly used, because it reduces the oil viscosity and is less expensive than LPG.

One difficulty with CO_2 injection is that petroleum companies must first secure, transport, and store an adequate supply of CO_2. However, CO_2 is a green-house gas produced in large volumes by many industrial factories. CO_2 capture and sequestration (CCS) by using it to stimulate oil production mitigates two important problems. Certain producers inject CO_2 injection into coal mines to produce coal-seam methane [12].

Not all locations and reservoirs are appropriate for this technique; Suitability depends on geology and fluid characteristics. Nitrogen gas can be used in combination with CO_2 flooding when complete CO_2 flooding is not economical. Nitrogen and flue gas are lower in cost and have shown success in re-pressuring reservoirs.

7.7.2 Chemical Injection

There are many chemicals that can improve oil mobility by reducing its viscosity and/or reducing the intermolecular interactions which hold oil onto rock. Chemical methods include polymer flooding, surfactant flooding and alkaline flooding. In all of the chemical injection methods, the chemicals are injected into several wells (injection wells) and the production occurs in other nearby wells (production wells).

7.7.2.1 Polymer Flooding

Polymer (augmented) flooding consists in mixing a dilute solution of a water-soluble polymer, such as polyacrylamides (PAM), partially hydrolyzed polyacrylamide (HPAM), or xanthan gum (a polysaccharide biopolymer), to increase the viscosity of the injected water and improve the mobility ratio between the injected and in-place fluids. The polymer solution affects the relative flow rates of oil and water and sweeps a larger fraction of the reservoir than water alone, thus increasing the amount of oil recovered in some formations. This method is suitable for heterogeneous reservoirs (fractures, layers with large permeability contrasts, impermeable layers) where the oil's viscosity is too high for normal extraction.

7.7.2.2 Surfactant (Micellar) Flooding

Surfactant flooding recovers additional oil left behind by waterflooding. Surfactants may be used in conjunction with polymers. Much of the remaining oil is trapped in capillaries as microscopic oil droplets, which usually constitute more than half the residual oil. By the injection of surfactant solution, the oil is mobilized through a large reduction in the interfacial tension between oil and water. The decrease in oil/water surface tension reduces residual oil saturation and improves the macroscopic efficiency of the process.

Dilute solutions of surfactants such as sulfonates or biosurfactants such as *rhamnolipids* may be injected. Primary surfactants usually are augmented with co-surfactants, activity boosters, and co-solvents to improve stability of the formulation. Figure 7.13 depicts a two-step enhanced oil recovery process, in which a surfactant slug is injected into the well followed by a larger slug of water containing a high-molecular-weight polymer, which acts like a piston to propel the chemicals.

7.7.2.3 Caustic Flooding (Alkaline Flooding)

Caustic flooding, also called alkaline flooding, is accomplished by adding sodium hydroxide to injection water. Alkaline flooding is a very complex process. It improves oil recovery by using in situ surfactants (soap) produced from the reaction of alkali

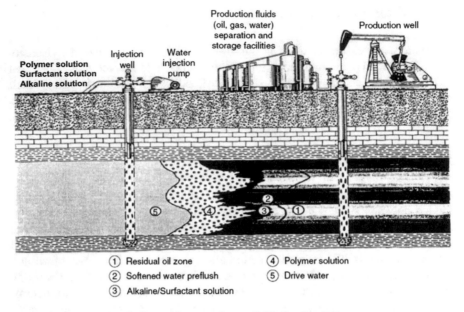

Fig. 7.13 Recovery by alkaline-surfactant-polymer (ASP) flooding [12]

and the organic acids naturally occurring in the oil. It enhances oil recovery by lowering the interfacial tension, decreasing the rock wettability, emulsification of the oil, mobilization of the oil, and helping to draw the oil out of the rock.

Caustic flooding is usually accompanied in conjunction with surfactant and polymer flooding. The combination is called alkaline-surfactant-polymer (ASP) flooding, shown in Fig. 7.13, in which the three slugs are used in sequence. Alternatively, the three fluids could be mixed together and injected as a single slug. The objective of the ASP flooding process is to reduce the chemical consumption per unit volume of oil, resulting in a reduction in cost.

7.7.3 Microbial Stimulation

Microbial injection is part of microbial enhanced oil recovery (MEOR) and is rarely used because of its higher cost and because the method is not widely accepted. These microbes function either by partially digesting long hydrocarbon molecules, by generating biosurfactants, or by emitting carbon dioxide which can be used in gas injection.

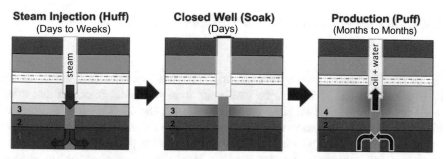

1. Heavy crude oil, 2. Heated zone, 3. Condensed steam zone, 4. Flowing oil and condensed steam

Fig. 7.14 Huff and puff cyclic steam injection [14]

7.8 Thermal Recovery

Various methods are used to heat the crude oil in the formation to reduce its viscosity and surface tension, thus, increasing the permeability of the oil. The heated oil may also vaporize and then condense forming improved oil. Methods include cyclic steam injection (Huff and Puff), steam flooding, and combustion.

7.8.1 Cyclic Steam Stimulation (Huff and Puff)

Cyclic steam stimulation (CSS) is also known as the Huff and Puff method, shown in Fig. 7.14. It requires only one wellbore and consists of 3 stages: injection (Huff), soaking, and production (Puff). First, steam at elevated pressure is injected into a well at a temperature of 300–340 °C for a period of weeks to months. Next, the well is allowed to sit for days to weeks to allow heat from the elevated pressure steam to soak into the formation and reduce the viscosity of the oil around the well. Finally, the hot oil is pumped out of the well for a period of weeks or months. Once the production rate falls off, the well is put through another cycle of injection, soaking, and production. The process is used for thinner shallow production near bitumen reservoirs. High reservoir porosity and oil saturation is ideal for this method of recovery. This process is repeated until the cost of injecting steam becomes higher than the money made from producing oil [13]. This process can typically remove about 25% of the total reserve.

Steam degrades (weakens) the formation by removing alkali- and alkaline-earth ions from reservoir constituents, such as particulates, thereby causing the formation to swell. This decreases permeability and slows down production. Watkins and Kalfayian [15] invented methods for decreasing swelling by injecting ammonium salts and ammonia precursors with the steam, replacing the removed K^+, Mg^{2+}, etc., with NH_4^+.

7.8.2   Steam Flooding (Steam Stimulation or Steam Drive Injection)

Steam flooding is also known as steam drive injection or steam stimulation, shown in Fig. 7.15. In this method, some wells are used for steam injection and other wells for oil recovery. Two steps are involved: first, the oil is heated by steam to higher temperatures to decrease its viscosity so that it more easily flows through the formation toward the producing wells; second, the oil is pushed to the production wells in a similar manner as water flooding. More steam is needed for this method than for the cyclic method.

7.8.3   In Situ Combustion (Toe-to-Heel Air Injection)

The steam injection methods mentioned above are applicable for oils with viscosities in the range of 100–10,000 cP. Heat provided by the steam reduces the viscosity of the oil, thereby making it mobile. For heavy oils and bitumen with viscosities greater than 10,000 cP, combustion becomes the method of choice.

An effective and revolutionary in situ combustion method for producing heavy oil is THAI (toe-to-heel air injection) [16], shown in Fig. 7.16. THAI combines a

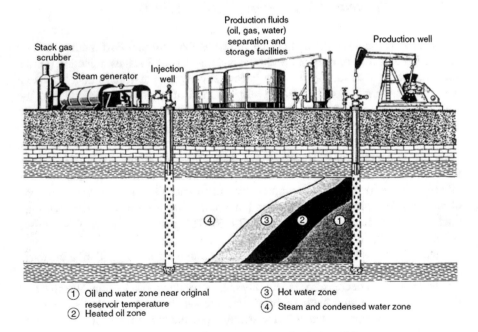

Fig. 7.15 Steam drive injection (steam flooding or steam stimulation) [12]

vertical air injection well with a horizontal production well. THAI is also applicable
to horizontal wells, which due to their geometry are not suitable for steam injection
methods.

For the first few months, steam is injected in the vertical well to preheat the horizontal well and condition the reservoir around the vertical well. Then air is injected
in the vertical well and combustion initiated. The combustion raises temperatures
to approximately 400–600 °C (750–1110 °F). At these temperatures, both thermal
cracking and coking occurs. In this process, about 10% of the oil is lost to coke. The
oil from thermal cracking is of higher quality than reservoir oil. The mobilized oil
flows by gravity as a **toe** to the horizontal section of the L-shape production well.
The combustion front sweeps the oil from the toe to the **heel** of the horizontal producing well, recovering the original oil-in-place while partially upgrading the crude
oil in situ. The combustion gasses bring the mobilized oil and vaporized water to the
surface, so no pumps are needed.

Another good feature of this process is the minimal amount of water supply
needed. Once the first few months of steam injection has been completed, no more
water or even natural gas is used. Combustion continues as long as air is injected.
With less equipment also comes a smaller footprint at the surface and less land is
required. The combustion is also self-limiting, so as soon as air injection is stopped
the flames burn out. With this method, the area at and near the combustion zone will
turn into coke Eventually, the reservoir no longer retains any oil or natural gas.

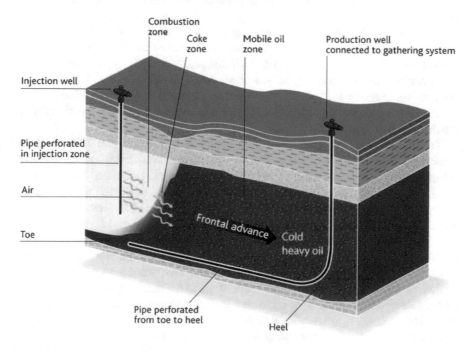

Fig. 7.16 Toe-to-heel air injection (THAI) for oil recovery [17]

7.9 Recovery of Heavy Oils and Bitumen

Heavy crude oil or extra heavy crude oil is highly viscous, and cannot easily flow under normal reservoir conditions. Heavy crude oil has been defined as any liquid petroleum with an API gravity less than 20°. In 2010, the World Energy Council defined extra heavy oil as crude oil having a gravity of less than 10° and a reservoir viscosity of no more than 10,000 cP. Compared to the lighter crude oils, heavy crude oils have higher viscosity and specific gravity, as well as heavier molecular composition with significant contents of nitrogen, oxygen, and sulfur compounds and heavy-metal contaminants.

Natural bitumen, including tar sands or oil sands, shares the attributes of heavy oil but is yet more dense and viscous, having a viscosity greater than 10,000 cP. Natural bitumen and heavy oil resemble the resids from the refining of light crude oil. They are thought to be the residue of formerly light oil that has lost its light-molecular-weight components through degradation by bacteria, water-washing, and evaporation (weathering). Conventional heavy oils and bitumens differ in the degree by which they have been degraded from the original crude oil.

According to World Resources Institute, remarkable quantities of heavy oil and oil sands are found in Canada and Venezuela. The largest reserves of heavy crude oil in the world were located north of the Orinoco Basin in eastern Venezuela. It was estimated that there were 270 billion barrels of recoverable heavy or extra-heavy oil reserves in the area, similar to the amount of conventional oil reserves in Saudi Arabia.

Heavy oil and bitumen are recovered in several ways. They can be dug out with conventional mining techniques, or they can be liquefied by the injection of high-pressure steam, for example in cyclic "huff and puff" operations (see above).

At surface facilities, notably in Venezuela and Canada, recovered heavy oil and bitumen are diluted with lighter hydrocarbons. The resulting "Dilbit" (diluted bitumen) flows under ambient conditions, so it can be transported conventionally in pipelines and oil tankers.

Kerogen is recovered from oil shale by several methods. From 1985 to 1990, Unocal recovered some 4.6 million barrels of synthetic crude oil from oil shale in a complex mining and upgrading venture at Parachute Creek, Colorado [18]. The plant yielded roughly 40 gallons of oil per ton of rock. In the vertical-shaft retort, crushed kerogen-containing shale was pumped up from the bottom of the retort vessel. Hot recycle gas flowed counter-currently downward, decomposing the rock and releasing hydrocarbons. Condensed shale oil was removed from the retort at the bottom. Part of the hot gas was recycled. The rest either was used to produce heat and hydrogen, or recovered as product. The spent shale was removed from the top of the retort, cooled, and stored in pits or returned to the mine. In the reducing environment of the retort, sulfur and nitrogen were converted to H_2S and NH_3, which were recovered from product gases by conventional means. The plant yielded high-quality synthetic crude oil suitable for further refining in conventional facilities.

More discussions on diluted bitumen and synthetic crude oil for transportation can be found in Chap. 17, which covers mid-stream operations.

Other oil shale processes involve partial combustion, either underground, at the surface, or in shafts drilled horizontally into kerogen-rich formations. From 1972 to 1991, Occidental Petroleum developed an in situ process, in which explosives were used to create underground chambers of fractured oil shale. The oil shale was ignited with external fuel, and air and steam were injected to control combustion. The hot rock fractured and released shale oil, which was pumped to the surface from a separation sump and collecting well.

Bitumens derived from oil shale and many tar sands contain small but significant amounts of arsenic, which are severe poisons for catalysts in refineries.

7.9.1 Canadian Oil Sand (Tar Sand) and Bitumen

As mentioned earlier, the Western Canadian Sedimentary Basin contains world's largest reserve of oil sands. Total natural bitumen reserves in the world are estimated at 249.67 billion barrels, with 70.8% in Canada. The largest fields are located in northern Alberta, Athabasca, Cold Lake and Peace River, shown in Fig. 7.17.

The oil sands are loose sand or partially consolidated sandstone containing naturally occurring mixtures of sand, clay, and water, saturated with a dense and extremely viscous form of bitumen. Bitumen is so heavy and viscous (thick) that it will not flow unless heated or diluted with lighter hydrocarbons. The viscosity of bitumens is greater than 10,000 centipoises under reservoir conditions and their API gravities are less than 10° API.

7.9.2 Recovery of Oil Sand Bitumen

Oil sands are recovered using two main methods: open-pit mining and in situ drilling. The method depends on how deep the reserves are deposited. In Alberta, 97% of the total surface area of the oil sands region could be developed in situ.

7.9.2.1 Open Pit Mining

Approximately 20% of the oil sands lie close enough to the earth's surface to be mined, which impacts 3% of the surface area of the oil sands region.

Open-pit mining is similar to many coal-mining operations. Large shovels scoop the oil sands into trucks, which take it to crushers, where the large clumps are broken down. The oil sand is then mixed with water and transported by pipeline to a separation plant. A combination of very hot water, agitation, and other processing methods are required to increase bitumen separation to about 75%. The resulting bitumen is

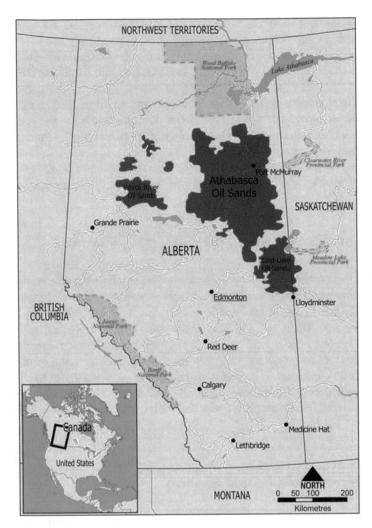

Fig. 7.17 Oil sand fields in northern Alberta, Canada [19]

not very usable, though, and has to be upgraded. Roughly two tons of sand must be processed to produce just one barrel of useable oil, so the return on investment for this method is extremely low; not to mention all of the negative environmental impacts. Compared to other methods, it has the largest land-use footprint, produces the most greenhouse gases. It negatively affects surrounding wildlife, and pollutes a vast amount of water. So companies generally try to avoid this if possible.

Tailings ponds are an operating facility common to all types of surface mining. In open-pit oil sands processing, tailings consisting of water, sand, clay and residual oil are pumped to these ponds, where settling occurs. According to industry sources, 78–86% of the water is recovered and reused [20]. Supposedly, when the ponds are no longer required, the land will be reclaimed. Reclamation could be troublesome, because the tailings contain significant amounts of arsenic in concentrations high enough to be toxic but too low to be of any commercially value.

7.9.2.2 Cold Heavy Oil Production with Sand (CHOPS)

One of the simpler methods of heavy oil extraction is the Cold Heavy Oil Production with Sand method or CHOPS. This method is simpler because it does not require sand filtration, allowing sand and oil to be extracted together. The method works because removing sand creates more space within the reservoir, allowing larger pockets of liquid oil to form. It's called "Cold" because there no heat is injected into the reservoir, so it's also relatively energy efficient. This method isn't super effective and can only recover about 5–6% of the oil within the deposit, but since there is no heat injection it is pretty cheap to employ. The disposition of the oily sand, after it has been separated from the extracted liquid, is challenging. Some is used in the construction of asphalt roads, but there are some problems with that too. Most is currently deposited in underground salt caverns.

7.9.2.3 In Situ Drilling/Steam Assisted Gravity Drainage (SAGD) [21]

The majority of the oil sands lie more than 70 m (200 ft) below the ground and are too deep to be mined. These reserves can be recovered through wells with thermal stimulation, as discussed above.

A form of steam flooding that has become popular since 1996 in the Alberta oil sands is steam assisted gravity drainage (SAGD), in which two horizontal wells are drilled, one a few meters above the other. Steam is injected into the upper well as shown Fig. 7.18. The steam injection can also be performed by multiple vertical injection wells above the horizontal production well. The steam increases the temperature of the oil sand around it. The viscosity of the oil sand decreases as the temperature increases. The oil sand and condensed steam trickle down due to gravity. The condensed steam and less viscous bitumen from the oil sands flow into the lower horizontal production well. Pressure pushes the bitumen and water into the storage tanks above ground. The bitumen-water mixture is separated into bitumen and water. The bitumen goes off to be refined while part of the water is recovered and recycled within the process. SAGD is a unique process that has taken precedence, due to its higher recovery, lower cost, and greater efficiency. The development of SAGD is fairly new in comparison to other methods of oil recovery. The progress of drilling techniques overlapped with the development of SAGD, so the use of horizontal wells became less expensive and more efficient. SAGD recovery ranges from 50 to 70% of oil-in-place.

In SAGD, some of the lighter contaminants will rise with the steam and won't have to be separated later. This doesn't happen with a lot of other heavy oil extraction methods. However, it has some drawbacks, one being just the cost of generating the

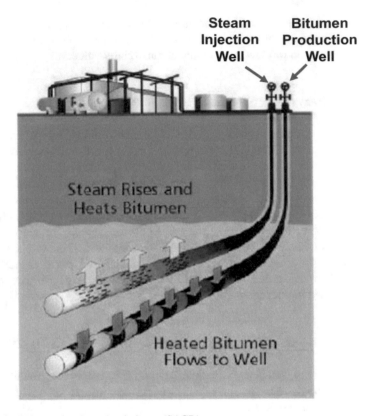

Fig. 7.18 Steam assisted gravity drainage (SAGD)

large amount of steam required. Variations on this method employ solvents instead of steam to increase the liquidity of the oil, making the process more energy efficient. Another drawback is high water use. There is a concern that it relies too much on nearby water supplies and will deplete lakes and river streams. There have been instances of contamination due to leaking wells. Treating the contaminated water is costly and not always done [22].

7.10 Hydraulic Fracturing (Fracking) [23]

Natural gas is available as associated gas in crude oil reservoirs and as non-associated gas in natural gas reservoirs and in the form of condensates. Recently, unconventional sources, such as shale gas, tight gas and coal-bed methane, are commercially exploited at large scales [24]. Unconventional resources are defined as those that cannot be produced commercially without altering rock permeability or fluid viscosity.

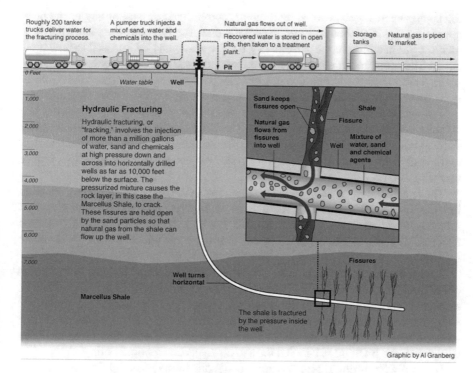

Fig. 7.19 Hydraulic fracturing (fracking) [25]

Tight gas refers to natural gas reservoirs produced from reservoir rocks with such low permeability, having less than 0.1 millidarcy (mD) matrix permeability and less than 10% matrix porosity, that considerable hydraulic fracturing is required to harvest the well at economic rates. These reservoirs do not have depth constraints. They can be developed deep or shallow, at high or low pressure and temperature, in stacked multilayered or single layer configurations, and in homogeneous or naturally fractured formations.

It has been estimated that shale contains as much as 30% of today's oil reserves. Oil and gas are trapped in fine fissures and cannot be recovered conventionally. Hydraulic fracturing (Fig. 7.19), a well stimulation technique also known as fracking, is a process used in nine out of 10 natural gas wells in the United States. High quality crude oils are also produced. Fracking involves injecting a specially designed fluid under controlled pressure intermittently over a short period (three to five days) to create fractures in a targeted deep-rock formation. The fractures permit oil, natural gas and brine to flow to the wellbore.

A well drilled for hydro-fracking goes vertically down until it hits shale. It then is drilled horizontally through the shale for up to 5000 ft. The next step is injection of fluid, which has a wide variety of potential compositions. The "fracking fluid" is typically a slurry of water (90%), proppant (9.5%) and chemical additives, such as thickening agents. gels, foams, light diesel, compressed gasses, or any of the other approximately 750 chemical additives registered for this purpose. The goal is to increase permeability of the surrounding rock to allow any oil released to flow more easily. The proppants are small grains of sand, treated like resin-coated sand, man-made ceramic beads, aluminum oxide or other particulates used to hold the fractures open when the hydraulic pressure is removed from the well to recover gas and oil.

The actual fracturing part, though, is caused by extremely high pressure and velocity, sometimes as much as 15,000 psi and 265 L/min. The resulting cracks free fluid hydrocarbons, including gases, light oils, and heavy oils, allowing them to flow more easily into the well. Typically, the additives decrease surface tension, reduce viscosity, and prevent emulsions.

The injected fluids are somewhat recovered with the oil and disposed by injection into deep wells. This so-called "frack water" is highly contaminated with dissolved minerals, toxic trace metals, and heavy oil. It could be recovered, but only at great expense. The disposition of frack water is a question which causes great public concern.

To check on the status of fracturing and to possibly catch any leakage before it gets out of hand, seismic monitoring is used. The length and depth of the fractures can be measured and compared with expectations. If problems are discovered, especially those that may cause environmental problems, fracturing can be stopped or at least reduced. Monitoring also can reveal close-by reservoirs. After fracturing has been completed, the frack oil is handled the same way as conventional oil.

Hydraulic fracturing is highly controversial due to the potential for adverse environmental impact. Responsible operators apply fracking well below water tables and cement their wells using best practices. When such practices are applied, the likelihood of loss of containment is minimal. However, disreputable companies have fractured with explosives instead hydraulic pressure and have done so at shallow depths, where problems are more likely.

Fracking affects the quality of air due to the amount of fuel being burned to run the pumps at such high pressure. Some reports also say that up to 90% of the fracturing liquid remains underground and cannot be reclaimed; over time, this may affect nearby water resources. In some areas, fracking increases seismic activity substantially. Prior to injection into disposal wells, wastewater is stored at the surface in open vats, from which degassing contaminates the air with methane and other greenhouse gases. The U.S. congress exempted some aspects of fracking from the Safe Drinking Water Act (SDWA). The EPA has strict regulations on what oil companies can do at their fracturing sites, but nobody can know in advance exactly what may happen underground, where most of the problems can occur.

The oil and gas that cannot be recovered by conventional means, such as primary to tertiary recoveries, are classified as unconventional oil and gas (UCOG). UCOG includes oil sand, bitumen, extra-heavy oils, gas hydrates, coalbed methane, gas in tight sand, etc. The oil and gas recovered by hydraulic fracturing are known as shale oil and shale gas. The shale oil from fracking can be confused with the oil obtained from retorting oil shale. Shale oils from fracking are high quality oils because the heavy metals and asphaltenes remain in the reservoir or formation. But shale oils from oil shale retorting contain the heavy fractions and toxic inorganic materials, such as arsenic and mercury. It would probably be better to refer the shale oil from fracking as "fracking oil" to differentiate from the "shale oil" from oil shale retorting. Shale gas from fracking is natural gas, mainly methane. The gas from oil shale retorting contains methane, too, but it also contains significant C_2-to-C_4 gases.

A primary consideration when extracting unconventional petroleum, such as from SAGD and hydraulic fracturing, is the net-energy gain, or NEG. The NEG of a process is the amount of energy available in the oil recovered minus the amount of energy that had to be put into retrieve it. If this value is not high enough then it is not worth recovering the petroleum.

7.11 Wellsite Pretreatment

Crude oil comes from the ground mixed with a variety of substances: gases, water, salt, and dirt. These must be removed before the crude can be transported effectively and refined without undue fouling and corrosion. Some cleanup occurs in oil fields and midstream processes such as the preparation of syncrudes. Natural gas is mostly methane, but it may contain hydrogen sulfide, CO_2, water, higher hydrocarbons, mercury compounds, and noble gases (He and Ar).

Natural gas and crude oil need to be transported to a gas plant or a refinery, usually a considerable distance away from the fields, for processing into useful products. For large-scale transportation, pipelines and tankers are commonly used. For smaller scales, especially for the distribution of petroleum products, railroad cars, barges and tank trucks are used to a large extent. Prior to transportation, most gas and oil require some form of pretreatment near the reservoir to meet transportation requirements and safety specifications/regulations.

Considerable planning, including possible trading, is involved in determining how and where to ship to the oil and gas. Transportation and trading, even storage, can be considered as "midstream" operations between the field production (upstream) and refinery or gas plant (downstream).

In certain cases, pretreatment of natural gas can be conveniently performed at the wellhead. The produced oil, on the other hand, is usually collected from several wells and sent to a central facility for separation of gas, oil, water and sand.

A field separator at well site is often no more than a large covered vessel that provides enough residence time for gravity separation into four phases: gases, crude oil, water, and solids. Generally, the crude oil floats on the water. The water is withdrawn from the bottom and is disposed of at the well site. Gases are withdrawn from the top and piped to a natural-gas processing plant or reinjected into the reservoir to maintain well pressure. Crude oil is pumped either to a refinery through a pipeline or to storage to await transportation by other means.

Low density natural gas is mostly moved by pipelines. It is also transported by seagoing tankers at ambient or atmospheric pressure, with the cargo under refrigeration; special alloys to resist brittleness at temperatures as low as $-160\ °C$ became available in the late 1960s, enabling operators to cryogenically liquefy natural gas (LNG) for shipping.

More discussions in transportation can be found in Chap. 17.

References

1. http://science.howstuffworks.com/environmental/energy/oil-drilling4.htm. Accessed 15 Aug 2014
2. Mueller MD, Jones FL, Quintana JM, Ruddy KE, Mims MG (1993) Well casing flotation device and method. US Patent 5,181,571, 26 Jan 1993
3. Unocal claims extended reach records. Oil Gas J. 16 Sept 1991. http://www.ogj.com/articles/print/volume-89/issue-37/in-this-issue/drilling/unocal-claims-extended-reach-records.html. Retrieved 1 Nov 2012
4. http://www.kingwelloilfield.com/topics/features/Blowout-Preventer.html. Retrieved 15 Aug 2016
5. Tarr BA, Flak L Underground blowouts. http://www.jwco.com/technical-litterature/p06.htm
6. Speight JG (2006) The chemistry and technology of petroleum. 4th edn. Revised and Expanded, Marcel Dekker
7. Maksimova EV, Cooper CK (2017) Offshore production. In: Hsu CS, Robinson PR (eds) Springer Handbook of Petroleum Technology. Springer, New York
8. https://en.wikipedia.org/wiki/Oil_platform. Accessed 23 Nov 2016
9. https://info.drillinginfo.com/offshore-rigs-primer-offshore-drilling/ Accessed 9 Oct 2018
10. Tzimas E (2005). Enhanced oil recovery using carbon dioxide in the European energy system (PDF). European Commission Joint Research Center. Retrieved 01 Nov 2012
11. http://petrowiki.org/Polymer_waterflooding_design_and_implementation. Retrieved 10 Sept 2016
12. Oudinot AY, Riestenberg DE, Koperna GJ Jr (2017) Enhanced gas recovery and CO_2 storage in coal bed methane reservoirs with N_2 co-injection. Energy Proc 114:5356–5376
13. Wikipedia: Steam injection (Oil Industry). https://en.wikipedia.org/wiki/Steam_injection_%28oil_industry%29
14. https://en.wikipedia.org/wiki/File:Cyclic_steam_stimulation,_oil_well.pdf
15. Watkins DR, Kalfayan LJ (1985) Methods for maintaining the permeability of fines-containing formations. US Patent 4,498,538, 12 Feb 1985
16. Greaves M, Xia TX, Ayasse C Underground upgrading of heavy oil using THAI-'toe-to-heel air injection'. In: SPE international thermal operations and heavy oil symposium, Calgary, Alberta, Canada, 1–3 Nov 2015. https://www.onepetro.org/conference-paper/SPE-97728-MS. Retrieved 15 Sept 2016
17. Naqvi SAA Enhanced oil recovery of heavy oil by using thermal and non-thermal method. https://cdn.dal.ca/content/dam/dalhousie/pdf/faculty/engineering/peas/MEngProjects/2012MEngProjects/Syed%20Ata%20Abbas%20Naqvi%2C%20Enhanced%20Oil%20Recovery%20of%20Heavy%20Oil%20by%20Using%20Thermal%20and%20Non-Thermal%20Methods.pdf
18. Duir JH, Randle AC, Reeg CP (1990) Unocal's parachute creek oil shale project. Fuel Process Technol 25(2):101–117
19. Athabasca oil sands: https://en.wikipedia.org/wiki/Athabasca_oil_sands. Retrieved on September 15, 2016
20. Tailing Ponds. https://www.canadasoilsands.ca/en/explore-topics/tailings-ponds. Retrieved 9 Sept 2018
21. Wikipedia: Steam-assisted gravity drainage. https://en.wikipedia.org/wiki/Steam-assisted_gravity_drainage. Retrieved 15 Sept 2016

22. Canada's 500,000 Leaky Energy Wells: "Threat to Public". https://thetyee.ca/News/2014/06/05/Canada-Leaky-Energy-Wells/. Retrieved 9 Oct 2018
23. https://en.wikipedia.org/wiki/Hydraulic_fracturing. Retrieved 15 Sept 2016
24. https://www.sciencedirect.com/topics/earth-and-planetary-sciences/natural-gas-reservoirs. Retrieved 9 Oct 2018
25. ProPublica: What is hydraulic fracturing. https://www.propublica.org/special/hydraulic-fracturing-national. Retrieved 15 Sept 2016

Part III
Petroleum Refining - Processes and Products

Chapter 8
Petroleum Processing and Refineries

8.1 Significant Events in Petroleum Processing

Bitumen was used in ancient times for many reasons—for water-proofing boats and canals, as mortar for bricks, as a medicinal ointment, for making spears and other weapons, and in mummification. Liquid petroleum and natural gas were used mainly for heating, lighting, cooking, and warfare. Ancient writers referred to flaming arrows and firepots based on pitch, turpentine, naphtha, sulfur, and charcoal. How the naphtha was obtained is a subject of debate, but by 300 AD, it was being recovered with simple distillation equipment such as that described by the 3rd century Greek alchemist Zosimos of Panopolis, shown in Fig. 8.1. Only small amounts of light fractions were recovered from petroleum for separate use. The ancient people wanted bitumen, mostly, so they allowed petroleum to "age" by storing it in open pits, allowing the light ends evaporate.

Naphtha was an essential component of Greek fire, which was invented during the reign of Constantine IV Pogonatus (668–685) by Callinicus of Heliopolis, a Jewish refugee from Syria. The composition of Greek fire is still debated, but suggestions include combinations pitch, naphtha, quicklime, sulfur, and saltpeter. The substance could be thrown in pots or discharged thorough syphon tubes. It ignited on contact with air and could not be extinguished with water. In 673, Greek ships defended Constantinople with this formidable weapon, crippling an Arab fleet that was trying to attack the city.

The demand for petroleum started increasing after the invention of kerosene lamps in the mid- and late-19th century. Since then, several historical events influenced the evolution of today's sophisticated technology for refining petroleum into useful fractions. The progress in development is essentially market driven. The market is driven, of course, by supply and demand. From about 1970 to 2014, the Organization of Petroleum Exporting Countries (OPEC) largely succeeded in controlling prices by controlling supply. Starting in 2011, OPEC's influence diminished due to oil production by hydraulic fracturing in the United States, which dropped the spot price of bench-mark crudes from >$110 per barrel in mid-2014 to <$30 per barrel

© Springer Nature Switzerland AG 2019
C. S. Hsu and P. R. Robinson, *Petroleum Science and Technology*,
https://doi.org/10.1007/978-3-030-16275-7_8

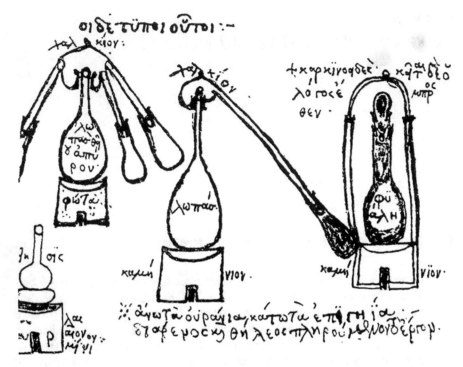

Fig. 8.1 Distillation equipment described by 3rd Century Greek alchemist Zosimos of Panopolis [1]

in early 2016 (in 18 months). Since then, the price gradually recovered to above $70 per barrel today (3Q, 2018). Figure 8.2 shows how oil prices have varied since 1859, and how prices have corresponded to certain world events.

Before they can be used, kerosene, gasoline, and other fractions must be separated from crude petroleum or other sources in a refinery. The earliest known refinery was built in 1745, at the behest of the Empress Elisabeth of Russia. The plant was built near an oil well in Ukhta by Fiodor Priadunov to distill a kerosene-like substance from "rock oil" for use in oil lamps by Russian churches and monasteries.

For illumination, kerosene was superior to vegetable oils and most animal fats. Kerosene burned cleanly and steadily, while most animal fats tended to burn with smoky flames. The best animal fats for lamps came from whales. In the 1850s, the over-hunting of whales was threatening the survival of several species. Prices were fluctuating wildly. Experts agree that petroleum saved the whales [3].

In 1847, James Oakes built a "rock oil" refinery in Jacksdale, England. The unit processed 300 gallons per day to make "paraffin oil" for lamps. Around the same time, James Young built a coal-oil refinery in Whitburn, Scotland.

Samuel M. Kier of Pennsylvania is credited to give birth to the U.S. refining industry. Kier recovered crude oil as a by-product of his brine well, which was the basis of his salt business. Originally, he sold the crude oil as a medicine. In 1849,

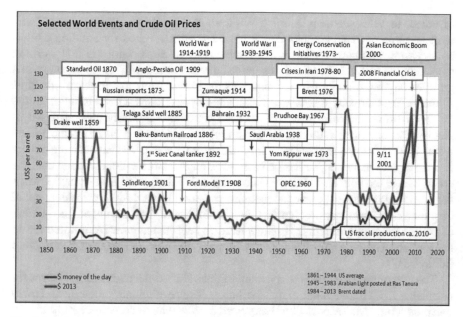

Fig. 8.2 Selected world events and crude oil prices, 1859–2015. Prices are annual averages. Items in purple boxes indicate the start dates for major new production. The underlying format comes from the BP Statistical Review of World Energy—2015. Recent prices come from the US Energy Information Administration [2]

when the medicine fad faded, he successful refined the liquid petroleum to produce "Carbon Oil" for kerosene lamps and started building the first U.S. petroleum refinery in Pittsburgh, Pennsylvania.

The market for kerosene boomed—supply drove demand, in this case—and the demand for whale oil plummeted after Edwin L. Drake struck oil near Titusville, Pennsylvania in 1859, triggering the Pennsylvania Oil Boom. By the end of the 1860s, there were 58 refineries operating in Pittsburgh alone.

In 1870, John Rockefeller and five partners established the Standard Oil Company and began consolidating the U.S. refining business. Less than 10 years later, Standard Oil controlled 90% of US refineries and pipelines. Critics accused Rockefeller of engaging in unethical practices, such as predatory pricing and colluding with railroads to eliminate his competitors, in order to gain a monopoly in the industry. In 1911, the U.S. Supreme Court ruled Standard Oil to be in violation of anti-trust laws and ordered it to dissolve. Although Rockefeller is vilified as the ultimate "robber baron," he probably saved the industry from itself. In other parts of the world, notably Austrian Galicia, unrestrained over-production, even when prices were very low, irreparably depleted one of the world's largest oil fields and damaged the nearby environment; evidence of that damage lingers even today, more than 100 years later [4].

The use of kerosene lamps started to decline after the invention of light bulbs by Thomas Edison in 1878. At the same time, several internal combustion engines were invented. In 1889, Gottlieb Daimler, Wilhelm Mayback and separately Karl Benz built gasoline powered automobiles. Rudolf Diesel is credited for building a high-compression prototype engine in 1897 after Akroyd Stuart built the first working diesel engine in 1892. Such engines burned fuels that are difficult to vaporize in a gasoline engine. A major development was the switch from coal to oil in warships. In the 1890s, Germany was talking about fueling new ships with petroleum. In the 1910s, Winston Churchill, in his role as First Lord of the Admiralty, pushed the Royal Navy in this direction. With coal, he explained, storage bunkers closest to the boilers were emptied first. As the coal inventory decreased, the distance between the boilers and the supply increased, so it took more time and/or more sailors to bring coal via conveyors to the boilers. This could handicap a ship during the heat of battle, when the need for both steam power and manpower were highest. Liquid petroleum could be pumped, so inventory didn't affect the availability of fuel or combatants—unless of course the supply ran out. Britain's need to fuel its navy far from home led to the formation of the Anglo-Persian Oil Company, the precursor to British Petroleum (today's BP).

The invention of internal combustion engines shifted the main market driver for petroleum from illumination to transportation. The revolutionary moving assembly line production of the Model T by Henry Ford in 1908 made automobiles affordable to the general public. Demand for gasoline jumped, and continued to rise steadily—with a pause during the Great Depression—until the 1970s, when unrest in the Middle East triggered events that dampened supply.

Steady improvements in design and manufacturing technology have decreased transportation costs and increased the maximum size of crude distillation units. Unit operating costs are lower in large industrial areas (zones) such as Rotterdam; Singapore; Chiba/Yokohama; South Korea; Sarnia, Ontario; Greater Los Angeles; Greater Chicago; Greater Seattle; the San Francisco Bay Area; and largest of all, the U.S. Gulf Coast between Corpus Christi, Texas and Mobile, Alabama. Through the Colonial and Plantation Pipelines, the Gulf Coast Basin supplies a significant percentage of the fuel consumed in the Northeast United States. In basins, several refineries and petrochemical plants can share offsites (power, steam, industrial gases) and have ready access to nearby suppliers of chemicals and catalysts, engineer service companies, not to mention a large, concentrated pool of trained personnel.

These improvements in technology, coupled with basin economics, led to considerable consolidation—the merging of two or more adjacent facilities—and to the shutdown of smaller, isolated, less efficient plants. From 1981 to 2014, the number of operating U.S. refineries went from 321 to 123. At the same time, the average U.S. refinery capacity rose from about 55,000 barrels per stream day (BPSD) to about 146,500 BPSD.

Table 8.1 summarizes the significant events in petroleum processing from 1860 to 2000 [5].

Table 8.1 Significant events in petroleum refining 1861–2000

Date	Description
1878	Thomas Edison invents the light bulb. The use of kerosene lamps starts to decline
1889	Gottlieb Daimler, Wilhelm Mayback and (separately) Karl Benz build gasoline-powered automobiles
1901	Ransom E. Olds begins assembly-line production of the Curved Dash Oldsmobile
1908	Ford Motor Company offers Model T's for US$950 each
1912	William Burton and Robert Humphreys develop thermal cracking
1913	Gulf Oil builds the world's first drive-in filling station in Pittsburgh, Pennsylvania
1919	UOP commercializes the Dubbs thermal cracking
1929	Standard Oil of Indiana (now BP) commercializes the Burton process for delayed coking at Whiting, Indiana
1933	UOP introduces the catalytic polymerization of olefins to form gasoline
1934	Eugene Houdry, working for Sun Oil, patents Houdry Catalytic Cracking (HCC)
1938	A consortium of refiners develops sulfuric acid alkylation, which is first commercialized at the Humble (now ExxonMobil) refinery in Baytown, Texas
1940	Phillips develops HF alkylation
1942	Standard Oil of New Jersey (now ExxonMobil) commercializes the FCC process at Baton Rouge, Louisiana
1949	Old Dutch Refining in Muskegon, Michigan starts the world's first catalytic reformer based on the UOP Platforming processes
1950	Catalytic hydrotreating is patented by Raymond Fleck and Paul Nahin of Union Oil
1960s	UOP introduces C_4 and C_5/C_6 isomerization processes
1961	Standard Oil of California (now Chevron) introduces catalytic hydrocracking
1970	The world celebrates Earth Day. The newly created U.S. Environmental Protection Agency passes the Clean Air Act, which requires a 90% reduction in auto emissions by 1975. The European Union issues similar requirements
1972	Mobil invents ZSM-5. During the next three decades, this shape-selective catalyst finds uses in numerous processes, including FCC, catalytic dewaxing and the conversion of methanol to gasoline
1975	The catalytic converter goes commercial. The phase-out of tetraethyl lead begins
1990	The U.S. Congress issues the Clean Air Act Amendments of 1990, which lay the framework for reformulated gasoline and low-sulfur diesel
1990s	Several processes are developed to remove sulfur from gasoline. These include SCANfining(Exxon), OCTGAIN (Mobil), Prime G (Axens), and S Zorb (Phillips)
1993	Chevron commercializes Isodewaxingfor converting waxy paraffins into high-quality lube base stock
2000	The European Commission issues the Auto Oil II report, which includes a timetable for low-sulfur gasoline and ultra-low-sulfur diesel

8.2 Process Development Driven by Market Demand

As mentioned, refining process development is driven by changes in market demand. The demand for more and more kerosene began in the mid-19th century. Russians began using fuel oil to power ships and locomotives in the early 1880s. By 1890, oil was displacing coal in American locomotives, especially in Southwestern states, where coal was rare and oil was abundant. After the advent of the Model T in 1908, the market demand for automobile fuels began to outstrip the demand for kerosene. Refiners were pressed to develop technology for converting kerosene and heavier fractions, such as thermal cracking, polymerization, catalytic cracking, alkylation, isomerization, and catalytic reforming. These processes greatly increased gasoline yields. Various hydrotreating processes were developed, first to protect catalysts in catalytic reforming, isomerization, and fixed-bed hydrocracking units, and later for environmental protection.

Lead-based fuel additives were developed to meet the need of more powerful aircraft engines in the late 1930s. During World War II, high octane aviation gasoline enabled Allied combat aircraft to fly at higher altitude with more power and speed than enemy aircraft. The need for even better gasoline performance spurred the development of alkylation, polymerization, and isomerization processes.

The invention of jet planes during World War II created a demand for jet fuels, which have the same boiling range as kerosene. The demand increased rapidly during the 1950s and 1960s, due to increased military demand during the Cold War and to the rapidly expanding use of jets by commercial airlines. High quality lubricant oils were also in great demand.

Catalytic reforming, catalytic isomerization and hydrocracking were developed to meet the requirements of high-compression engines. Many other processes for petroleum-derived materials were also developed. The process development from 1862 to 1975 is shown in Table 8.2.

In addition to increasing overall crude oil demand, the automobile increased demand for naphtha, on which gasoline is based. Gasoline still accounts for most petroleum consumption in the Unites States. In North America, gasoline demand is still high, but in Europe and Asia, diesel demand is higher than gasoline demand.

In 2013, United States consumed, on average, 79% of its petroleum as naphtha, kerosene and gas oil (NKG); major destinations for NKG are gasoline, jet fuel and diesel, respectively. But straight-run yields of NKG are lower than the demand. In Table 8.3, the NKG gaps range from 17.2 to 45.2% and the naphtha gaps range from 11.9 to 26.2%.

These gaps create a strong economic incentive to convert "Other" into NKG. Here's a way to estimate the size of the incentive:

- The crack spread is the difference between the price of a crude oil and the wholesale prices of petroleum products extracted from it. Crack spread calculations are used to estimate refinery profit margins. Figure 8.3 shows that the crack spread for Brent crude varied between $11 and $22 per barrel between June 2012 and May 2013.
- Assume the United States consumes 18 million barrels per day of oil.

Table 8.2 Process development from 1862 to 1975 [6]

Year	Process	Purpose	By-products
1862	Atmospheric distillation	Produce kerosene	Naphtha, Tar, etc.
1870	Vacuum distillation	Produce lubricant oils originally, after 1930's cracking feedstocks	Asphalt, residual coker feedstocks
1913	Thermal cracking	Increase gasoline yields	Olefins, residual fuels
1913	Thermal resid hydrocracking	Produce fuels and syncrude	Heavy residuals
1916	Solvent Sweetening (not Hydrotreating)	Reduce sulfur and odor	none
1930	Thermal reforming	Improve Gasoline Octane Number	Olefins, residuals
1932	Hydrogenation (first-generation HDS)	Remove sulfur	Hydrogen sulfide
1932	Coking	Produce fuels	Coke
1933	Solvent extraction	Improve lubricant viscosity index	Aromatics
1935	Solvent dewaxing	Improve lubricant pour point	Wax
1935	Catalytic polymerization	Improve gasoline yield and octane number	Petrochemical feedstocks
1937	Catalytic cracking	Produce high octane gasoline	Petrochemical feedstocks
1939	Visbreaking	Reduce viscosity of fuel oils	Distillates, Tar
1940	Isomerization	Produce alkylation feedstock	Naphtha
1940	Alkylation	Increase gasoline yield and octane number	Aviation gasoline
1942	Fluid Catalytic Cracking	Increase gasoline yield and octane number	Petrochemical feedstocks
1949	Catalytic reforming	Convert low octane naphtha	Aromatics. gasoline
1950	Deasphalting	Remove asphalt from cracking feedstocks and resids	Asphalt
1954	Catalytic hydrodesulfurization (HDS)	Remove sulfur	Hydrogen sulfide
1956	Mercaptan oxidation (Merox)	Convert mercaptans to sulfides	Sulfides
1957	Catalytic isomerization	Convert n-butane to isobutane and low octane n-paraffins to isoparaffins	Gasoline, alkylation feeds

(continued)

Table 8.2 (continued)

Year	Process	Purpose	By-products
1960	Hydrocracking	Produce fuels from heavy oil and remove contaminants such as sulfur	Alkylation feedstocks, lube base stocks, olefin plant feeds
1971	CCR platforming	Convert low octane naphtha with greater yields	Aromatics, gasoline
1974	Catalytic dewaxing	Improve lubricant pour point	Distillate fuels
1975	Modern resid hydrocracking	Increase fuel yields from residual oil	Heavy residuals

Table 8.3 Conversion gaps for four crudes

Product cut	Naphtha	Kerosene	Gas oil	Other	Gaps	
Becomes	Gasoline	Jet	Diesel		NKG	Naphtha
U.S. consumption, %	42	9	28	21	–	–
Amount in crude, %						
Brent	30.1	9.9	15.1	44.9	23.9	11.9
Bonny Light	27.7	12.5	21.6	38.2	17.2	14.3
Green Canyon	16.9	8.5	14.1	60.5	39.5	25.1
Ratawi	15.8	7.4	10.6	66.2	45.2	26.2

Gaps are the difference between U.S. consumption of a given product and the amount of that product in the crude before conversion

NKG gap = U.S. consumption of NKG − NKG in crudes Naphtha gap = U.S. consumption of naphtha − naphtha in crudes

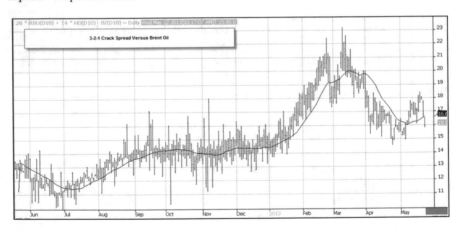

Fig. 8.3 The 3:2:1 crack spread for Brent crude between June 2012 and May 2013 [7]

U.S. Refinery Yields, vol% [1,2]	Summer Jul 2014	Winter Dec 2014
Liquified Petroleum Gases (LPG) [3]	5.3	2.4
Motor Gasoline [4]	44	46.5
Aviation Gasoline	0.1	0.1
Kerosene-Type Jet Fuel	9.7	9.7
Kerosene	0.1	0.2
Distillate Fuel Oil	29.1	30.7
Residual Fuel Oil	2.4	2.3
Petrochemical Feed (naphtha)	1.3	1.4
Petrochemical Feed (other)	0.7	0.5
Special Naphthas	0.3	0.2
Lubricants	1	0.9
Waxes	0.1	0
Petroleum Coke	5.5	5.3
Asphalt and Road Oil	2.3	1.7
Still Gas [3]	4.2	4
Miscellaneous Products	0.6	0.5
Volume Gain [5]	6.7	6.4

(1) After removal of sulfur, nitrogen, and other contaminants.
(2) Volumes of solids are expressed in fuel-oil equivalents.
(3) Still gas is largely methane but can also contain C2-C4 compounds
(4) Does not include ethanol.
(5) The volume swells due to cracking. Light products have lower densities than crude oil.

Fig. 8.4 Industry-average volume-percent yields from petroleum refineries in the United States. Sulfur and other contaminants are excluded. Motor gasoline yields do not include ethanol. In Europe, yields of motor gasoline and distillate fuel oil would be approximately reversed, and yields of petroleum coke would be lower. About 90 vol.% of the products are fuels—gasoline, kerosene, jet and fuel oils. Most refinery gases and petroleum coke also are used as fuels

- Assume the NKG gap is 23.9%—the same as for Brent in Table 8.3.
- Assume the crack spread is $11 per barrel.
- On this basis, the U.S. conversion incentive is $47.3 million dollars per day.

Today's modern refineries waste almost nothing. It is possible to convert more than 99 wt% of a conventional crude into something other than ash, and the ash can be blended into clinker for cement manufacturing.

Figure 8.4 presents a quantitative petroleum product breakdown in the United States for July and December 2014 [8]. About 83 wt% (90 vol.%) went to fuels. In Europe and Asia, yields of motor gasoline and distillate fuel oil (diesel) are approximately reversed.

Table 8.4 Non-fuel products from petroleum refineries

Product		Uses
Lubricant base stocks, greases, industrial oils		
	$\gg$800 products—from "baby oil" to axle grease—all with stringent specifications	
Construction materials		
	Asphalt	Paved roads, water-proofed buildings. Four major classes
	Waxes	Coated paper packaging, water-proofing, candles
Carbon		
	Petroleum coke	In addition to fuel use: anodes for making steel
	Needle coke	Anodes for making aluminum and high-grade steel
	Graphite	Metallurgy, specialty fibers, paints, pencils
Chemicals		
	BTX, cyclohexane	Solvents
	Ammonia	Metal cleaning, fertilizer
	Sulfur	Sulfuric acid, fertilizers
	Carbon dioxide	Dry ice, carbonated soft drinks
Raw materials for petrochemical plants		
	Aromatics	
	C_3/C_4 olefins	Olefin production plants
	Straight-run naphtha	Olefin production plants
	UCO[a]	Olefin production plants

[a]UCO is highly upgraded unconverted oil from hydrocrackers

Petroleum provides mostly transportation fuels, but it also supplies hundreds of non-fuel products. Table 8.4 lists some non-fuel products produced inside refineries, often without further processing in a petrochemical plant. The last section of the table lists some petrochemical precursors.

8.3 Refinery Capacities in the United States and in the World

According to the U.S. Energy Information Administration (EIA), the top ten refineries in the U.S. have the capacities are listed in Table 8.5. Many of the largest refineries are no longer owned by large integrated oil companies. In fact, many large plants previously owned by big oil companies have been sold to smaller refining companies. The location of the refineries stays the same as refineries cannot be easily relocated.

Table 8.5 Top 10 U.S. refineries[a] operable capacity (EIA. Jan. 2018 database) [9]

Name of refinery	Location	Barrel per calendar day (BPCD)
Port Arthur Refinery (Motiva Enterprises)	Port Arthur, Texas	636,500
Galveston Bay Refinery (Marathon Petroleum)	Texas City, Texas	571,000
Baytown Refinery (ExxonMobil)	Baytown, Texas	560,500
Garyville Refinery (Marathon Petroleum)	Garyville, Louisiana	556,000
Baton Rouge Refinery (ExxonMobil)	Baton Rouge, Louisiana	502,500
Whiting Refinery (BP)	Whiting, Indiana	430,000
Lake Charles Refinery (Citgo)	Lake Charles, Louisiana	427,800
Beaumont Refinery (ExxonMobil)	Beaumont, Texas	344,600
Port Arthur Refinery (Valero)	Port Arthur, Texas	335,000
Pascagoula (Chevron)	Pascagoula, Mississippi	330,000

[a]Only refineries with atmospheric crude oil distillation capacity
BPCD—barrels per calendar day

Table 8.6 lists the ten U.S. states with the largest refining capacity, comparing data for 2005 with 2018. Note how capacities decreased during this time frame in California, Pennsylvania, and New Jersey while they increased in the other seven. In contrast, Wyoming has 4 refineries, their total capacity is only 105,000 barrels per calendar day.

The majority of world's 10 largest refineries are located in Asia, led by the 1.24 million bbl/day capacity Jamnagar refinery complex in India, followed by one each in Venezuela, United Arab Emeritus (UAE) and Singapore, and three each in South Korea and the United States. Table 8.7 lists the top 10 large oil refineries in the world, based on processing capacity.

Many companies own several refineries, large and small. Based on 2015 data, the five largest oil refining companies are: ExxonMobil Refining and Supply Co. (5,580,000 BPCD), Royal Dutch Shell Group (4,109,000 BPCD), Sinopec (3,971,000 BPCD), BP PLC (2,859,000 BPCD) and Saudi Arabian Oil Co. (2855 BPCD), where BPCD stands for barrel per calendar day.

Table 8.8 compares the refining capacity in thousand barrels per calendar day (MBPCD) of the top 13 countries in the world in 2010 and 2014. Some countries do not have any refineries. On the other hand, it can be seen that the capacity does not necessarily reflect the demand inside the country. The refined products are imported by the countries that need them. The United States, Canada, and certain other countries import products at some locations and export products from others.

Table 8.6 Ten U.S. states with the largest refining capacity [9]

State	Number of operating refineries		Crude distillation capacity (1000 BPCD)	
	2005	2018	2005	2018
Texas	27	29	4628	5701
Louisiana	17	17	2773	3303
California	21	16	2027	1892
Illinois	4	4	896	999
Washington	5	5	616	638
Pennsylvania	5	4	770	601
Ohio	4	4	551	598
Oklahoma	5	5	485	522
Indiana	2	2	433	442
New Jersey	5	3	666	418
Top 10 total	93	91	13,845	15,114
U.S. total	144	140	17,125	18,598

BPCD—barrels per calendar day

Table 8.7 Ten largest refineries in the world (2018) (in MBPCD: thousand barrels per calendar day) [10]

Name of refinery	Location	Country	Crude distillation capacity (MBPCD)
Jamnagar Refinery (Reliance)	Jamnagar, Gujarat	India	1240
Paraguana Refinery complex (PDVSA)	Punto Fijo, Falcon	Venezuela	940
SK Energy Ulsan Refinery (SK energy)	Ulsan	South Korea	840
Ruwais Refinery (Abu Dhuba Natinal Oil Company)	Ruwais	UAE	817
Yeosu Refinery (GS Caltex)	Yoesu	South Korea	730
Ornan Refinery (S-oil)	Onsan	South Korea	669
Jurong Island Refinery (ExxonMobil)	Jurong Island	Singapore	605
Port Arthur Refinery (Saudi Aramco)	Port Arthur, Texas	United States	603
Galveston Bay Refinery (Marathon petroleum)	Texas City, Texas	United States	571
Baytown Refinery (ExxonMobil)	Baytown, Texas	United States	560

Table 8.8 Refining capacity by country (in MBPCD: thousand barrels per calendar day) [11]

Country	2010	2014
United States	17,736	17,791
China	10,302	14,098
Russia	5511	6338
India	3703	4319
Japan	4291	3749
South Korea	2712	2887
Saudi Arabia	2107	2822
Brazil	2093	2235
Germany	2091	2060
Iran	186p	1985
Italy	2396	1984
Canada	1951	1965
Mexico	1702	1375
United Kingdom	1757	1368

Although China was among the first countries to discover and use oil and natural gas, the development of the modern Chinese oil industry was inhibited by relatively low reserves and lack of demand from the transportation sector in earlier years. Between 1966 and 1976, when petroleum use in Western countries was growing exponentially, the Cultural Revolution quashed Chinese economic growth and, consequently, the growth of energy demand. China began to drill wells in 1878, but until 1949 only a few small oil fields were exploited. Since 1949, several large fields have been discovered and exploited. Refining capacity also increased steadily. Today, China has the second largest refining capacity in the world. Many Chinese refineries are now as advanced as any in the world.

The United States and Europe contain 45% of the world's refineries and account for 46% of world's crude distillation capacity. The high usage of FCC in North and South America reflects the higher demand for gasoline in those regions. Note the low use of thermal cracking and the high use of hydrotreating in the United States and the Asia Dragons (South Korea, Singapore, Hong-Kong and Taiwan).

8.4 Basic Operation Categories for Refining Processing

The goals of modern refinery operations are to be safe, economically feasible, and environmentally acceptable. They can be divided into the following categories:

- **Separation processes** involve no change in the size and basic structure of molecules. In other words, they are only physical separations, no chemistry is involved. Separation processes include distillation, absorption/adsorption, deasphalting, extraction and settling. Settling is important at wellsites when preparing crude oil for shipment, and it is also used in refineries.

- **Conversion processes** change the size and structure of the molecules: they involve changing chemical bonds. They convert less valuable fractions into more valuable fractions and products. When refiners talk about conversion, they generally mean boiling point reduction, especially breaking C–C bonds by catalytic cracking, coking, thermal and hydrocracking. But conversion also entails making valuable products from light molecules. Such processes include alkylation, isomerization, polymerization, and reforming. Boiling point reduction is also accomplished without breaking C–C bonds in the processes that saturate aromatics and remove sulfur, nitrogen, or oxygen.
- **Treating/Finishing processes** involve the removal of undesirable components. Such processes include desalting; hydrotreating to achieve hydrodesulfurization (HDS), hydrodenitrogenation (HDN), and hydrodemetalation (HDM); and solvent scrubbing for sweetening. Mercaptan oxidation to remove or reduce the odor of crude oils for shipping is also considered a treating process.
- **Solvent refining** processes include solvent extraction, dewaxing and deoiling, usually to prepare lubricants and high-quality waxes.
- **Blending** is applied to both crude oils and the products. The blending (mixing) of various process streams is common in the refinery operations. In feedstock blending, crude oils of different grades and containing different concentrations of heteroatom-containing compounds are mixed in a proper ratio to suit refinery requirements and constraints. In product blending, hydrocarbon fractions, additives and other components are mixed to produce finished products that meet product specifications, which are established based on performance and environmental requirements.
- **Other operations** include hydrogen production, waste and process water treatment, tail gas treatment, sulfur recovery, storage and handling (including tank farm maintenance/management,), product movements, etc.

8.5 Conversion Strategy

For refiners, the goal of a conversion process is to reduce boiling point, primarily by transforming low-valued heavy fractions into high-valued light products. This is accomplished by increasing hydrogen-to-carbon (H/C) ratio through **hydrogen addition** or **carbon (coke) rejection**. The most common hydrogen addition processes are hydrotreating and hydrocracking, which increase the H/C ratio by adding hydrogen to the molecules. Taken together, these two processes are referred to as hydroprocessing. When discussing hydroprocessing, some authors include catalytic reformers (which generates hydrogen) and isomerization units, in which hydrogen is not consumed but is needed to reduce catalyst coking. The flow schemes for fixed-bed hydrotreating and hydrocracking units are similar. Hydrotreating is an essential pre-

Table 8.9 Feeds and products for hydroprocessing units

Feeds	Products from hydrotreating	Products from hydrocracking
Heavy naphtha	Catalytic reformer feed	LPG
Straight-run light gas oil	Kerosene, jet fuel	Naphtha
Straight-run heavy gas oil	Diesel fuel, heating oil, fuel oil	Naphtha
Atmospheric residue	Lube base stock, low-sulfur fuel oil, RFCC[a] feed	Naphtha, middle distillates, FCC feed, lube base stock
Vacuum gas oil	FCC feed, lube base stock	Naphtha, middle distillates, FCC feed, lube base stock, olefin plant feed
Vacuum residue	RFCC[a] feed	See note[b]
FCC light cycle oil	Diesel blend stocks, fuel oil	Naphtha
FCC heavy cycle oil	Blend stock for fuel oil	Naphtha, middle distillates
Visbreaker gas oil	Diesel blend stocks, fuel oil	Naphtha, middle distillates
Coker gas oil	FCC feed	Naphtha, middle distillates, FCC feed, lube base stock, olefin plant feed
Deasphalted oil	Lube base stock, FCC feed	Naphtha, middle distillates, FCC feed, lube base stock

[a]RFCC = "residue FCC" or "reduced crude FCC," processes designed for feeds that contain high concentrations carbon-forming compounds
[b]Traditional fixed-bed hydrocrackers cannot process vacuum residue. However, ebullated-bed and slurry-phase hydrocrackers can. Products from the latter include naphtha, middle distillates, and FCC feed

cursor to fixed-bed gasoil hydrocracking. Feeds and products for typical hydrotreaters and hydrocrackers are shown in Table 8.9.

Other conversion processes—FCC, thermal cracking, and delayed coking – often are called coke rejection processes. In fact, they yield both lighter molecules with higher H/C and heavier products with lower H/C ratios at the same time. In coking and FCC, the heavier product is coke, in which the H/C ratio is <1.0. It is common to hear "carbon rejection" instead of coke rejection, but "carbon rejection" is a misleading phrase; elemental carbon per se is not removed.

Refiners define conversion as any means of transforming material that boils above a particular cutpoint, known as the conversion cutpoint, into material that boils below the conversion cutpoint. A typical conversion cutpoint for naphtha-oriented operations is 400 °F (204 °C), while a typical conversion cutpoint for diesel-oriented operation is 650 °F (343 °C). The equation for conversion is simple:

$$\text{Conversion} = P/FF * 100\%$$

Table 8.10 Molecular weight, H/C and boiling points for selected hydrocarbons

Compound	Molecular Weight	Formula	H/C	Boiling point °C	°F
Paraffins					
Methane	16.04	CH_4	4.0	−164	−263.2
Ethane	30.07	C_2H_6	3.0	−88.6	−127.5
Propane	44.10	C_3H_8	2.67	−42.1	−43.7
Butane (iso)	58.12	C_4H_{10}	2.50	−6.9	19.6
Octane (iso)	114.23	C_8H_{18}	2.25	99.2	210.6
Cetane (n)	226.44	$C_{16}H_{34}$	2.13	287	548.6
Aromatics					
Benzene	78.11	C_6H_6	1.0	80.1	176.2
Naphthalene	128.17	$C_{10}H_8$	0.8	218	424.4
Benzopyrene	252.32	$C_{20}H_{12}$	0.6	–	–

where P = the amount of material in the product that boils below the conversion cutpoint and FF = the total amount of fresh feed. In "true conversion" calculations, the amount of P in the FF prior to conversion is subtracted from the denominator.

Table 8.10 compares compounds with different hydrogen content. For any given class of hydrocarbons, "light" means a lower molecular weight, lower boiling point, lower density, and a higher hydrogen-to-carbon ratio (H/C). Methane, the lightest hydrocarbon, has an H/C ratio of 4.0. Benzopyrene has an H/C ratio of 0.6. The H/C ratio of common crude oils ranges from 1.5 to 2.0, while the H/C ratio for asphaltenes is about 1.15.

8.6 Refinery Categories

A refinery is an industrial complex that manufactures petroleum products, such as gasoline, from crude oil and other feedstocks. Many different types of refineries exist around the world.

Refinery configurations evolve to meet market demands. Such demands are both economic and geographic. A few refineries were built to supply one customer with a specific set of products, for example jet fuel for a nearby airport or air force base. Some refineries in the U.S. Midwest mostly supply diesel for farm equipment. Refineries with ocean access near major seaports, such as Rotterdam, Singapore, and Houston, tend to be exceptionally flexible, able to process feeds from just about anywhere in the world, and capable of supplying intermediate s to other refineries and nearby chemical plants.

The refineries are categorized according to their capacities and the types of processing units used to produce the petroleum products: topping, hydroskimming, con-

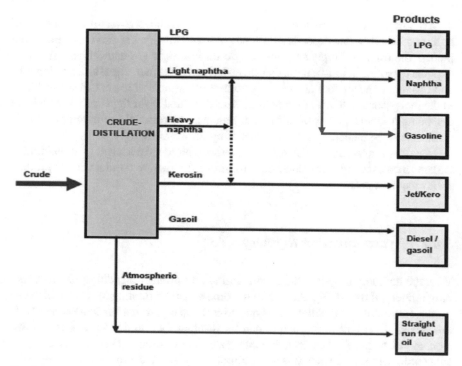

Fig. 8.5   A topping refinery [13]

version and integrated. Almost all U.S. and European refineries are either *conversion* or *integrated* refineries, as are the newer refineries in Asia, the Middle East, South America, and other areas experiencing rapid growth in demand for light products. By contrast, much of the refining capacity in Europe and Japan is in *hydroskimming* and *conversion* refineries [12]. Due to shutdowns of many isolated refineries in Japan, refining in that country is largely integrated with chemical plants within major industrial zones near Chiba, Yokohama, and Osaka/Mizushima. The same could be said for Europe, where refineries in Rotterdam/Europoort, La Mede/Laverra, Austria and Italy are closely associated with chemical plants.

8.6.1   Topping (Simple) Refinery

The simplest refinery configuration is topping refinery, shown in Fig. 8.5, which is designed for petrochemical manufacture or production of industrial fuels for local markets in remote oil-producing areas to reduce transportation costs. Only the desired products are shown on the right side of the figure. Gas oil product can be heating oil, fuel oil and lubricant feedstock.

A topping refinery consists of tankage, a distillation unit, recovery units for gas and light hydrocarbons and necessary utility systems (steam, power and water treatment plant). The refinery simply separates crude oil into light gas and refinery fuel gas, naphtha (gasoline boiling range), distillates kerosene, jet fuel, diesel and heating oils, and residual or heavy fuel oil. There is no hydrotreating unit to control sulfur levels in the fuel products. Hence, topping refineries are suitable for light crudes or streams that contain small amounts of sulfur components. The residual sulfur components in products can be treated by solvent sweetening for removal.

In modern times, except for small refineries which make asphalt or in extremely isolated locales in 3rd world countries, topping refineries are obsolete due to product sulfur requirements.

8.6.2   Hydroskimming Refinery

With the addition of hydrotreating, reforming and product blending units to a topping refinery, shown in Fig. 8.6, it will become a hydroskimming refinery which can produce desulfurized distillate fuels and high-octane gasoline for market needs. In this design, over half of the output can be residual fuel oil due to lack of conversion. N-butane in LPG can be added into gasoline to increase RVP for winter use in cars (shown as a dotted arrow as an optional flow). A typical hydroskimming refinery would include an atmospheric distillation tower (pipestill), a catalytic naphtha reforming unit, light-ends recovery and fractionation, treating, and blending. The pipestill performs initial distillation of crude oil into gas, naphtha, distillates and residual fuel. The naphtha may be separated into gasoline blending stocks, solvents and reformer feed. The distillates include kerosene, jet fuel, diesel and heating oil. The residual fuel is blended for use as a bunker fuel oil. Hydroskimming refineries are commonplace in regions with low need for conversion, either due to light feedstocks (crudes) or market demand. In other words, there is no need to alter the natural yield patterns of the crudes they process.

8.6.3   Conversion Refinery

A conversion refinery has the most complex, versatile and efficient configuration. A conversion refinery incorporates all of the basic units of topping and hydoskimming refineries, plus gas oil conversion plants (catalytic cracking and hydrocracking), olefin conversion plants (alkylation and polymerization) and residue conversion units (coking, RFCC, etc.). Many such refineries also incorporate solvent extraction processes for manufacturing lubricants and petrochemical units to recover ethylene, propylene, benzene, toluene and xylenes (BTX) for further processing into solvents and high polymers.

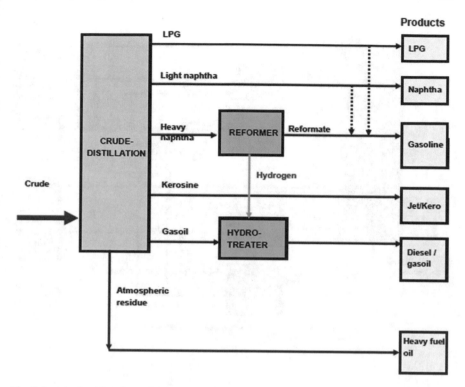

Fig. 8.6 A hydroskimming refinery [13]

Conversion refineries produce a whole range of gas, fuel and lubricant base oil products. They improve the natural yield patterns of the crudes they process as needed to meet market demands for light products, but they still produce some heavy, low-value products, such as residual fuel and asphalt.

In the recent past, one could say that there were two types of conversion refineries: the Cracking Refinery, shown in Fig. 8.7 for catalytic cracking, and the Coking Refinery, shown in Fig. 8.8. Cracking refineries transform heavy crude oils to gasoline, jet fuel, diesel fuel, and petrochemical feedstocks through catalytic cracking or hydrocracking or both. Coking refineries include coking in addition to catalytic cracking and hydrocracking to convert gas oil and residuum fractions. Coking units "destroy" the heaviest and least valuable crude oil fraction (*residual oil*) by producing cokes and lighter streams that serve as additional feed to other conversion processes (e.g., catalytic cracking) and to upgrading processes (e.g., catalytic reforming) that produce the more valuable light products. In general, all product streams are hydrotreated to remove residual sulfur and nitrogen (hydrotreaters are not shown in the figure).

Refinery size is usually measured in terms of atmospheric distillation capacity, as discussed in Sect. 8.3. In 1960, Wilbur L. Nelson developed Complexity Index relative to the atmospheric (initial) distillation to quantify the relative costs of the

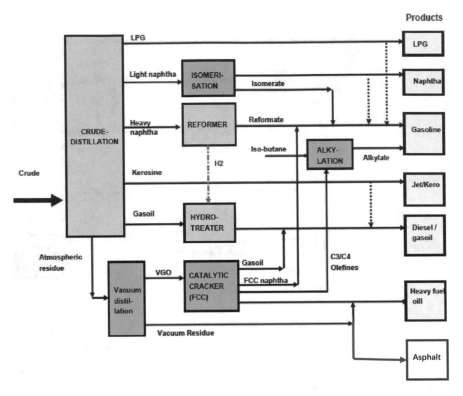

Fig. 8.7 A catalytic cracking refinery [13]

components that constitute the refinery (refinery units), listed in Table 8.11. The index is used not only as an indicator of the investment cost but also the value addition potential of a refinery. The sum of the products of the Nelson complexity index with the capacity of each unit (or the percentage of crude oil it processes) in the refinery is Equivalent Distillation Capacity (EDC) of the refinery in total as the benchmark of investment cost and manpower requirement.

8.6.4   Integrated Refinery

An integrated refinery includes all units of a conversion refinery as well as a lubricant and/or petrochemical manufacturing plant, making fuels, lubricants, olefins, and benzene/toluene/xylenes (BTX). It can also include chemical plants for producing solvents and high polymers. An integrated refinery except for the petrochemical plant units is shown in Fig. 8.9. Lubricant and petrochemical plants can be separated from the main refinery complex, but are located next to the complex through pipeline connections for feed transfer.

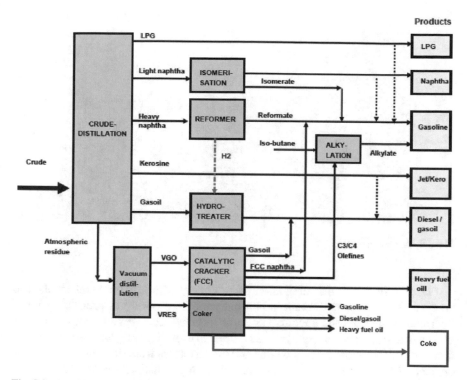

Fig. 8.8 A coking refinery [13]

Table 8.11 Equivalent distillation capacity index of refining units

Unit	Index	Unit	Index
Atmospheric distillation	1.0	Lubes	60.0
Vacuum distillation	2.0	Asphalt	1.5
Thermal processes	5.0	Hydrogen (Mcfd)	1.0
Catalytic cracking	6.0	Oxygenates (MTBE/TAME)	10.0
Thermal cracking	3.0	Catalytic reforming	5.0
Catalytic hydrocracking	6.0	Visbreaking	2.5
Catalytic hydrofining	3.0	Fluid coking	6.0
Alkylation/polymerization	10.0	Delayed coking	6.0
Isomerization/aromatization	15.0	Others	6.0

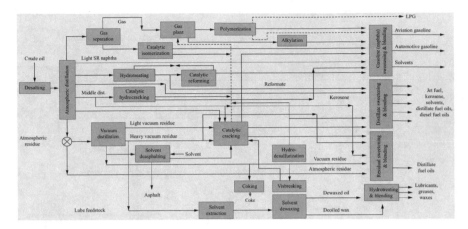

Fig. 8.9 Integrated refinery

Figure 8.10 shows the integration of a refinery with a petrochemical plant as an example of integrated refining. Normal paraffins from a gas processing plant, hydrotreated naphtha and upgraded bottoms from hydrocracker are sent to an olefin plant where steam cracking or pyrolysis is applied to produce ethylene, propylene and a mixture of C_4 olefins. The C_4 olefins and the C_4 stream from the fluidized catalytic cracker (FCC) are sent to C_4 processing plant to produce butadiene and butylene with butane as a co-product. The C_6–C_8 aromatics produced in the olefin plant and catalytic reformer are sent to BTX (benzene, toluene and xylenes) recovery units to separate into individual aromatic products. Benzene can be further hydrotreated (hydrogenated) to produce cyclohexane.

8.7   Refinery Configuration

The process flow schemes for a representative high-conversion refinery are shown in Figs. 8.11 and 8.12. We say "representative" because no two commercial refineries are exactly the same. They contain different types of process units, and the units are arranged differently. The configuration of a high-conversion refinery is very complex. Many units are connected to each other, as some product streams of some units are the feeds for the subsequent units. The processes for distillation cuts other than residua (atmospheric and vacuum) are shown in Fig. 8.11, and the processes for residua are shown in Fig. 8.12.

Prior to being subjected to refinery operations, crude oils are stored in storage tanks (tank farm) within or adjacent to the refinery. Different grades of crude oils are usually blended to meet the refinery feed requirement or specifications and refinery constraints. For example, high acid crudes are blended with low acid crudes according to the corrosion tolerance of the refinery equipment.

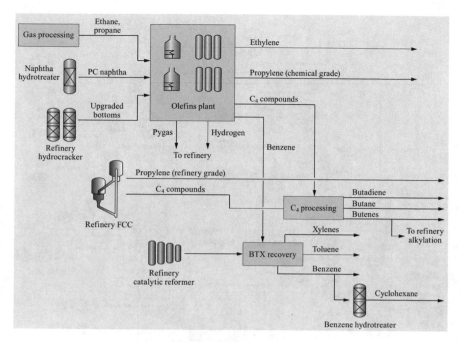

Fig. 8.10 Integration of a refinery and petrochemical plant

The first major process unit in a refinery is the **desalter**, which removes dirt, water, salt, and other water-soluble contaminants from blended crude oils. From the desalter, the crude(s) go to an **atmospheric distillation unit (ADU)**, which produces several streams. These include refinery fuel gas containing mostly methane and ethane, petroleum gas (C_3 and C_4) containing mostly propane and butanes, one or two straight-run naphtha streams, one or two straight-run middle distillates streams, and atmospheric residue (AR). All ADU products contain sulfur, either as H_2S, inorganic sulfur, or organic sulfur compounds. The heavier fractions also contain nitrogen, oxygen, and trace contaminants. At some point, with one major exception, **hydrotreating** removes organic sulfur and other contaminants from C_5+ liquid streams. Sulfur can be removed from relatively sweet naphtha with **mercaptan oxidation** (Merox) technology.

Here are the main destinations for individual streams from atmospheric distillation units, from the bottom of the tower to the top:

- The AR goes to a **vacuum distillation unit (VDU)**, which produces vacuum gas oil (VGO) and vacuum residue (VR). Refiners allocate VGO based on business requirements. Common VGO destinations include a fluid catalytic cracker (**FCC**), a **hydrocracker,** a **lube base stock facility**, and **fuel-oil blending**. VR can be a coker feed.
- Either before or after hydrotreating, straight-run middle distillates are divided into light and heavy fractions, which become kerosene or gas oil, respectively. Both

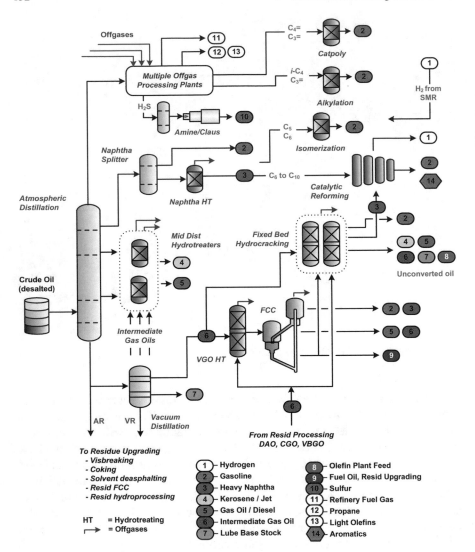

Fig. 8.11 Oil refinery process units: upgrading processes for top-of-the-barrel cuts—VGO and lighter—which boiling below about 1050 °F (556 °C). Please refer to the key for stream descriptions. Almost all units produce some kind of offgas; offgas processing is described above. No single drawing can show all the complexities of a high-conversion refinery. Polymerization unit for poly gasoline and a visbreaker are not shown in this figure. Please refer to the text for additional details

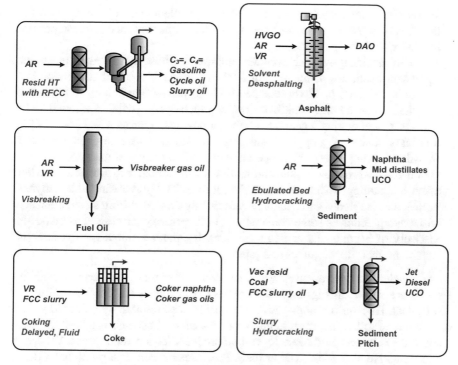

Fig. 8.12 Oil refinery conversion processes for bottom-of-the-barrel cuts, which can include either atmospheric residue (AR) and/or vacuum residue (VR). Commercially, RFCC processes mostly AR Ebullated-bed hydrocracking processes either AR or AR/VR blends. Slurry hydroprocessing handles VR, coal, FCC slurry oil, and/or coal tar. Delayed coking processes VR. Please refer to the text for additional details

can serve as fuel oil. If it meets specifications, the kerosene can be blended into jet fuel and the gas oil can be blended into diesel. Certain refineries produce just one middle distillate stream, typically wide-range gas oil.

- Naphtha is split into light and heavy fractions. Desulfurized light naphtha (LN) includes C_5 molecules along with differing amounts of C_4 and C_6. LN can be blended into gasoline or separated into constituent components. Small amounts of mixed butanes can be blended into gasoline. N-butane is sold as such or isomerized into i-butane. I-butane goes to **alkylation (ALKY)** units or is dehydrogenated to C_4 olefins for chemical plants. Without excess isobutane, C_3–C_5 olefins can go to a catalytic polymerization unit to produce gasoline. Heavy naphtha (HN) includes C_6–C_{12} molecules. **Catalytic reforming (CRU)** and **isomerization (ISOM)** units are common HN destinations. C_6 compounds may be excluded from CRU feeds if a refiner wants to minimize benzene production. C_9–C_{12} molecules can be blended into aviation gasoline.
- H_2S is removed from refinery fuel gas by adsorption in alkanolamine units. The H_2S is converted to elemental sulfur in **Claus process** units. Sweet refinery fuel

gas serves as fuel gas throughout the refinery. Petroleum gas (C_3 and C_4) is sold as liquefied (LPG) or exported to a chemical plant, either before or after segregation into individual components.

- Thermal cracking units, such as **cokers** and **slurry-phase hydrocrackers**, produce significant quantities of methane and ethane. In catalyst-based conversion units—FCCs and fixed-bed hydrocrackers—C_1/C_2 production is low, and for paraffins, the iso/normal ratio is high. Processes which operate without external hydrogen—FCC and coking—generate olefins, aromatics, hydrogen and coke. FCC is a primary source of propylene and butylenes for downstream chemical plants. In hydrocracking units, which operate with large amounts of external hydrogen, olefins and coke are low, product aromatics are lower than feed aromatics and light alkanes including isobutane for alkylation feed. The polyolefin feeds, ethylene, propylenes, butylenes and butadiene, are mainly produced from steam cracking.
- Visbreaking achieves some conversion, but its primary purpose is to reduce the viscosity of heating oils and fuel oils, making them suitable for furnaces and improving their flow in tubes and pipes.

As much as anything else, VR conversion determines refinery profitability. Options are shown in Fig. 8.12. The cost of crudes depends on their sulfur and VR content. High-sulfur, high-VR crudes cost less, so refineries configured to handle them tend to be more profitable. VR is a source of asphalt, bright-stock lubricants, greases, waxes, and specialty carbon products. To a certain extent, VR can be blended into No. 6 fuel oil (bunker fuel). Processes for converting AR and VR into lighter products include **delayed coking**, **visbreaking**, modified versions **residue FCC (RFCC)**, and **residue** hydrocracking with either ebullated bed or slurry-phase technology.

Individual process units can be significantly different, but they have several things in common:

- **Heat**. All units require heat, which is supplied by fuel gas or fuel oil, but sometimes just by steam. Even if external heat isn't necessary during normal operation of a particular unit, heat is always required for startup. In high-conversion refineries, off-gases from operating units might provide all of a refinery's fuel gas needs. Others refineries are "fuel gas long," meaning that they generate more gas than they can legally burn in a flare.
- **Power and Steam**. All units require electric power and most use one or more grades of steam. FCC units and sulfur plants are net producers of high-grade steam, but they usually import low-grade steam from elsewhere in the refinery.
- **Amine Adsorption: Recovery of Sulfur and CO_2**. H_2S-containing "sour" off-gases, both olefinic and non-olefinic, are treated in multiple plantwide amine systems, which remove the H_2S (and any CO_2 which happens to be present) by acid/base adsorption and transport them the refinery sulfur plant. Sulfur plants employ the Claus process, in which H_2S is converted to elemental sulfur via combustion. Claus tail-gas plants boost recovery to >99.9%.
- **Offgas Processing**. Offgas is a general term for gases produced by refinery units. Depending on their source, offgases can include H_2, N_2, H_2S, light paraffins

(methane, ethane, propane and butanes), and light olefins (ethylene, propylene, and butenes.) As mentioned, H_2S is removed from all gases by amine adsorption and converted into elemental sulfur. The purified "sweet" gases go to either to a **saturated gas plant** (sat gas plant) or an **unsaturated gas plant** (unsat gas plant). The sat gas plant treats offgases from hydrotreaters and catalytic hydrocrackers, and it can also co-process natural gas. Feeds to the unsat gas plant contain ethylene, propylene, and butylenes; most olefins are produced by FCC and coking units. Products from the gas plants become refinery fuel gas, which contains H_2, N_2, ethane, and/or ethylene; **LPG**—liquid petroleum gas, which contains mostly propane with limited amounts of butanes; and one or more **refinery olefin streams.** Refinery fuel gas is burned to supply process heat. LPG can be sold as-is or fractionated to purify the propane. Isobutane goes to refinery alkylation units, where it reacts with propylene and butylenes to form high-value C_7–C_8 alkylate, excellent gasoline blend stock. Purified ethylene and propylene go to polyolefin plants. Refinery propylene is a primary source of polypropylene.

- **Hydrogen** is required by hydrotreaters, hydrocrackers, isomerization units, catalytic dewaxing units, and others. Hydrogen is produced by catalytic reformers, steam/methane reformers, and nearby olefin plants. FCC and delayed coking units also produce significant amounts of low-purity hydrogen, but it is costly to recover and use that hydrogen.
- **Over-the-fence gases**. It is common in certain areas for refineries to purchase "over-the-fence" hydrogen and nitrogen from industrial gas suppliers. Nitrogen is required during shutdowns and startups as an inert gas for protection of the interior of the equipment.
- **Instrument Air**. Valves are opened and closed with actuators driven by "instrument air." It is important to keep the air clean and dry to prevent the fouling of lines.

8.8 Refinery Fuel Products

Figure 8.13 illustrates how streams from individual process units might be blended into four major groups of finished fuel products. (A fifth major group—lubricants—is discussed separately.) As shown, the gasoline blender prepares finished gasoline from 12 different streams. Oxygenates (ethanol) are added separately to the blender product, called Reformulated Gasoline Blendstock for Oxygen Blending (RBOB gasoline), commonly known as reformulated gasoline (RFG). In some regions, such as Southern California, detergent additives are added at blending and distribution terminals, some of which are located hundreds of miles away from any refinery. The figure over-simplifies middle distillate blending.

Kerosene and jet fuel are shown in the same box, because their boiling ranges are similar. The cut points of certain streams are changed based on market demand, and some streams can go to two or three places. Depending on how it is cut, straight-run light gas oil can become jet fuel, kerosene, or diesel.

Refineries can produce several grades of fuel oil. The bottom line is: A material isn't gasoline unless it meets a specification, such as ASTM D-4814 in the United States and many other countries, or EN 228 in Europe.

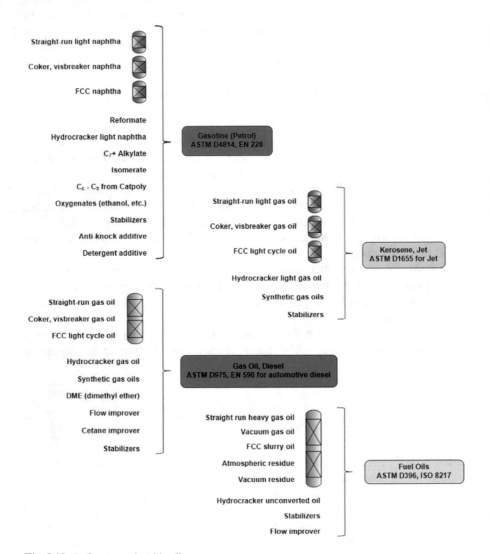

Fig. 8.13 Refinery product blending

References

1. Berthelot M Collection des anciens alchimistes grecs (3 vol, Paris, 1887–1888, p 161); F. Sherwood Taylor, "The Origins of Greek Alchemy," Ambix 1 (1937), 40
2. Energy Timelines. Energy Kids. http://www.eia.gov/petroleum/. Retrieved 3 Dec 2015
3. Pees ST (2004) Whale oil versus the others. In: Oil history, Samuel T. Pees, Meadville, PA
4. Frank F (2007) Oil Empire: visions of prosperity in Austrian Galicia. Harvard University Press
5. Hsu CS, Robinson PR (2006) Practical advances in petroleum processing. Springer, New York, NY
6. OSHA Technical Manual (2005) Section IV, Chapter 2, "petroleum refining processes". U.S. Department of Labor, Occupational Safety and Health Administration, Washington, DC
7. Falling crack spreads cloud near term for refiners. http://seekingalpha.com/article/1468771-falling-crack-spreads-cloud-near-term-for-refiners. Retrieved 3 Dec 2015
8. U.S. Energy Information Administration: Detailed breakdown. http://www.eia.gov/dnav/pet/pet_pnp_pct_dc_nus_pct_m.htm. Retrieved 3 Dec 2015
9. (a) U. S. Energy Information Administration: Top 10 US Refineries Operable Capacity. Jan. 1, 2018 database, published June 2018. http://www.eia.gov/petroleum/refinerycapacity, (b) https://www.eia.gov/petroleum/refinerycapacity/table4.pdf. Retrieved 30 Sept 2018
10. List of Oil Refineries. https://en.wikipedia.org/wiki/List_of_oil_refineries. Retrieved 1 Oct 2018
11. (a) Asia-Pacific refining primed for capacity growth. Gas Oil J 112(12), Dec. 1, 2014. pp 34–45; (b) https://www.statista.com/statistics/273579/countries-with-the-largest-oil-refinery-capacity/. Retrieved 1 Oct 2018
12. An Introduction to Petroleum Refining and the Production of Ultra Low Sulfur Gasoline and Diesel Fuel. The International Council on Clean Transportation, 24 Oct 2011
13. Hauge K (2015) Refining ABC. http://www.statoil.com/en/InvestorCentre/Presentations/Downloads/Refining.pdf. Retrieved 3 Dec 2015

Chapter 9
Crude Storage, Blending, Desalting, Distillation and Treating

Large quantities of crude oils are transported through pipelines or tankers. At the marine terminal, a cargo of crude oil may be routed through a pipeline directly to a storage tank in the refinery tank farm, or transferred to holding tank, where it is kept temporarily before going to the refinery. The marine terminal also has berths to load refined products. Storage tanks are containers that hold the crude oil or compressed gases (gas tanks), as well as intermediate stocks (partially refined), finished products and chemicals for the short- or long-term storage.

9.1 Crude Storage and Blending

Crude oils shipped to the refinery are stored in storage tanks. Liquid storage tanks are often cylindrical in shape, perpendicular to the ground with flat bottoms, with a fixed frangible or floating roof. Gas tanks for compressed or liquefied gas are in spherical shape.

Different grades of crude oils are stored in different tanks. Care must be taken to avoid mixing incompatible crudes, which can cause precipitation (due to deasphalting)., and fouling. Above ground storage tanks can also be used to hold blended crudes, refined products, water, waste matter, and hazardous materials, while meeting strict industry standards and regulations.

A refinery seldom refines a single crude. Prior to being charged to the refinery units, crude oils and imported stocks are unloaded to storage tanks, then blended and transferred to surge tanks for individual units. Proper blending improves distillation unit throughput and the performance of downstream units. It also can improve product quality and reduce energy cost [1]. Crude blending is based on computer models, which are used in conjunction with scheduling models and operations plans. The models and plans include timing, desired volumes, etc. They are unique for every refinery, because they depend on the refinery configuration, logistics constraints, tank inventory, feedstock composition, and local (and global) market forces. For operations planning, some refiners still use comparatively simple stand-alone LP

© Springer Nature Switzerland AG 2019
C. S. Hsu and P. R. Robinson, *Petroleum Science and Technology*,
https://doi.org/10.1007/978-3-030-16275-7_9

(linear program) models, which include just a limited number of parameters (5 or 6) for important process units. But the industry is moving toward sophisticated non-linear optimizers, which are linked to fully rigorous process models. The non-linear optimizers include site-specific scheduling and some are linked to corporate models which include multiple refineries and chemical plants. Important advanced features include back-casting (a comparison of previous plans with actual operation) and trader-assistance applications.

Several crudes with similar molecular characteristics are usually blended together. Incompatibilities between crude oils can occur, for example, between paraffinic crude oil and heavy asphaltic crude oil. Blending is used to maintain, as nearly as possible, constant feed quality for the crude distillation unit. For example, high acid crudes are mixed with low acid crudes, and high-sulfur crudes are mixed with low sulfur crudes to bring the acid or sulfur contents to a tolerable level for refinery operations.

9.2 Desalting

Crude oil introduced to refinery processing contains many undesirable impurities, such as sand, inorganic salts, drilling mud by-products, polymers, corrosion byproducts, etc. The salt content in the crude oil varies depending on source of the crude oil and how it is transported. When a mixture from many crude oil sources is processed in refinery, the salt content can vary greatly.

These undesirable impurities, especially salts and water, need to be removed prior to distillation. Two processes are used: dewatering and desalting. Dewatering removes water and constituents of brine that accompany the crude oil. Desalting is a water-washing operation that removes water-soluble minerals and entrained solids. Both operations can be performed at the production field, at the refinery, or both places.

The impurities of most concern in crude oil include:

- The inorganic salts, which can be decomposed to form acids or alkali in the crude oil pre-heat exchangers and heaters. Salts are most frequently present in crude oil are $CaCl_2$, $NaCl$ and $MgCl_2$, which can form hydrogen chloride when heated in the presence of water. Hydrogen chloride gas condenses to aqueous hydrochloric acid at the overhead systems of distillation columns, causing serious corrosion of equipment.
- Naphthenic and carboxylic acids, as measured by the total acid number (TAN) test, also induce corrosion.
- The sand, silt, or salt cause deposits and foul heat exchangers. They can also cause significant damage to pumps, pipelines, etc. due to abrasion or erosion.
- Inorganic and organic compounds of sodium, calcium nickel, vanadium, iron, arsenic, and other metals in the crude can poison and deactivate catalysts.

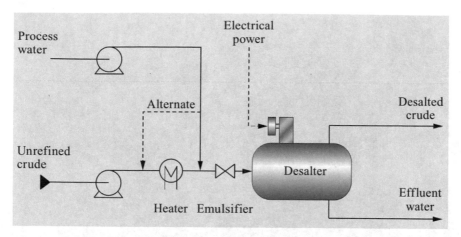

Fig. 9.1 Flow diagram of desalting involving an electrostatic desalter

- Organo-silicon compounds are added to crudes to improve flow rates through pipelines and pumps. These Si compounds degrade catalysts by weakening alumina or aluminosilicate supports. They are only partly removed by desalting.

At the refinery, the crude is treated with hot water in one or more desalters. Desalters employ either chemical or electrostatic precipitators to remove dissolved salts and collect remaining solids. In chemical desalting, water and surfactants are added to the crude, heated to dissolve salts and other impurities, then sent to a settling tank, where the water and oil separate.

Figure 9.1 shows a flow diagram of a typical desalting unit. The blended crude is mixed with 3–10 vol.% water at 200–300 °F to dissolve salts in the crude oil and to reduce viscosity and surface tension for easier mixing and subsequent separation of the water. The oil/water mixture is homogenously emulsified. Then the emulsion enters the desalter vessel, where it separates into oil and water phases under the influence of electrostatic coalescence, in which a high-potential field (typically 12–35 kV), is imposed across a pair of electrodes, shown in Fig. 9.2. The electrostatic field promotes coalescence with polarization. Polarization of water droplets pulls them out from the oil-water emulsion phase. The salt dissolved in these water droplets is also separated along the way. Coalescence is aided by passage through a tower packed with sand, gravel, and the like.

An emulsion between oil and water can also be broken by adding chemicals or treating agents, such as soaps, fatty acids, sulfonates, long-chain alcohols or other de-emulsifiers.

A desalting unit can be designed with single stage or two stages, shown in Fig. 9.3. Two-stage desalting system is normally applied, that consists of 2 electrostatic coalescers (desalter). For resid processing, 3-stage desalting is used for some crudes.

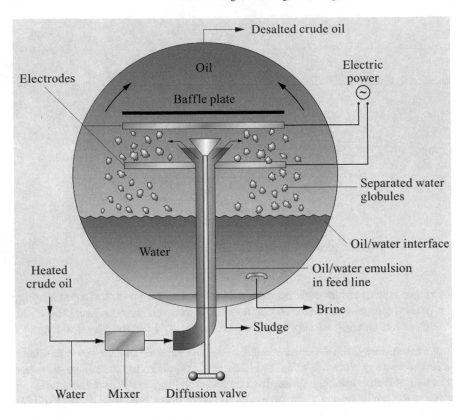

Fig. 9.2 Cross-section view of a desalter

If the crude isn't desalted, residual solids can clog downstream equipment and deposit on heat exchanger surfaces, thereby reducing heat-transfer efficiency. Salts can induce corrosion in major equipment and deactivate catalysts.

9.3 Distillation

Distillation can be considered as the heart of any refinery. Crude oils are made of numerous components of different boiling points. The simplest way to separate them is with continuous distillation into different fractions (distillates or cuts) of various boiling ranges. At just above atmospheric pressure, heavy molecules in most crude oils decompose above 650 °F (~350 °C). To achieve additional separation of heavy fractions, continuous distillation is carried out under vacuum (i.e., reduced pressure-typically at 40 mmHg) to boil out additional components. Vacuum distillation increases the yield of total distillates. The relationship between boing points under atmospheric pressure and under 40 mmHg vacuum is shown in Fig. 2.1. For

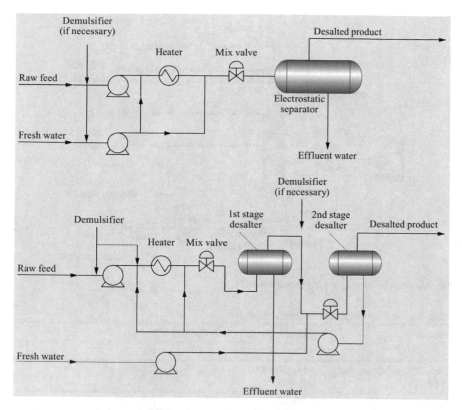

Fig. 9.3 Single- and Two-Stage electrostatic desalting systems

example, at 500 °F under 40 mmHg vacuum, the compounds with boiling point at 750 °F under atmospheric pressure can be boiled out. Hence, the atmospheric equivalent boiling point (AEBP) at 500 °F under 40 mmHg vacuum is 750 °F. The AEBP of 650 °F is ~900 °F. Hence, additional components that boil between 650 and 900 °F can be distilled out of the crude oil under 40 mmHg vacuum without severe decomposition. The upper temperature for vacuum distillation in refineries can be slightly higher, up to 1050 °F. At this temperature, there is a greater tendency for thermal decomposition, but the decomposition does not occur immediately; the feed flows out of the column and undergoes cooling before any damage is done.

Figure 9.4 shows a simplified diagram for crude oil distillation. There are many trays in a distillation tower (also called a pipestill or column), which will be discussed later. The desalted crude is introduced near the bottom of the atmospheric distillation tower. The lightest fractions, gases and naphtha, flow out of the tower at the top as an overhead stream, which can be further fractionated into separate gas and naphtha streams. The effluents from middle trays are heavy naphtha, kerosene, light gas oil and heavy gas oil. Kerosene and light gas oil are often referred also as middle distillates or distillate fuels that include kerosene, jet fuel and diesel. Steam is introduced to

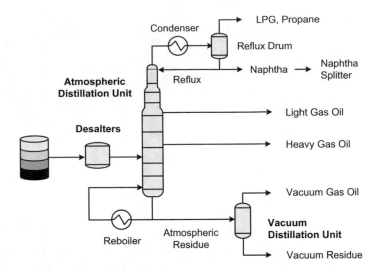

Fig. 9.4 Simplified diagram of crude distillation

Table 9.1 Boiling ranges of distillation fractions

Fraction	Boiling range (to the nearest 5°)	
	°C	°F
Liquefied petroleum gas (LPG)	−40 to 0	−40 to 30
Light naphtha	30–85	80–185
Gasoline	30–200	90–400
Heavy naphtha	85–200	185–390
Kerosene	170–270	340–520
Light gas oil	180–340	350–650
Heavy gas oil	315–425	600–800
Lubricant oil	>400	>750
Vacuum gas oil	340–565	650–1050
Residuum	>540	>1000

improve separation by reducing the partial pressures of hydrocarbons. The residue, or bottom, of the atmospheric pressure distillation tower, also called as reduced crude, is introduced into a vacuum distillation unit (tower) to be fractionated into vacuum gas oil and vacuum residue.

The boiling ranges of the distillation fractions are listed in Table 9.1. These numbers are only used as references. The actual cut points may vary at various refineries, depending on the season and market demands.

The distillation cuts obtained in refineries are not well-defined. There are always overlaps between adjacent cuts, as shown in Fig. 9.5 for a desired cut point of 315 °F

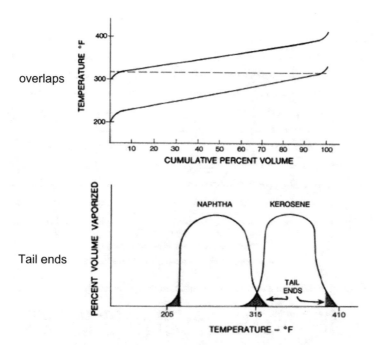

overlaps

Tail ends

Fig. 9.5 Overlaps and tail ends of distillation cuts [2]

between naphtha and kerosene. The leading edge of the kerosene distillation curve overlaps with the tailing edge of naphtha. The overlap between cuts can be "sharp" or "sloppy," depending on several factors, especially oil flow rates, steam flow rate, and heat balance. The designations are arbitrary, and often nothing can be done to decrease overlap at maximum flow rate without making hardware changes. Overlaps exist for all distillation cuts, in some commercial units, the overlaps are very large. Such "sloppy cuts" are more common in units running far above (or far below) their design feed rate.

Figure 9.6 demonstrates the application of boiling point overlaps. The desired cutpoints for light naphtha, heavy naphtha, kerosene, and heavy diesel are 90 °F (32 °C), 190 °F (88 °C), 300 °F (149 °C), and 525 °F (274 °C), respectively. Due to operational constraints, the observed effective cutpoints are 99 °F (37 °C), 188 °F (87 °C), 302 °F (150 °C), and 523 °F (273 °C), respectively. The overlap between kerosene and heavy diesel is considerable. If we move the initial boiling point (IBP) for the heavy diesel at 360 °F (182 °C) instead of 523 °F (274 °C), it would result in considerable entrainment of valuable kerosene into the far-less-valuable bottom product.

A petroleum refinery can adjust distillation yields to meet market demands, in part, just by adjusting cut points. The swing cut between 150 and 205 °C can go into any of the three products—naphtha (gasoline), kerosene (jet fuel) and gas oil (diesel and heating oil), shown in Fig. 9.7, depending on the seasonal and market demands.

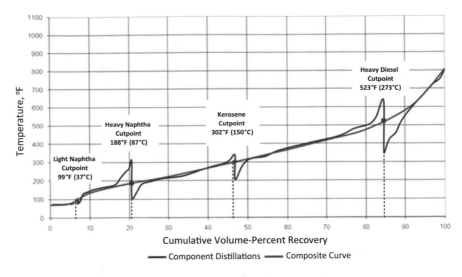

Fig. 9.6 Overlap in distillation curves from a commercial hydrocracker

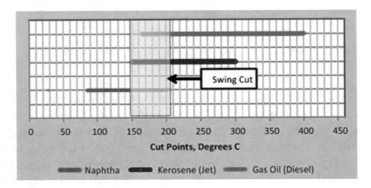

Fig. 9.7 Swing cut region for naphtha, kerosene and gas oil

In summer "driving season," the cutpoint is moved to higher temperatures to increase the yield of transportation fuels. In winter, the cutpoint is set at lower temperatures to increase the yield of heating oil (gas oil heavier than diesel).

9.3.1 Atmospheric Distillation

Figure 9.8 shows a flow diagram of an atmospheric distillation unit. The crude oil enters a desalter at 250 °F to remove salt and water, as described before. The desalted oil goes through a network of pre-heat exchangers to a fired heater, which brings the temperature up to 657–725 °F (347–385 °C). If the oil gets much hotter than this,

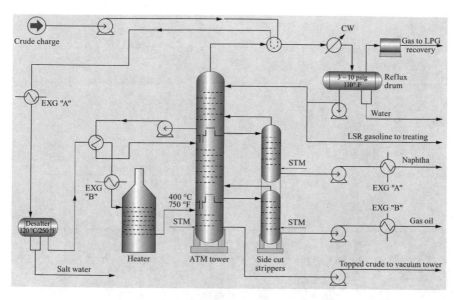

Fig. 9.8 Flow diagram of an atmospheric distillation tower

it starts to crack, generating carbon. The carbon would deposit inside the pipes and equipment through which the oil flows.

The hot crude enters the tower just above the bottom, as shown in Fig. 9.8. Steam is added at the bottom to enhance separation; it does so largely by decreasing the vapor pressure of hydrocarbons in the column. When it enters the tower, most of the oil vaporizes. The steam flows upward with vaporized crude while the condensed liquid flows downward as in a countercurrent fashion. The hottest trays are in the bottom section with the coolest at the top section. Unvaporized oil drops to the bottom of the tower, where it is drawn off.

Products are collected from the top, bottom and side of the column. Side-draw products are taken from trays where the temperature corresponds to the cutpoints for a desired product (naphtha, kerosene, light gas oil and heavy gas oil). Some of the side-draws can be returned to the tower as a pump-around or pump-back stream to control tower temperatures and improve separation efficiency.

Two side cut strippers for naphtha and gas oil are shown as examples. There can be additional strippers for kerosene (jet fuel) and diesel (light gas oil). Also not shown is the reboiler at the bottom of the tower; this will be discussed later. An atmospheric distillation tower usually contains 30–50 fractionation trays, with 5–8 trays in each section. Product strippers for cut streams also have 5–8 trays. The strippers remove entrained light components from liquids. Stripper bottom streams can be drawn off as products of a specific boiling range or returned to the distillation tower.

Inside the distillation tower (also called pipestill or column), the vapors rise through the distillation trays, which contain perforations, bubble caps, downcomers, and/or modifications thereof, shown in Fig. 9.9 (perforations on the trays not

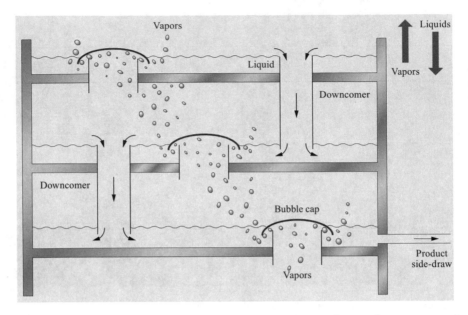

Fig. 9.9 Bubble cap and downcomer on a distillation tray. Intermediate products are removed through side-draw trays

shown). Vapors and liquids flow counter-currently. Each tray permits vapors from below to bubble up through the relatively cool condensed liquid on top of the tray. This vapor/liquid contact knocks heavy material out of the vapor. Condensed liquid flows down through a pipe (downcomer) to the hotter tray below, where the higher temperature causes re-evaporation. A given molecule evaporates and condenses many times before finally leaving the tower.

Figure 9.10 is another drawing for an atmospheric distillation tower. The section above the feed tray is called enriching, or rectification, section and the section below the feed trap is stripping section. Gas and naphtha are withdrawn from the top tray as overhead. After condensation, a portion of the liquid is introduced back to the top tray as reflux. Reflux also controls temperature in the enriching section. It also controls entrainment of heavier components in the naphtha (the lightest distillate). Reflux ratio is the amount of reflux liquid returning to distillation tower divided by the amount of liquid withdrawn as product per unit time. With a higher reflux ratio, fewer theoretical plates are required.

At the bottom of the tower, the liquid (atmospheric residue) passes through a reboiler to recover light components from the heavy liquid. The reboiler helps control temperatures in the stripping section.

The bottom stream from the main fractionator (atmospheric distillation tower) is called atmospheric bottoms, atmospheric residue, reduced crude, topped crude, or long resid.

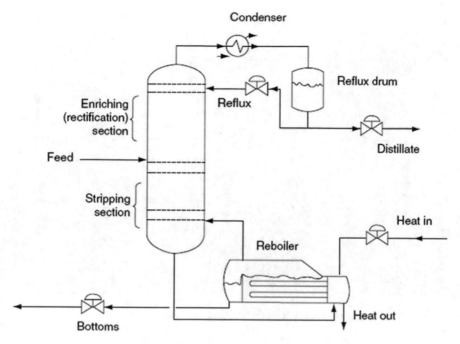

Fig. 9.10 Atmospheric distillation (Fractionation) tower with a reflux for overhead and a reboiler for bottoms

9.3.2 Vacuum Distillation

Figure 9.11 shows a flow diagram of a vacuum distillation unit. The atmospheric residue goes to a fired heater, where the typical outlet temperature is about 730–850 °F (390–450 °C). From the heater, the atmospheric residue goes to a vacuum distillation tower. Steam ejectors reduce the absolute pressure to 25–50 mmHg vacuum, or about 7.0 psia (0.5 bara). Under vacuum, hydrocarbons vaporize at lower temperatures than atmospheric boiling points. For example, the equivalent atmospheric boiling point of 800 °F under 40 mmHg vacuum is ~1050 °F. Thus, molecules with normal boiling points above 650 °F (343 °C) are less likely to undergo thermal cracking and can be vaporized at lower temperatures. There are fewer trays than the atmospheric distillation tower to fractionate the topped crude into light vacuum gas oil, heavy vacuum gas oil and vacuum residuum at the bottom. As in atmospheric distillation, some gas and light components entrained or decomposed during heating in furnace are carried out at the top of the tower.

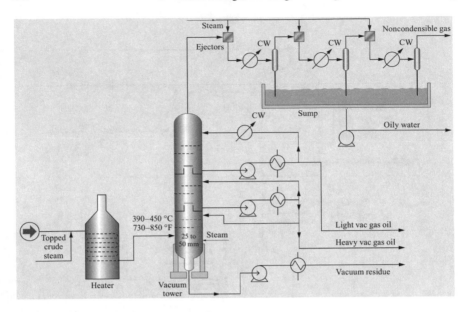

Fig. 9.11   Flow diagram of a vacuum distillation Unit

9.3.3   Distillation Yields of Straight Run Fractions

The products from distillation prior to upgrading are called straight-run products. At a given set of cutpoints, the yields of different fractions depend on the crude oil being processed. Figure 9.12 shows TBP of a light and a heavy crude for kerosene yield (cut points between 315 and 450 °F). The light crude yields more kerosene than the heavy crude, and hence has a higher value.

Table 9.2 lists a few crude oils and their typical straight-run yields. Total naphtha includes light, medium and heavy naphtha, and the middle distillates include kerosene and atmospheric gas oil. Naphtha is used for making gasoline and aromatics, kerosene for jet fuel and atmospheric gas oil for diesel. Table 9.3 shows that the demand for transportation fuels exceeds the straight-run yields for the crudes in Table 9.2. Obviously, crudes containing less heavy material—VGO and vacuum residue are more valuable.

The higher-valued crude oils, such as Brent and Bonny Light, have higher API gravity with higher naphtha and middle distillate yields. They tend to have less sulfur. The oil having high sulfur content increases processing costs because the sulfur must be removed. Hence, the oils, such as Green Canyon and Ratawi, have lower values. Since sulfur is not removed during distillation, the straight-run distillation products have to be treated for sulfur removal.

Products from the crude distillation unit, i.e., the straight-run distillates, go to other process units, as shown in Table 9.4. The lightest cuts are gas and light naphtha. The gas goes to a gas processing plant or is liquefied into liquefied petroleum gas (LPG).

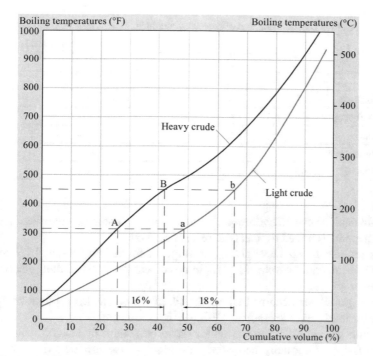

Fig. 9.12 Comparison of kerosene yields from a light and a heavy crude oil [2]

Table 9.2 Typical straight-run yields from various crudes

Source field	Brent	Bo liny Lt.	Green Canyon	Ratawl
Country	Norway	Nigeria	USA	Mid East
API gravity	38.3	35.4	30.1	24.6
Specific gravity	0.8333	0.8478	0.8752	0.9065
Sulfur, wt%	0.37	0.14	2.00	3.90
Yields, wt% feed				
Light ends	2.3	1.5	1.5	1.1
Light naphtha	6.3	3.9	2.8	2.8
Medium naphtha	14.4	14.4	8.5	8.0
Heavy naphtha	9.4	9.4	5.6	5.0
Kerosene	9.9	12.5	8.5	7.4
Atmospheric gas oil	15.1	21.6	14.1	10.6
Light VGO	17.6	20.7	18.3	17.2
Heavy VGO	12.7	10.5	14.6	15.0
Vacuum residue	12.3	5.5	26.1	32.9
Total naphtha	30.1	27.7	16.9	15.8
Total middle distillate	25.0	34.1	22.6	18.0

Table 9.3 Average U.S. consumption of petroleum products, 1991–2003

Product	Consumption (barrels/dav)	Percent of total (%)
Gasoline	8,032	43.6
Jet Fuel	1,576	8.6
Total distillates	3,440	18.7
Residual fuel oil	867	4.7
Other oils	4,501	24.4
Total consumption	18,416	100
Sum of gasoline, Jet and distillates	13,048	70.8

The light naphtha can be hydrotreated and sent to the motor gasoline blending pool. Heavy naphtha is a feed for catalytic reforming units.

Kerosene can be used for lighting, heating, and for making jet fuel. In either case, it must first undergo hydrotreating. Light gas oil can go to diesel fuel (distillate fuel oil) blending.

Heavy gas oil can become fuel oil, diesel, lube base stock, or a light component of feed for fluid catalytic cracking (FCC) or hydrocracking.

Vacuum gas oil (VGO) and vacuum resid (residuum or residue, VR) are low valued. They are normally converted into higher-value products through various upgrading processes, as in a conversion refinery. VGO can become fuel oil or lube base stock, but its primary destinations are FCC and hydrocracking units, which are discussed in subsequent chapters. Figure 8.12 gives more details of the possible destinations of vacuum resid which is also known as "bottom of the barrel".

9.4   Hydrogenation, Dehydrogenation and Condensation

The saturation of aromatics and polyaromatics (Fig. 9.13) is common to several petroleum refining processes. It is especially important in catalytic reforming, hydrotreating, hydrocracking, and cyclohexane production. Aromatics saturation is reversible, i.e., hydrogenation is accompanied by dehydrogenation, except when saturated C-C bonds are cracked or when polyaromatics condense to form larger polyaromatics. Carbon-carbon single bonds in polynaphthenes and naphthenoaromatics can be broken irreversibly, but bonds within aromatic rings are more stable and difficult to break. Condensation is a reaction in which two ring molecules are combined into one. Aromatic ring condensation is accompanied by dehydrogenation.

Figure 9.14 summarizes thermodynamic calculations for the competition between the saturation of naphthalene and the condensation of naphthalene with o-xylene to form chrysene. At high pressures and low temperatures, equilibrium favors saturation. At low pressures and high temperatures, equilibrium favors dehydrogenation. At high-enough temperatures, equilibrium favors condensation. Fixed-bed catalytic

Table 9.4 Destinations for straight-run distillates

Fraction	°C	°F	Next Destination	Ultimate product(s) or Subsequent destination
LPG	−40 to 0	−40 to 31	• Sweetener	• Propane fuel
Light naphtha	39–85	80–185	• Hydrotreater	• Gasoline
Heavy naphtha	85–200	185–390	• Cat. reformer	• Gasoline, aromatics
Kerosene	170–270	340–515	• Hydrotreater	• Jet fuel, No. 1 diesel
Gas oil	180–340	350–650	• Hydrotreater	• Heating Oil, No. 2 diesel
Atmospheric resid	340+	650+	• Visbreaker	• FCC or hydrocracker feed, low-viscosity resid
			• Resid hydrotreater	• Resid FCC
			• Ebullated bed hydrocracker	• Naphtha, gas oils, FCC
Vacuum gas Oil	340–566	650–1050	• FCC	• Gasoline, LCO, gases including C_3/C_4 olefins
			• Hydrotreater	• Fuel oil, FCC, lubes
			• Hydrocracker	• Naphtha, jet, diesel, FCC, olefins, lubes
			• Solvent refining	• DAO, asphalt
Vacuum resid	540+	1000+	• Coker	• Coke, coker gas oil, coker naphtha, gases
			• Solvent refining	• DAO, asphalt
			• Slurry-phase hydrocracker	• Traditional hydrotreater or hydrocracker

Resid is an abbreviation for residua, and 340+ (etc.) means everything that boils above 340 °C (etc.)
LCO: light cycle oil; *FCC:* fluidized catalytic cracking

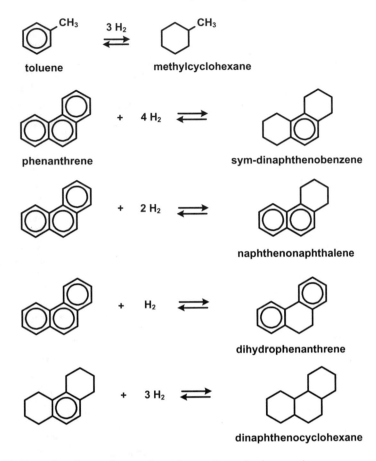

Fig. 9.13 Examples of saturation reactions of aromatics and polyaromatics

hydrotreating and hydrocracking units operate in the aromatics crossover region between 315 and 425 °C (600 and 800 °F).

Figure 9.15 shows the so-called zig-zag mechanism for the production of large polyaromatics (aromatic ring growth) by adding 2-carbon (*cata*-condensation) and 4-carbon (*peri*-condensation) species [3]. The condensation of large polyaromatics via the Scholl reaction can lead eventually to coke formation and deactivation of the catalysts used in FCC, hydrotreating and hydrocracking.

Figure 9.16 shows a mechanism for the one-at-time buildup of rings on a nucleus of coke, or ring growth of aromatics. The mechanism includes the following steps, all of which are to some extent reversible:

- Hydrogen abstraction by a radical
- Alkylation (olefin addition)
- Cyclization
- Dehydrogenation

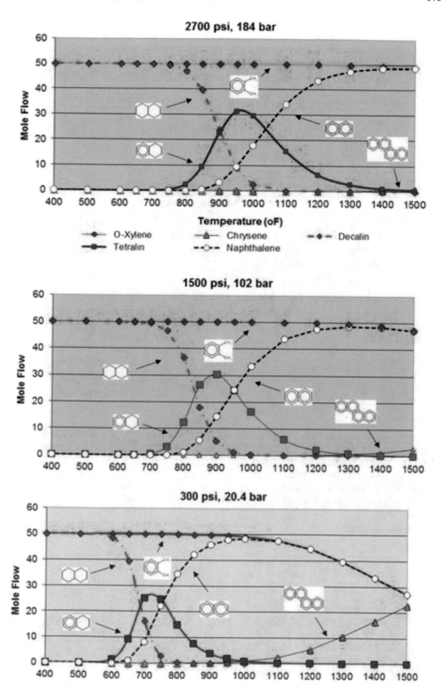

Fig. 9.14 Thermodynamic calculations illustrating the competition between the saturation (hydrogenation) and the condensation (dehydrogenation) of polyaromatics. Data for the graphs were generated by Aspen Plus for a six component system comprising naphthalene ($C_{10}H_8$), tetralin ($C_{10}H_{12}$), decalin ($C_{10}H_{18}$), o-xylene (C_8H_{10}), chrysene ($C_{18}H_{12}$) and hydrogen (not shown)

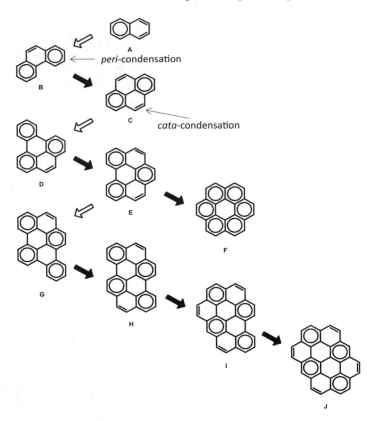

Fig. 9.15 Zig-zag mechanism for the condensation of polyaromatics by sequential addition of 2-carbon and 4-carbon units. The isomers shown are **a** naphthalene, $C_{10}H_8$; **b** phenanthrene, $C_{14}H_{10}$; **c** pyrene, $C_{16}H_{10}$; **d** benzo[e]pyrene, $C_{20}H_{12}$; **e** benzo[ghi]perylene, $C_{22}H_{12}$; **f** coronene, $C_{24}H_{12}$; **g** dibenzo[b,pqr]perylene, $C_{26}H_{14}$; **h** benzo(pqr)naphtho(8,1,2-bcd)perylene, $C_{28}H_{14}$; **i** naphtho[2'.8',2.4]coronene, $C_{30}H_{14}$; and **j** ovalene, $C_{32}H_{14}$. Note how the H/C ratio goes down as condensation increases, from 0.8 for naphthalene to 0.4375 for ovalene

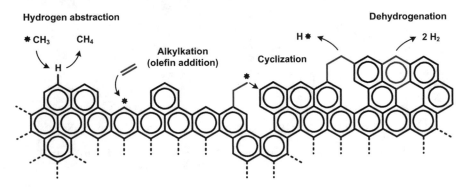

Fig. 9.16 Mechanism for the addition of rings to an existing layer of coke

9.5 Treating/Sweetening

Gases and straight-run distillates need to be treated for removal of undesirable impurities before they can be used as products.

Several treating processes entail seemingly simple acid-base reactions, such as alkanolamine treating and caustic scrubbing. Alkanolamine treating removes acid gases—H_2S and CO_2—from fuel gas and off-gas streams. In some hydrotreaters and hydrocrackers, high-pressure amine units remove H_2S from the recycle gas. Caustic scrubbers are used in several ways, including removing the last traces of H_2S from the hydrogen used for processes in which the catalysts are highly-sulfur sensitive.

9.6 Hydrotreating

Hydrotreating is used in the pretreatment of process streams or the finishing of products in hydrofining or hydrofinishing. It removes sulfur, nitrogen, oxygen and trace elements which are detrimental to subsequent processes or affect the quality and appearance of the products. If the contaminants are not removed, products will not meet specifications.

Hydrotreater feeds range from naphtha to vacuum residues. Generally, each fraction is treated separately. Materials with higher boiling points require more severe treatment conditions. Table 9.5 lists feeds and products for hydrotreating and hydroprocessing units.

The main hydrotreating reactions are summarized below:

Olefin saturation: olefin + H_2 → paraffin + heat
Aromatic saturation: aromatics + H_2 → naphthenes + heat
Dehydrogenation: naphthenes + heat → aromatics + H_2
HDS: sulfur compounds + H_2 → hydrocarbons + H_2S + heat
HDN: nitrogen compounds + H_2 → hydrocarbons + NH_3 + heat
HDO: oxygen compounds + H_2 → hydrocarbons + H_2O + heat
HDM: organometallics + H_2 → hydrocarbons + metal + adsorbed contaminants

As mentioned in Sect. 9.4, the saturation of aromatics and the dehydrogenation of the corresponding naphthenes are in equilibrium. It is important to note that, except for dehydrogenation, all reactions are exothermic.

HDS is hydrodesulfurization. HDN is hydrodenitrogenation. HDO is hydrodeoxygenation, and HDM is hydrodemetallation [4]. HDM is especially relevant to the hydrotreating of heavier feeds, such as VGO and residue. From such streams, HDM removes trace elements, such as nickel, vanadium, iron, silicon, arsenic, etc., which are poisons to downstream catalysts.

Arsenic and mercury are especially challenging. They form volatile alkyl compounds, which can appear in all distillate fractions.

Table 9.5 Feeds and products for hydroprocessing units

Feeds	Products from hydrotreating	Products from hydrocracking
Heavy naphtha	Catalytic reforming feed	LPG
Straight-run light gas oil	Kerosene, jet fuel	Naphtha
Straight-run heavy gas oil	Diesel fuel	Naphtha
Atmospheric residue	Lubricant base stock, low sulfur fuel oil, RFCC feed	Naphtha, middle distillates, FCC feed, lubricant base stocks.
Vacuum gas oil	FCC feed, lubricant base stock	Naphtha, middle distillates, FCC feed, lubricant base stocks, olefin plant feed
Vacuum residue	RFCC feed	See note (b)
FCC light cycle oil	Diesel blend stock, fuel oil	Naphtha
FCC heavy cycle oil	Fuel oil blend stock	Naphtha, middle distillates
Visbreaker gas oil	Diesel blend stocks, fuel oil	Naphtha, middle distillates
Coker gas oil	FCC feed	Naphtha, middle distillates, FCC feed, lubricant base stock, olefin plant feed
Deasphalted oil	Lubricant base stock, FCC feed	Naphtha, middle distillates, FCC feed, lubricant base stock

Note (a) FCC = fluid catalytic cracking, RFCC = residue FCC
(b) Traditional fixed-bed hydrocrackers cannot process vacuum residue. The use of ebullated-bed and slurry-phase hydrocrackers produces naphtha, middle distillates and FCC feed

Organosilicon compounds can be present in naphtha-range cuts from delayed coking units. Heavier fractions may be contaminated by silicon-containing flow improvers.

All reactions occur at the same time, to one extent or another. Obviously, HDO and HDM occur only when oxygen and trace elements are present.

9.6.1 Hydrotreating Process Flow

Figure 9.17 presents a process flow scheme for a one-reactor fixed-bed hydrotreater with four catalyst beds. Reaction conditions depend on feed quality and process objectives.

The feed is warmed with heat exchange, mixed with hydrogen-rich gas and passed through a furnace, where it is heated to the desired reactor inlet temperature, typically 600–780 °F. The temperature depends on process objectives and catalyst activity.

The heated mixture flows down through reactors loaded with catalysts. In many diesel hydrotreaters and almost all VGO hydrotreaters, the reactors have multiple beds, separated by quench decks. As the hydrotreating reactions occur, they con-

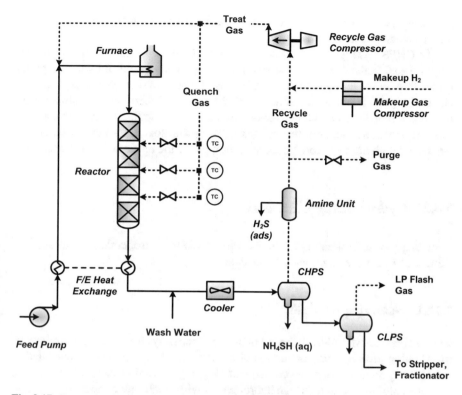

Fig. 9.17 Representative hydrotreating unit with one reactor, four catalyst beds, and two flash drums. Naphtha hydrotreaters have one bed in one reactor, and may employ hydrogen once-through (no recycle). F/E exchanger is the feed/effluent heat exchanger. Temperature controllers (TC) control temperature by manipulating the flow of quench gas. CHPS is the cold high-pressure separator. CLPS is the cold low-pressure separator. Wash water is injected to remove ammonia as aqueous ammonium bisulfide, which otherwise would precipitate in cold spots in or downstream from the CHPS, blocking flow and inducing corrosion

sume hydrogen and generate heat. The heat is controlled by bringing relatively cool recycle gas into the quench decks, where it mixes with reaction fluids from the bed above. Makeup gas comes into replace consumed hydrogen. Gas flow can be once-through in naphtha hydrotreaters, but in distillate and VGO hydrotreaters, unconsumed hydrogen is recycled.

Fluids exiting the reactor are cooled with heat exchange before going to an arrangement of separation towers, which include a stripper and, for high-pressure units, two or more flash drums.

Hydrotreating produces both H_2S and NH_3. Under reaction conditions, these remain in the gas phase. But at lower temperatures, they combine to form solid ammonium bisulfide (NH_4SH). Ammonia also reacts with chlorides to form NH_4Cl; chloride can come with makeup gas, feed, or wash water. These salts can deposit in air coolers and heat exchangers, blocking flow and—even worse—inducing corrosion.

Fortunately, they are water-soluble, so they can be controlled by injecting wash water into the reactor effluent for removal.

The CHPS flash gas is recycled. An optional amine unit removes H_2S from the recycled gas. Some of the recycle gas might be purged to control hydrogen purity. Depending on the unit and the feed, oils from the separation towers can go various places. Stripped naphtha, kerosene or diesel might go directly to downstream units or product blenders. In VGO or residue hydrotreaters, liquids from the separation section go to a fractionator. Other off-gases, along with sour high-pressure purge gas, are treated with amine before going to other units or the refinery fuel-gas system.

9.6.2   Hydrotreating Chemistry

A summary of hydrotreating reactions appeared in the introduction to this section. Here, we discuss the reactions in more detail.

9.6.2.1   Saturation of Olefins

Olefins are rare in straight-run feeds, but they are relatively abundant in cracked stocks from coking or FCC units. Saturation of olefins occurs rapidly. If not controlled, it can lead to polymerization and consequent plugging of catalyst beds. The best ways to control olefins are (a) to design hydrotreaters with small low-temperature beds up front and/or (b) to employ activity grading. In activity grading, catalysts are layered, with low activity catalysts on top, followed by successively more-active catalysts.

9.6.2.2   Saturation of Aromatics/Dehydrogenation of Naphthenes

Saturation of aromatics and dehydrogenation of naphthenes are described in Sect. 9.4. Saturation reactions play a significant role in hydrotreating and hydrocracking. In hydrotreating, the aromatics crossover effect increases the difficulty of removing nitrogen compounds and large sulfur compounds. The condensation of aromatics leads to coke formation and consequent catalyst deactivation.

9.6.2.3   Hydrodesulfurization (HDS) for Sulfur Removal

Figure 9.18 shows representative hydrodesulfurization (HDS) reactions. For sulfur removal from the first four reactants, the mechanism is straightforward. That is, the sulfur-containing molecule interacts with an active site on the catalyst, which removes the sulfur atom and replaces it with two hydrogen atoms. Additional hydrogen converts the sulfur atom into H_2S, which desorbs from the catalyst, leaving behind a regenerated active site.

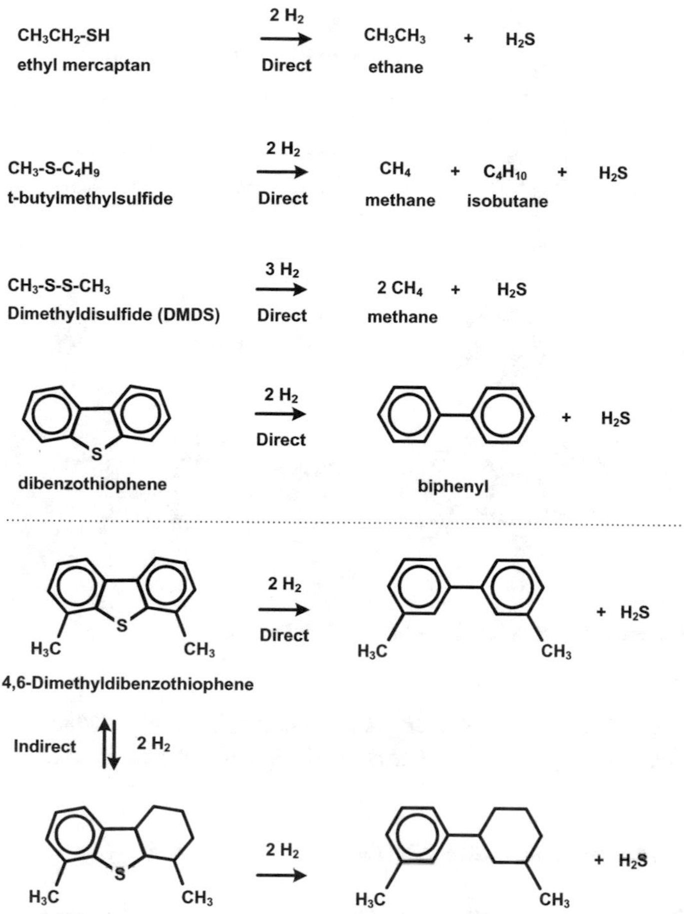

Fig. 9.18 Representative hydrodesulfurization (HDS) reactions

For the fifth reactant—4,6-dimethyldibenzothiophene (4,6-DMDBT) which has two methyl groups near the vicinity of the sulfur atom—HDS proceeds via both a direct and an indirect route. Overall, the indirect route is considerably faster. The 4,6-DMDBT molecule is planar, and the two methyl groups block the access of the sulfur atom to the catalyst surface as "hindered sulfur", thereby inhibiting direct HDS. In the indirect route, saturating one of the aromatic rings that flank the sulfur atom converts the planar structure into a puckered configuration with tetrahedral C-C bonds. This puckering rotates one of the inhibiting methyl groups away from the sulfur atom, giving its better access to the catalyst.

Making ultra-low-sulfur diesel (ULSD), in which the sulfur content is less than 10–15 wppm, requires severe hydrotreating, after which the only remaining sulfur compounds are the above-mentioned 4,6-DMDBT and other di- and trimethyl dibenzothiophenes. Because the removal of sulfur from these compounds is more facile

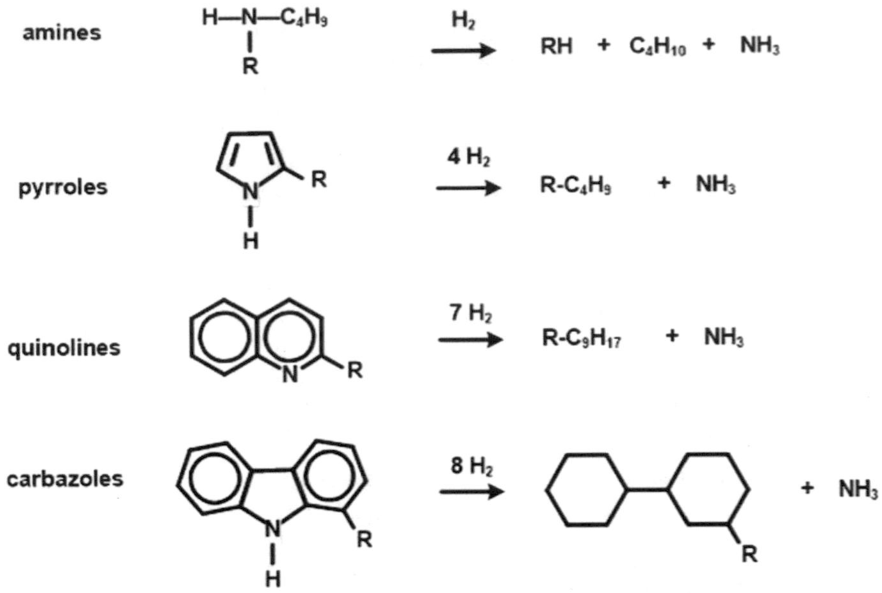

Fig. 9.19   Representative hydrodenitrogenation (HDN) reactions

after prior saturation, the crossover phenomenon affects the production of ULSD significantly, so much so that it governs the design of commercial units.

9.6.2.4   Hydrodenitrogenation (HDN) for Nitrogen Removal

Figure 9.19 shows representative HDN reactions, and Fig. 9.20 presents the mechanism for the HDN of quinoline. As with sulfur removal from hindered DMDBTs, the aromatics crossover phenomenon is important for deep HDN because nitrogen removal requires prior saturation of an aromatic ring adjacent to the nitrogen atom.

9.6.2.5   Hydrodeoxygenation (HDO) for Oxygen Removal

In feeds to industrial hydroprocessing units, oxygen is contained in furans, organic acids, ethers, peroxides, and other compounds. Some are formed by reaction with air during transportation and storage of crude oil, distillates, and cracked stocks generated by delayed coking and FCC units (see below). Phenols and cresols and quite stable, but some hydrodeoxygenation (HDO) reactions proceed so rapidly that they can cause problems with excessive heat release. Also, oxygen compounds can form gums and polymers, which inhibit flow and increase pressure drop. Feedstocks from biomass contain considerably more oxygen than conventional crude oil and petroleum distillates. While a conventional vacuum gas oil may contain 0.5 wt%

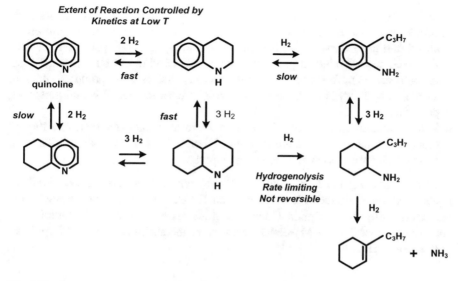

Fig. 9.20 Mechanism for the HDN of quinolone

oxygen, a bio-derived oil in the same boiling range might contain >40% oxygen. Transportation fuels must meet tight specifications, or they can't be used—at least not for very long. Producing fuels from bio-derived oils in conventional oil refineries requires extra hydrogen and presents processing and storage challenges. Large-scale hydrogen production generates large quantities of CO_2. When assessing the environmental impact of replacing conventional crude oil fractions with bio-derived oils, it is crucial to consider the hydrogen required for upgrading.

9.6.2.6 Hydrodemetallation (HDM) for Trace Element Removal

Metals and other trace elements poison catalysts in hydrotreaters and downstream process units. The following are the most troublesome:

- Nickel and vanadium are present in high-boiling fractions, mostly in asphaltenes. Asphaltenes are mixtures of waxy solids with porphyrins.
- Corrosion generates soluble iron.
- Entrained salt brings in alkali and alkaline earth salts, primarily carbonates and bicarbonates of sodium and calcium.
- Arsenic is present in crudes from West Africa, the Ukraine, Canada, and elsewhere. Synthetic crudes from oil sands and oil shale tend to contain significant amounts. Arsenic is one of the worst catalyst poisons—hundreds of times worse, on a weight basis, than Ni, V, or Fe. Arsenic forms organo arsines, which are relatively volatile. They tend to be most abundant in middle distillates.

- Mercury components, including elemental mercury, mercuric chloride, mercuric sulfide, mercuric selenide, dimethylmercury, diethylmercury, etc. are present in all oil and gas reservoirs in trace amounts, usually in a few ppb levels or less. However, several hundred ppb mercury can be found in crudes and natural gas in southeast Asia, Australia and South America, such as those produced offshore Thailand. Mercury is nearly as bad as arsenic, both for volatility and poisoning potential.
- Silicon comes in as silicones, which are added to crude oil to facilitate flow though pipelines. Silicones are also used to control foaming in delayed coking units. Most end up in light fractions, such as heavy naphtha.

In hydrotreating, trace elements are removed with graded guard material (adsorbents), including special wide-pore guard catalysts. Guard materials remove some contaminants by chemisorption. Other contaminants, such and Ni, V, and soluble Fe, are removed by HDM. HDM reactions convert the metals into sulfides, which adhere to the guard material.

9.6.3 Hydrotreating Catalysts

The earliest hydrodesulfurization catalysts were bauxite and fuller's earth. Later, catalysts containing cobalt molybdate on alumina and nickel tungstate on alumina substantially replaced the earlier catalyst and these catalysts are still used very extensively.

Most commonly used hydrotreating catalysts are MoS_2 promoted by Co_xS_y and/or NiS on gamma alumina. WS_2 has also been used commercially in place of MoS_2. Improved compositions and manufacturing methods have increased activity more than ten-fold since the process was invented in the 1950s, but despite intense ongoing efforts to find better catalysts, the original raw materials remain unsurpassed. Other metal sulfides, such as RuS_2, IrS_2, OsS_2, and RhS_2, are more active than MoS_2 and WS_2, but they either deactivate too quickly or they are too expensive.

When the catalysts are manufactured, the metals are oxides, which have relatively low activity. Before use, they must be activated by reductive sulfidation, a process commonly known as sulfiding.

For the low-pressure HDS of light feeds, CoMo catalysts are preferred. For high-pressure deep desulfurization and HDN, NiMo catalysts are used, either alone or in combination with CoMo. Some suppliers claim that "sandwich" configurations, with alternating layers of CoMo and NiMo catalysts, provide superior HDS performance. In the past, catalysts based on Ni-promoted WS_2 were employed for HDN, due to the high saturation activity of the WS_2. However, WS_2 catalysts are seldom if ever used now.

The catalyst support is just as important as the active metals. The support must be strong enough to endure considerable pressure and hydraulic stress, and it must have a high surface area. The supports are microporous, with 99+ percent of their surface areas inside the pores. The pore diameters must be wide enough to admit reactants but narrow enough to exclude very large residue molecules.

In fixed-bed units, hydrotreating catalysts last for 1–5 years, typically 2–3 years. During a catalyst cycle, the catalysts slowly deactivate and temperatures are raised to compensate. Start-of-run reactor average temperatures can range from 600 to 700 °F (315–370 °C), depending on hydrogen partial pressure, feed rate, feed quality, and desired product quality. When the required temperature reaches a limit set by metallurgy, the feed rate must be reduced, the product quality objective must be relaxed, or the catalyst must be changed. Typical end-of-run average temperatures range from 780 to 800 °F (416–427 °C), and typically end-of-run peak temperatures range from 800 to 825 °F (427–440 °C). A catalyst cycle can end for other reasons, such as a predetermined schedule, unacceptable pressure drop, or excessive production of low-value light gases.

Spent catalysts are fouled with coke. They are either regenerating offsite by careful combustion of the coke, or sold to catalyst reclamation companies. In-place regeneration is now very rare. Catalysts heavily contaminated with trace elements are not regenerated. It is not unusual to see uncontaminated catalysts restored to 85–95% of their initial activity.

9.7 Mercaptan Oxidation

Mercaptan oxidation processes, such as the two-step UOP Merox process, convert foul-smelling mercaptans into disulfides. For Merox, the overall reaction is as follows:

$$4\,R\text{-}SH + O_2 \rightarrow 2\,R\text{-}S\text{-}S\text{-}R + 2\,H_2O$$

In the equation, R represents an alkyl group, such as methyl ($-CH_3$), ethyl ($-CH_2CH_3$), and so on. Methyl mercaptan is CH_3SH. Feed mercaptans can include methyl mercaptan (methane thiol), ethyl mercaptan (ethanethiol), 1-propyl mercaptan, 2-propyl mercaptan, t-butyl mercaptan, pentyl mercaptan, etc. C_6+ mercaptans end up in heavy naphtha cuts and are moved, together with other sulfur compounds, with conventional hydrotreating. The alkyl groups in the disulfide product can be different. For example, one of the products could be methylethyldisulfide.

In the conventional Merox process, the mercaptan-containing feedstock is prewashed with dilute NaOH to remove hydrogen sulfide (H_2S). The prewashed feed reacts with a different sodium hydroxide solution, one which contains a proprietary catalytic additive:

$$2\,R\text{-}SH + 2\,NaOH \rightarrow 2\,NaSR + 2\,H_2O$$

The next step regenerates the sodium hydroxide while producing a disulfide:

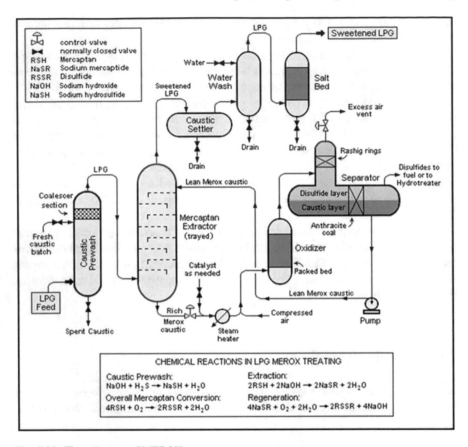

Fig. 9.21 Flow diagram of MEROX process

$$4\,\mathrm{NaSR} + 2\,\mathrm{O_2} + \mathrm{H_2O} \;\rightarrow\; 2\,\mathrm{R\text{-}S\text{-}S\text{-}R} + 4\,\mathrm{NaOH}$$

In the flow diagram, shown in Fig. 9.21, the mercaptan-containing feedstock enters the prewash vessel, which removes H_2S. A coalescer prevents caustic from being carried out of the vessel. The feed then enters the extractor, flowing up through several contact trays, which enhance the mixing of feedstock with caustic. The feed then goes to a settler, in which the gas separates from the caustic. The gas is water-washed to remove any residual caustic, and sent through a bed of rock salt to remove entrained water. The dry, sweet LPG exits the Merox unit. The hydrocarbon-rich caustic from the bottom of the extractor is mixed with a proprietary liquid catalyst, heated in an exchanger, then mixed with compressed air. The mixture goes to the oxidizer, where the extracted mercaptans are converted to into disulfides. The caustic/disulfide mixture flows to a separator, in which the mixture separates into a disulfide layer on top of an aqueous layer of "lean" caustic. The disulfides go to storage or to a hydrotreater. The caustic is recycled back to the extractor.

References

1. Foxboro: Crude oil blending. http://www.process-nmr.com/fox-app/pdf/crude-blend.pdf. Retrieved 6 Aug 2016
2. Hsu CS, Lobodin V, Rodgers RP, McKenna AM, Marshall AG (2011) Compositional boundaries for fossil hydrocarbons. Energ Fuels 25:2174–2178
3. Park J-I, Mochida I, Marafi AMJ, Al-Mutairi A (2017) Modern approaches to hydrotreating catalyst. Chapter 21 In: Hsu S, Robinson PR (eds) Springer handbook of petroleum technology. Springer, New York
4. Leffler WL (2008) Petroleum refining in nontechnical language, 4th Edition, PennWell

Chapter 10
Gasoline Production

10.1 Introduction

After Ford Model-T, there was a rapid growth in transportation vehicles. Motor gasoline ("mogas") became in high demand. Hence, many of early refining processes were designed for conversion to increase gasoline yields. The carbon number range of gasoline is between C_5 and C_{12}. This fraction can be obtained as straight run naphtha by distillation. However, not all naphtha molecules have good octane ratings suitable for automobile performance. Isomerization and reforming into structures with high octane ratings without altering the carbon numbers would be one approach. Other approaches include combining C_3 to C_5 molecules by alkylation and polymerization, or decreasing the size of $C_{12}+$ molecules by cracking and coking, which will be discussed in the following two chapters.

10.2 Isomerization [1]

Normal paraffins of the lighter naphtha fraction, butanes, pentanes and hexanes, have poor octane ratings for gasoline engines. Isomers of normal paraffins, on the other hand, have high octane ratings. During World War II, there was a demand of high-octane aviation gasoline, and isomerization became an important process. However, the majority of butane isomerization units were shut down after World War II due to lower demand. Demand was reduced mainly because tetraethyl lead (TEL) was being added to gasoline to boost octane. TEL was banned after research revealed environmental and health hazards. The phase-out of TEL created an octane gap, which resulted in new butane isomerization units being installed, with $AlCl_3$ as a catalyst. Subsequently, n-pentane and n-hexane isomerization processes were developed, using supported acid catalysts that can stand high temperature (370–480 °C) and high pressure (300–750 psi). A liquid-phase process employs a dissolved catalyst for C_5/C_6 isomerization.

© Springer Nature Switzerland AG 2019
C. S. Hsu and P. R. Robinson, *Petroleum Science and Technology*,
https://doi.org/10.1007/978-3-030-16275-7_10

Isomerization of larger *n*-paraffins occurs in many conversion units in the refinery. It is especially important in catalytic cracking where linear olefins can be produced, both in the presence and absence of excess hydrogen. The isomerization reaction is believed to occur through the formation and rearrangement of carbonium ions. Isomerization yield is increased by higher temperature, lower space velocity and lower pressure. However, in this section, we only focus on isomerization for the production of gasoline components. Butane isomerization provides isobutane as the major feed for subsequent alkylation. Pentane/hexane isomerization improves the octane number of light naphtha for gasoline blending. Carbon number is not supposed to change in the isomerization process.

Due to the advent of the alkylation process, the demand for isobutane increased. This deficiency was remedied by isomerization of the more abundant normal butane into isobutane. Modern isomerization catalysts are $AlCl_3$ supported on alumina and promoted by hydrogen chloride gas. In butane isomerization, *n*-butane is converted to isobutane normally in a two-stage process. The feed contains *n*-butane, typically mixed with other butanes. As shown in Fig. 10.1, mixed butane feed enters a deisobutanizer to separate isobutane followed by debutanizer to separate butanes from C_{5+} impurities. The purified butanes are mixed with hydrogen and heated to 230–340 °F (110–170 °C) under a pressure of 200–300 psi (14–20 bar). The catalyst is highly sensitive to water, so the feed must be thoroughly dried. In the low-temperature 1st stage, the catalyst comprises $AlCl_3$ promoted by HCl. Hydrogen gas is added to inhibit olefin formation and to control side reactions, such as disproportionation and cracking. In the high-temperature 2nd stage, the catalyst contains a noble metal such as platinum. The reactor effluent goes to a flash drum, from which hydrogen is recovered and recycled. Make up gas is added to compensate for the loss, which is low

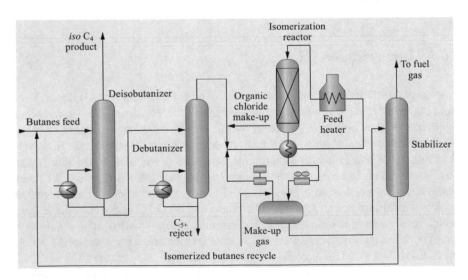

Fig. 10.1 Butane Isomerization

because the isomerization reaction itself does not consume hydrogen. HCl is removed in a stripper column. The liquids go to a stabilizer which separates C_4-(compounds with carbon numbers less than 4), which can be used as a fuel gas inside refinery, from isobutene and unconverted n-butane. The C_4 compounds are recycled and mixed with fresh feed. Product isobutane is collected at the top of deisobutanizer. Isobutane is the key feedstock for alkylation units, which produce high octane alkylates, a valuable component of gasoline.

Commercial processes have also been developed for the isomerization of low-octane normal pentane and normal hexane to higher-octane isoparaffins (see below). Here the reaction is usually catalyzed with supported platinum. As in catalytic reforming, the reactions are carried out in the presence of hydrogen. Hydrogen is neither produced nor consumed in the process but is employed to inhibit undesirable side reactions. The reactor step is usually followed by molecular sieve extraction and distillation. Though this process is an attractive way to exclude low-octane components from the gasoline blending pool, it does not produce a final product of sufficiently high octane to contribute much to the manufacture of unleaded gasoline.

Linear Isomers	RON
n-butane	93.8
n-pentane	61.7
n-hexane	24.8
Branched Isomers	
i-butane	110.1
i-pentane	92.3 (30.6 number increase)
2-methyl pentane (i-C6)	73.4 (48.5 number increase)
2,3-dimethylbutane (i-C6)	100.3 (75.5 number increase)

The C_5/C_6 isomerization process is shown in Fig. 10.2. The mixed C_5/C_6 feed enters a fractionator to separate n-C_5/C_6 from iso-C_5/C_6, which are the desired isomerization products. The purified n-C_5/C_6 is then mixed with hydrogen and hydrogen chloride and heated to 240–500 °C under 300–1000 psi pressure before entering the reactor, which is packed with HCl-promoted $AlCl_3$ catalyst or a Pt-containing catalyst. The residence time is 10–40 min. Again, hydrogen is used to control side reactions (disproportionation and cracking). The product stream goes through a flash tank to recover hydrogen for reuse, then enters a stabilizer to separate C_4 and lighter impurities before a C_5 and C_6 splitter. The C_5 components are recycled back to the feed fractionator to recover i-C_5 into isomerate, a valuable component of gasoline.

A two-stage reactor scheme is shown in Fig. 10.3. The first reactor contains a Pt-catalyst, which serves to hydrogenate any olefins and aromatics in the feed. It also converts organic chloride into HCl removes any other impurities carried by the feed stream. The second reactor is packed with $AlCl_3$ which is promoted by HCl to perform isomerization.

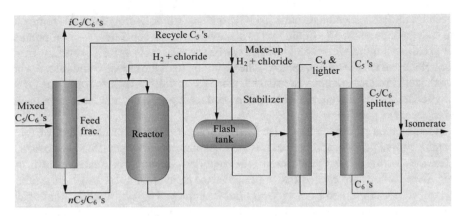

Fig. 10.2 C_5/C_6 Isomerization

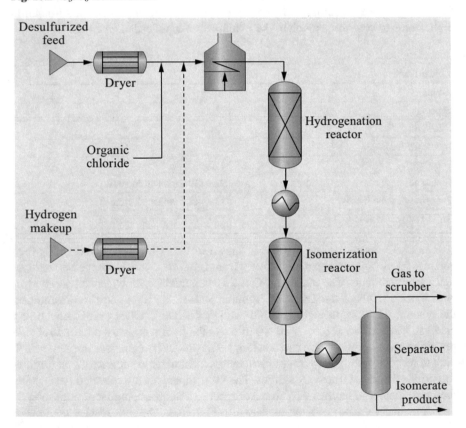

Fig. 10.3 C_5/C_6 Isomerization with two reactors and once-through hydrogen

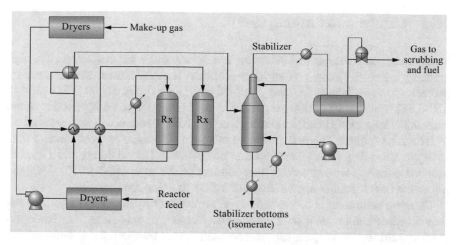

Fig. 10.4 UOP Penex Isomerization Unit [2]

Figure 10.4 shows the UOP Penex isomerization unit with a similar design [1]. Penex uses a platinum-containing catalyst. The process improves octane ratings from 50–60 to 82–86 or higher.

In the Shell Hysomer process for pentane/hexane isomerization [3], the feed is combined with hydrogen-rich gas, heated to 445–545 °F (230–285 °C) and routed to the Hysomer reactor at 190–440 psi (13–30 bar). As with fixed-bed hydrotreating and hydrocracking, the process fluids flow down through the catalyst bed, where a part of the n-paraffins are converted into branched paraffins. The catalyst is comprised of a noble metal on a zeolite-containing support. The reactions are exothermic, and temperature rise is controlled by injecting relatively cold quench gas. The reactor effluent is cooled by heat exchange and sent to a flash drum, which separates hydrogen from the liquid product. The hydrogen is recycled. The liquid is fractionated, and the n-paraffins are recycled. The net conversion of n-paraffins into branched products can be as high as 97%, and the octane can be boosted by 8–10 numbers.

Often, the heat-exchanger and fractionation systems of isomerization units are integrated with those of other process units, such as catalytic reformers. In the Union Carbide total isomerization process (TIP) [4], C_5/C_6 isomerization was integrated with molecular sieve separation, which provided complete conversion of n-paraffins.

The CDTech Isomplus process that was jointly developed with Lyondell Petrochemical achieves near-equilibrium conversion of n-butenes into isobutylene, and n-pentenes into isoamylene over a highly selective zeolite catalyst [5]. It was developed when methyl t-butylether (MTBE) was in high demand for reformulated gasolines. Isobutylene is also an important feedstock for polyisobutylenes.

Olefin isomerization converts straight-chain C_4–C_6 olefins into corresponding iso-olefins which can be used as alkylation feeds with excess an amount of isobutane for producing high octane gasoline.

10.3   Catalytic Reforming [6–9]

Straight-run gasoline from distillation has a very low octane number, i.e., it has a high tendency to knock. Thermal reforming was initially developed from thermal cracking processes with naphtha as the feed. The feed with an end point of 400 °F is heated to 950–1100 °F under 400–1000 psi pressure. The higher octane number is due to the cracking of long-chain paraffins into high-octane (65–80) olefins. Gases, residual oil or tar can be formed. Thermal reforming is less effective and less economical than catalytic reforming. Catalytic reforming was commercialized during the 1950s to produce reformate with research octane number on the order of 90–95 [6]. Catalytic reforming now furnishes approx. 30–40% of US gasoline requirements.

Catalytic reforming is conducted in the presence of hydrogen over hydrogenation-dehydrogenation catalysts supported on alumina or silica-alumina. The operation conditions for modern catalytic reforming are: temperature at 840–965 °F, pressure at 100–600 psi. Continuous catalyst regeneration (CCR) reformers operate at the low end of the pressure range, 100–150 psig, where production of aromatics is more favored by thermodynamics. The first reforming catalysts were based on molybdena-alumina or chromia-alumina, but since the commercialization of the Platforming process in 1949, noble metal catalysts have been used, either with supported platinum alone or with supported platinum-rhenium, platinum-rhenium-tin, or other tri-metallics on a silica-alumina or alumina base [7]. Hydrogen chloride is a co-catalyst.

Naphtha containing seven or more carbons is used as a reformer feed. Pentane and hexane are removed from desulfurized naphtha to be used as isomerization feeds, as discussed above. If the feed contains C_6 hydrocarbons, carcinogenic benzene is hydrogenated into cyclohexane to avoid its release into atmosphere upon incomplete combustion or when drivers fill their automobile tanks with gasoline. There are four main hydrocarbon types in naphtha: paraffins, olefins (mainly from catalytic cracking and coking units), naphthenes, and aromatics. Organic sulfur and most olefins are removed by prior hydrotreating; the sulfur content must be <1 ppmw. During the process, paraffins are cyclized into naphthenes, and naphthenes undergo dehydrogenation to form aromatics. Aromatics are left essentially unchanged. The preferred conditions for dehydrogenation are: high temperature, low pressure, and low space velocity.

An important co-product is hydrogen, which is required by other processes, especially hydrotreating and hydrocracking. The dehydrogenation reaction in reforming

Table 10.1 Typical feed and product distributions in reforming

Component	Feed (%)	Product (%)
Paraffins	30–70	30–50
Naphthenes	20–60	0–3
Aromatics	7–20	45–60

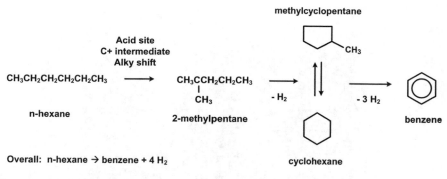

Fig. 10.5 Catalytic reforming reactions

is endothermic. Heat is required. Table 10.1 lists typical feed and product composi-tions in a reforming process.

Figure 10.5 summarizes the reactions of catalytic reforming, the purpose of which is to transform C_6 to C_{11} naphthenes and paraffins into aromatic compounds. The aromatics can go to chemical plants or be used as high-octane gasoline blend stocks. For the paraffins in the feed, the acidic sites of the catalyst isomerize n-paraffins to iso-paraffins. This is a key step for cyclization. It is followed by dehydrogenation to form aromatic compounds. The process yields considerable amounts of hydrogen; in Fig. 10.5, four moles of hydrogen are produced from one mole of hexane. The hydrogen is used in hydrotreaters, hydrocrackers, isomerization units, and others.

In contrast to hydrocracking, which operates at high pressure, catalytic reforming operates at low pressure and high temperature, which favors production of aromat-ics. The isomerization of alkylcyclopentanes into cyclohexanes is a key step in the production of benzene and alkylbenzenes.

The three major process flows for catalytic reforming are:

- Semi-regenerative
- Cyclic (fully regenerative)
- Continuous catalyst regeneration (CCR, moving bed).

Figure 10.6 shows a reactor for a fixed-bed semi-regenerative reformer. The feed is introduced through the top and flows down along the outer wall of the reactor and then flows radially inward through the thin annular catalyst bed. Screens are used to contain the catalyst and allow reactants to enter from the wall, pass through the catalyst bed, enter the central collection tube and exit the vessel at the bottom. This design allows the intimate contact between the feed and the catalyst.

The process flow is shown in Fig. 10.7. Catalyst cycles last from 6 to 12 months. A cycle ends when the unit is unable to meet its process objectives—typically C_5-plus yields >80 wt% when the octane number target is 100 RON. At the end of a cycle, the entire unit is brought down and coke is burned off the catalyst. The depentanized/dehexanized naphtha is hydrotreated to bring sulfur down to <5 ppm, then mixed with hydrogen and heated to >900 °F (>480 °C). The hot mixture passes

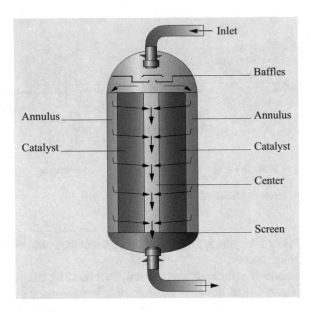

Fig. 10.6 Downflow fixed-bed catalytic Reactor [1]

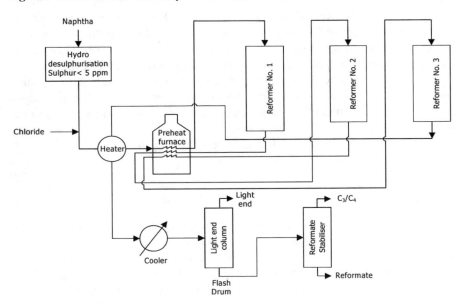

Fig. 10.7 Semi-regenerative catalytic reforming process flow

through a series of fixed-bed reactors. The feed is spiked with an organic chloride, which converts to hydrogen chloride in the reactors. This provides the required catalyst acidity and helps minimize catalyst coking. There are three reactors shown.

The major chemical reactions—dehydrogenation and dehydrocyclization—are endothermic and the reactors are adiabatic. Consequently, the temperature drops as reactants flow through a reactor. Between reactors, fired heaters bring the process fluids back to desired reactor inlet temperatures (RIT). The product stream of the final reactor is cooled by heat exchange and enters a high-pressure gas separator (light end column) at 38 °C where hydrogen is recovered. A portion of the hydrogen is recycled and the remainder is transferred to hydroprocessing units in the refinery. The liquid is then sent to a stabilizer to remove any remaining light ends, including petroleum gas (propane and butane). The final reformate product can have octane numbers (RON) ranging from 90 to 105.

A cyclic reformer has more reactors, typically six, and catalyst cycles in each reactor are much shorter—20 to 40 h. Reactor shutdowns are staggered so that only one reactor is down at a given time. A significant difference between the two processes is the length of the cycles.

While semi-regenerative Platforming went on-stream in 1949, [8, 11] the continuous catalyst regeneration (CCR) Platforming process was commercialized by UOP in 1971 [10, 11]. Hydrotreated feed mixes with recycle hydrogen and goes to a series of adiabatic, radial-flow reactors arranged in a vertical stack. Catalyst flows down the stack, while the reaction fluids flow radially through the catalyst beds. Heaters are used between reactors to reheat the reaction fluids to the required temperature. Flue gas from the fired heaters is typically used to generate steam. A CCR can operate at very low pressure (100 psig, 791 kPa). This improves yields of aromatics and hydrogen, but it accelerates catalyst deactivation by increasing the rate of coke formation. But faster coke formation is okay in a CCR reformer, because the catalyst is continuously being regenerated. The hydrogen recovery is superior to that for semi-regen Platforming, and correspondingly, product aromatics are higher. There are two licensed CCR configurations: side-by-side by Axens under the trade names Octanizer and Aromatizing, and staggered (shown in Fig. 10.8) by UOP under the trade name CCR Platforming. In CCR Platforming, catalyst flows from top to bottom by gravity flow.

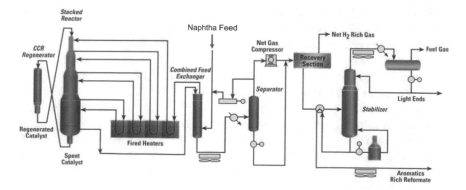

Fig. 10.8 UOP CCR platforming process

There are several commercial catalytic reforming processes.

- Magnaforming was developed by Engelhard and Atlantic Richfield (now a part of BP). Magnaforming employs a manganese-containing catalyst for improved sulfur tolerance along with a platinum-containing catalyst for reforming.
- Powerforming [12] was developed by Esso (now ExxonMobil) in 1954. It is a fixed-bed catalytic reforming process employing a platinum or a bimetallic catalyst. The process includes four reactors in series with intermediate reheat furnaces. A swing reactor enables each catalyst charge to be regenerated without loss of production.
- Ultraforming was developed by Standard Oil Indiana (Amoco, now a part of BP) in 1954. Ultraforming is a cyclic semi-regenerative process.
- Rheniforming was developed by Chevron in 1968. Rheniforming employs platinum-rhenium bimetallic catalysts where rhenium is combined with platinum to form more stable catalysts operated at low pressures. The amounts of metal are very low, about 0.6 wt%, but they are highly dispersed. Acidity is provided by the support, and dispersion is maintained by HCl.
- Houdriforming was developed by Houdry Division of Air Products and Chemicals, Inc. The catalyst used is usually Pt/Al_2O_3 or may be bimetallic. A small "guard case" hydrogenation pretreater can be used to prevent catalyst poisons in the naphtha feedstock from reaching the catalyst in the reforming reactors.

10.4 Alkylation [1, 13]

In alkylation, isobutane reacts with C_3 to C_5 olefins, which come mainly from cracking units, in the presence of strong acids to produce branched chain hydrocarbons. Without the presence of an excess amount of isobutane, C_3 to C_5 olefins can undergo undesired polymerization into sludge. These hydrocarbons, often referred to as alkylate, have high octane values (motor octane number from 88 to 95 and research octane number from 90 to 98) and low Reid vapor pressures, making them an excellent contributor to the gasoline blending pool.

Figure 10.9 shows the mechanism for the alkylation reaction of butenes with isobutane [1, 14]. The process uses an excess of isobutane to control heat and minimize olefin polymerization. As an initiation step exemplified in Reaction 1, 2-butene or its isomer is protonated by the acid catalyst to form a 2-butenium ion, which isomerizes to stable t-butyl ion. Reactions 2a, 2b and 2c show how different isomers lead to different C_8 carbocations. These C_8 carbocations will isomerize into more stable 2,2,4-trimethyl pentyl ion. The isomerization reactions will not reach completion due to thermodynamic equilibrium of reversible reactions, evidenced by the presence of small amounts of 2,2,3-trimethyl pentane and 2,2-dimethyl hexane in the product. The C_8 carbocations ions will react with predominant amount of isobutane through hydride transfer to be neutralized and leave t-butyl ion for propagating the alkylation reaction. However, some C_8—carbocations (C_4 dimer ions) can proceed to react with olefins to form C_{12}-carbocations (C_4 trimer ion) and continue poly-

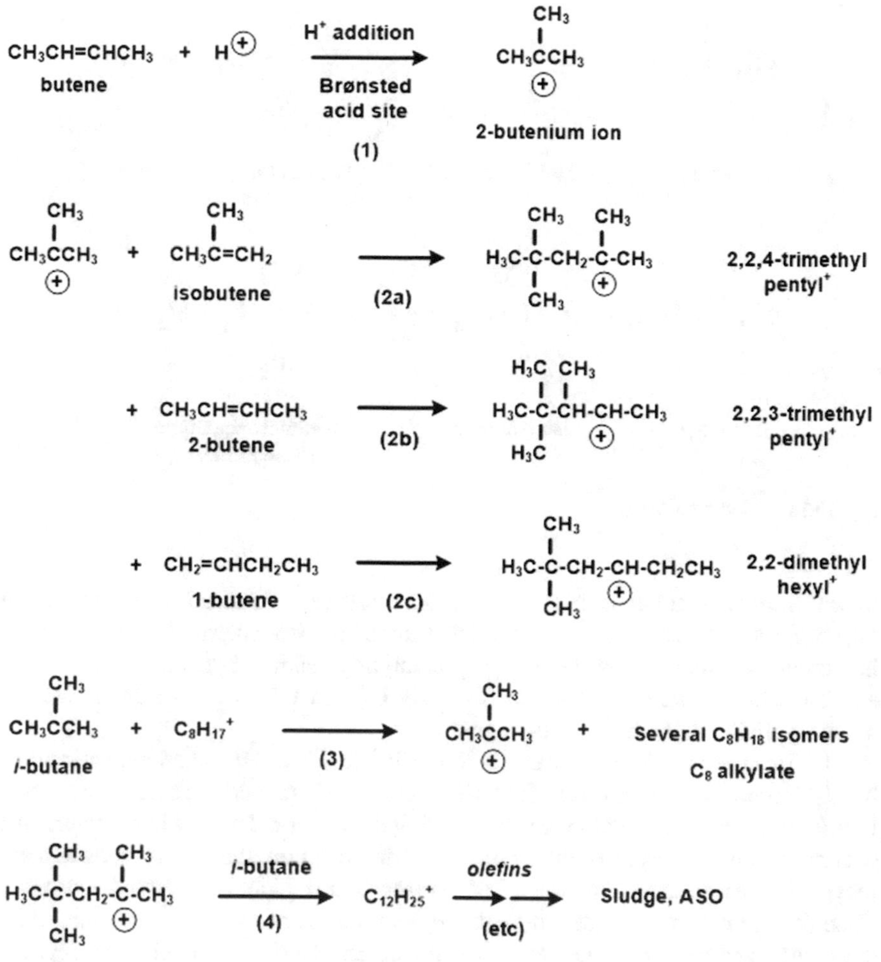

$CH_3CH=CHCH_3$ + $H^{\oplus}$ $\xrightarrow[\text{acid site}]{\substack{\text{H}^+ \text{ addition} \\ \text{Brønsted}}}$

$$\underset{\text{2-butenium ion}}{\overset{\displaystyle CH_3}{\underset{\oplus}{CH_3CCH_3}}}$$

butene

(1)

$\underset{\oplus}{\overset{\displaystyle CH_3}{CH_3CCH_3}}$ + $\underset{\text{isobutene}}{\overset{\displaystyle CH_3}{CH_3C=CH_2}}$ $\xrightarrow{\quad\quad}$ $\underset{\overset{\displaystyle CH_3}{}}{\overset{\displaystyle CH_3\quad CH_3}{H_3C\text{-}C\text{-}CH_2\text{-}\underset{\oplus}{C}\text{-}CH_3}}$ 2,2,4-trimethyl pentyl⁺

(2a)

+ $\underset{\text{2-butene}}{CH_3CH=CHCH_3}$ $\xrightarrow{\quad\quad}$ $\underset{\overset{\displaystyle H_3C}{}}{\overset{\displaystyle H_3C\quad CH_3}{H_3C\text{-}C\text{-}\underset{\oplus}{CH}\text{-}CH\text{-}CH_3}}$ 2,2,3-trimethyl pentyl⁺

(2b)

+ $\underset{\text{1-butene}}{CH_2=CHCH_2CH_3}$ $\xrightarrow{\quad\quad}$ $\underset{\overset{\displaystyle CH_3}{}}{\overset{\displaystyle CH_3}{H_3C\text{-}C\text{-}CH_2\text{-}\underset{\oplus}{CH}\text{-}CH_2CH_3}}$ 2,2-dimethyl hexyl⁺

(2c)

$\underset{\textit{i}\text{-butane}}{\overset{\displaystyle CH_3}{CH_3CCH_3}}$ + $C_8H_{17}{}^+$ $\xrightarrow{\quad\quad}$ $\underset{\oplus}{\overset{\displaystyle CH_3}{CH_3CCH_3}}$ + Several C_8H_{18} isomers C_8 alkylate

(3)

$\underset{\overset{\displaystyle CH_3}{}\oplus}{\overset{\displaystyle CH_3\quad CH_3}{H_3C\text{-}C\text{-}CH_2\text{-}C\text{-}CH_3}}$ $\xrightarrow[\text{(4)}]{\textit{i}\text{-butane}}$ $C_{12}H_{25}{}^+$ $\xrightarrow[\text{(etc)}]{\textit{olefins}}$ Sludge, ASO

Fig. 10.9 Alkylation mechanism for reaction of butenes with isobutane

merization to form sludge or acid soluble organics (ASO), shown in Reaction 4. Hence, temperature and space velocity control are important for limiting the extent of reaction ensuring maximum production of alkylate in the gasoline range (C_5 to C_{12}).

Figure 10.10 summarizes the alkylation reaction. 2,2,4-Trimethylpentane, commonly known as isooctane, is a reference compound for gasoline octane rating defined as 100.

Figure 10.11 shows the refinery streams that flow into a typical alkylation unit and product streams out of the unit. Isobutane feed comes from a hydrocracker, an FCC unit, or a butane isomerization unit as discussed earlier. All offgases from the refinery units are sent to a gas plant for treating (removing H_2S and NH_3) and to separate light

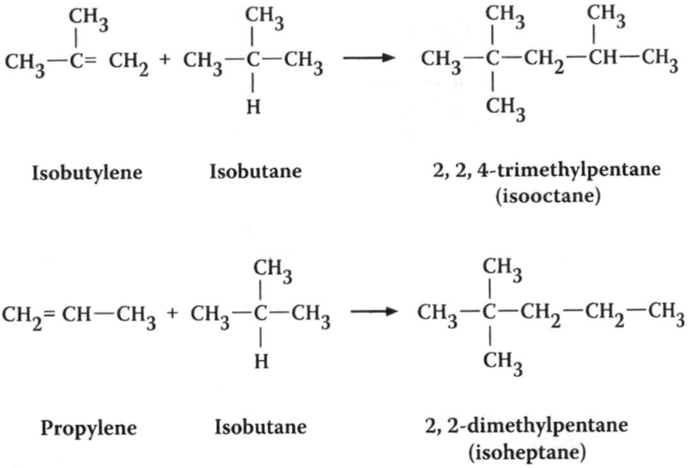

Isobutylene   Isobutane   2, 2, 4-trimethylpentane
                                                    (isooctane)

Propylene   Isobutane   2, 2-dimethylpentane
                                                    (isoheptane)

Fig. 10.10 Alkylation reactions

hydrocarbons, including isobutane. Olefins are not present in crude oils. Hence, they cannot come from distillation directly. They are mainly from thermal conversion units in the refinery, such as thermal cracking, catalytic cracking, hydrocracking, coking, etc. The main source of olefins, particularly $C_3^=$ and $C_4^=$, is from a fluid catalytic cracking (FCC) unit, shown in the figure.

The flow inside an alkylation unit is shown in Fig. 10.12. The alkylation reaction is highly exothermic, so the mixed feed of isobutene and cracked gas containing olefins is chilled to a low temperature before entering the reactor. The low temperature also decreases polymerization of the olefins. After the reaction, the acid is recovered and recycled. The product is caustic washed to remove the residual acid from the product. Then the product stream goes through the depropanizer to separate propane, to the deisobutanizer to recover isobutane for reuse, and to the debutanizer to separate butane from the desired product, alkylate.

Alkylate is an ideal gasoline component. It has a octane numbers exceeding 90. During World War II, alkylation became the main process for improving the octane of aviation gasoline (avgas). Alkylate has zero sulfur, zero benzene, and zero olefin content, so it is friendly to the environment. Its low Reid Vapor Pressure (RVP) reduces hydrocarbon emissions. In contrast, blending n-butane into gasoline increases RVP. With 10% ethanol in reformulated gasoline, RVP is higher by 1.1 to 1.25 psi [15]. Hence, there is no more need to add n-butane during winter or very cold weather for startup of gasoline engines. This causes a seasonal glut of n-butane.

Alkylation employs either hydrofluoric acid, as in a Hydrofluoric Acid Alkylation Unit (HFAU), or sulfuric acid, as in a Sulfuric Acid Alkylation Unit (SAAU). The process usually runs at low temperatures to avoid polymerization of the olefins. Temperatures for HF catalyzed reactions are 70–100 °F (21–38 °C). For sulfuric acid they are 35–50 °F (2–10 °C), so SAAU alkylation requires feed refrigeration.

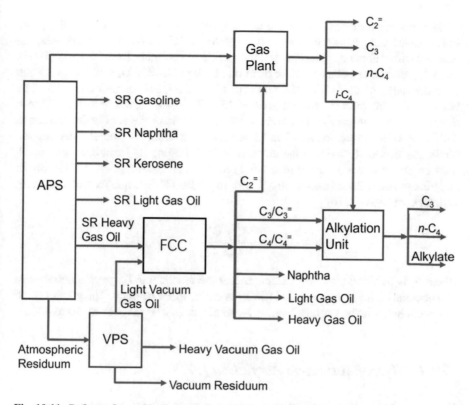

Fig. 10.11 Refinery flow with alkylation unit (APS: atmospheric pipe still or atmospheric distillation unit, VPS: vacuum pipe still or vacuum distillation unit: SR: straight run; FCC: fluidized catalytic cracking unit)

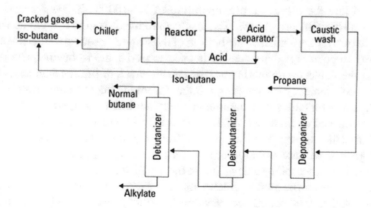

Fig. 10.12 Alkylation plant flow

The process variables affect the product quality in octane number and the yield. In the HFAU, increasing temperature from 60 to 125 °F (16–52 °C) will decrease the octane number about 3. In the SAAU, increasing temperature from 25 to 55 °F (−4 to 13 °C) decreases octane number from one to three. In HFAU, acid concentration is in the range of 86 to 90 wt%, while in the SAAU, between 93 and 95%. The isobutane:olefins ratio is between 5:1 and 15:1. The olefin space velocity, which is defined as the volume of olefins divided by the volume of the acid in the reactor, is 4–25 min. contact time in HFAU and 5–40 min. in SAAU. In general, lowering the olefin space velocity reduces the amount of high-boiling hydrocarbons produced, increases the product octane number, and reduces acid consumption. Mrstik, Smith and Pinkerton developed a correlation factor to predict alkylate quality with operating variables, defined as:[16]

$$F = \frac{I_E (I/O)_F}{100 (SV)_O}$$

where I_E is liquid volume % of isobutene in the reactor, $(I/O)_F$ is volumetric ratio of isobutene/olefin in the feed, and $(SV)_O$ is olefin space velocity. The greater the F value, the better is the alkylate quality. Normal values of F range from 10 to 40.

10.4.1   Hydrofluoric Acid Alkylation [17]

Figure 10.13 describes a HF alkylation unit or HFAU. HF is dangerous, and extra precautions are taken to ensure that it is always contained. The feed is mixed with recycled isobutene to give an isobutene/olefin ratio of 15:1. The combination is introduced into the reactor at a temperature of 70–100 °F. There it reacts in the presence of HF with purity >88%. The product mixture is sent to a settler and an acid stripper to recover the acid, HF. The bottom of the stripper is sent to the deisobutanizer to recover isobutene. A depropanizer is used to purify the isobutene. The product stream is then enters a defluorinator (for caustic wash of the residual acid) before it enters a debutanizer for separating n-butane from the final product, alkylate.

Phillips developed a single column HF alkylation reactor, shown in Fig. 10.14. The isobutane and olefin feed is mixed with acid in a reactor pipe at 70–80 °F which leads to a settler to separate the product from the acid. Acid flows down a side pipe by gravity, without a pump, for returning to the rector pipe after cooling. The hydrocarbon product mixture flows out at the top of settler.

A single tower can be used for fractionation of the product mixture. Propane comes off as an overhead. Isobutane recycles and is withdrawn at several trays above the feed tray. N-butane is taken off as a vapor several trays below the feed tray. Alkylate flows off at the bottom.

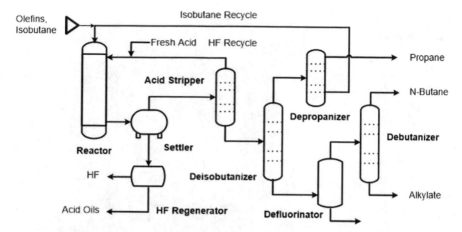

Fig. 10.13 HF alkylation process flow

10.4.2   Sulfuric Acid Alkylation [18]

Figure 10.15 shows an autorefrigeration sulfuric acid alklylation unit. The olefin feed is mixed with isobutane and acid, and chilled to 35–45 °F through a heat exchanger for minimizing redox reaction and preventing tar and SO_2 formation. A pressure at 5–15 psi is applied to prevent vaporization. Propane and lighter gases are withdrawn at the top of the reactor. The gases are compressed and liquefied. A portion of this liquid is vaporized in an economizer to cool the olefin feed, as autorefrigeration, before it is sent to the reactor or a depropanizer.

The C_4's and alkylate mixture is sent to deisobutanizer, after caustic wash to remove residual acid, to fractionate into isobutane for recycle with the isobutane feed, n-butane and alkylate.

A typical Stratco effluent refrigerated sulfuric acid alkylation unit is shown in Fig. 10.16 which uses acid settler effluent as the refrigerant to provide cooling in the contactor reactor. The Stratco contactor reactor is a horizontal pressure vessel containing an inner circulation tube, a tube bundle to remove the heat of reaction, and a mixing impeller. Hydrocarbon feed and recycle acid enter on the suction side of the impeller inside the circulation tube. As the feeds pass across the impeller, a fine emulsion of hydrocarbon and acid is formed by the extremely high shear forces induced by the impeller. Heat transfer from the reaction side of the tube bundle to the refrigeration side is aided by the high circulation rate, which also prevents any significant temperature differential within the contactor reactor. A portion of the circulating emulsion in the contactor reactor flows from the circulation tube, on the discharge side of the impeller, to the acid settler, where the hydrocarbon phase is separated from the circulating acid phase. The acid is recycled back to the contactor. Hydrocarbon leaves the settlers and is let down across a back-pressure control valve to the tube side of the contactor reactor bundle. The heat of reaction from the shell side

Fig. 10.14 Phillips HF
alkylation reactor

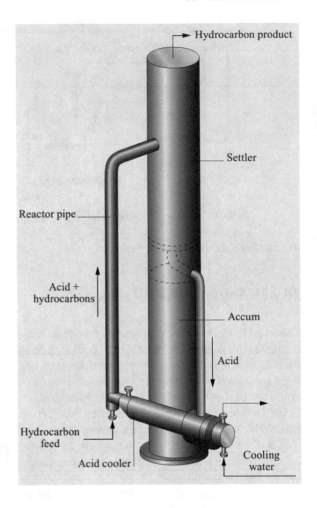

is removed by further vaporization of the hydrocarbon effluent as it passes through
the tubes before entering the condenser and separator.

10.4.3   HF Alkylation Process Versus H₂SO₄ Alkylation Process [19]

Currently, more than half of the world's approximately 700 refineries have alkylation
units that use HF or H_2SO_4. HF alkylation process has several advantages over
sulfuric acid alkylation process: (1) the reactors are smaller and simpler; (2) the HF
process is less sensitive to temperature than the H_2SO_4 process,cooling water can be
used instead of refrigeration; (3) it has a smaller settling device for emulsions; (4)

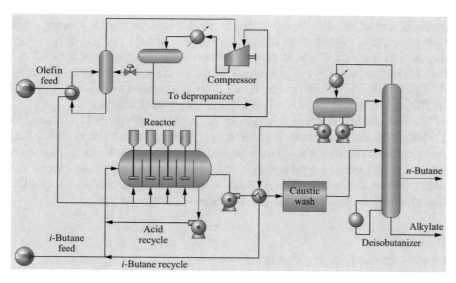

Fig. 10.15 Autorefrigeration sulfuric acid alkylation Unit [18]

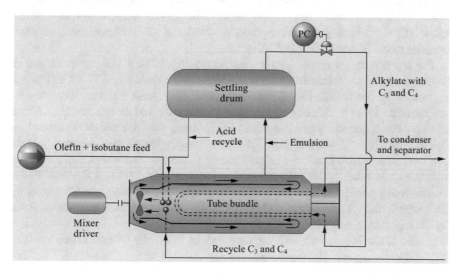

Fig. 10.16 Stratco effluent refrigeration process

HF consumption is low due to essentially complete regeneration: disposal of acid is not necessary; (5) operation is more tolerant to changes in temperature and isobutane to olefin ratio; and (6) there is less need in turbulence or agitation.

On the other hand, the advantages of sulfuric acid alkylation process over HF alkylation process include: (1) it does not require additional equipment, such as the HF stripper, HF regeneration tower, etc. (2) in the HF process, it is necessary to recover or neutralize various streams on-site. In H_2SO_4 process, the entire HC stream is neutralized off-site; (3) drying is not required for H_2SO_4 process, while drying down to a few ppm water is needed for the HF process. In the H_2SO_4 process, only feed coalescers are used to remove the free water that drops out of the chilled feed; (3) additional safety equipment is required for HF process, with greater costs; (4) capital costs for HF process are higher; (5) self-alkylation of isobutane occurs in the HF process; and (6) H_2SO_4 alkylation works better for butylene.

However, due to their strong acidity in the liquid phase, both HF and H_2SO_4 require relatively expensive corrosion-resistant vessels and equipment. Safety in handling and operations is a major concern for both, which also face disposal issues associated with spent acids and acid-soluble oils (ASO). These problems can be diminished with ionic-liquid alkylation and essentially eliminated by solid acid-catalyzed alkylation. These recently developed breakthrough processes are discussed below.

10.4.4 Ionic-Liquid Alkylation

Strongly acidic ionic liquid catalysts are safer and easier to handle than HF and H_2SO_4. The ionic liquid catalyst can be used in a lower volume at temperatures below 100 °C. The catalyst can be regenerated on-site, giving it a lower environmental footprint than HF and H_2SO_4 technologies.

The world's first composite ionic liquid alkylation (CILA) commercial unit of 120 kt/a was commissioned at Shandong Deyang Petrochemical Plant in Dongying, China in August, 2013, after a successful pilot plant test run. The process was retrofitted into an existing 65 kt/a sulfuric acid alkylation unit at a PetroChina Lanzhou Petrochemical plant starting in 2005, using the technology developed by China University of Petroleum [20]. The alkylation of isobutylene occurs in a strongly Lewis acidic ionic liquid based on aluminum (III) chloride. The retrofit of the existing sulfuric acid alkylation unit, called ionikylation, not only increased the yield of the process (compared to sulfuric acid), but also increased the process unit capacity by 40%, with attractive economics. This is by far the largest commercial usage of ionic liquids reported to date [21].

In the U.S., Honeywell UOP announced in 2016 the commercialization of ISOALKY™, a new alkylation technology developed by Chevron USA [22]. Chevron proved the technology in a small demonstration unit at its Salt Lake City refinery, where it operated for five years. Due to the success of the small unit, Chevron committed to convert its hydrofluoric acid (HF) alkylation unit in Salt Lake City to ISOALKY technology. The completed ISOALKY commercial unit will be operational in 2020 [23].

The catalyst for the process is a highly acidic ionic liquid—a non-aqueous liquid salt. The process operates at temperatures below 100 °C. The net reactions are the same as for other alkylation processes: isobutane + C_3–C_5 olefins → alkylate. The ionic liquid catalyst performs as well or better than HF and H_2SO_4, but with lower volatility. Due to the lower vapor pressure, the ionic liquid is easier to handle. HF alkylation units can be cost-effectively converted to ISOALKY technology. Other advantages include the ability to produce alkylate from a wider range of feedstocks using a lower volume of catalyst. The catalyst can be regenerated on-site, giving it a lower environmental footprint than HF and H_2SO_4 technologies.

10.4.5 Zeolite Catalyzed Alkylation

AlkyClean® is a solid-acid gasoline alkylation process developed by CB&I, Albemarle Catalysts, and Neste Oil [24]. The process employs a robust zeolite catalyst. According to the developers, the total installed cost of an AlkyClean unit is significantly lower than current HF and H_2SO_4 units. No product post-treatment or acid disposal is required. Due to the lack of corrosive acids and the relatively mild operating conditions, carbon steel can be used for construction. The catalyst is much more tolerant to water and other feed impurities, such as oxygenates, sulfur compounds, and butadiene. Deactivation from these impurities can be restored via gas phase regeneration with hydrogen at 250 °C. The product quality and from AlkyClean are comparable to that of liquid acid processes, but overall yields are higher. This is because there are no acid-soluble oils produced.

The world's first solid acid catalyst alkylation unit employing Albemarle's AlkyStar™ catalyst, together with CB&I's novel reactor scheme, was started up at the at Haiyi Fine Chemical's Zibo plant, a subsidiary of Shandong Wondull Petrochemical Group, in Shandong Province, China on August 18, 2015. Without the use of liquid acid catalysts, the solid-acid catalyzed AlkyClean technology is not only safer but also more environmentally friendly due to the reduction of waste streams, such as spent acids and acid-soluble oils. The alkylates produced consistently have high octane values between 96 and 98 [25].

10.5 Polymerization [1, 26, 27]

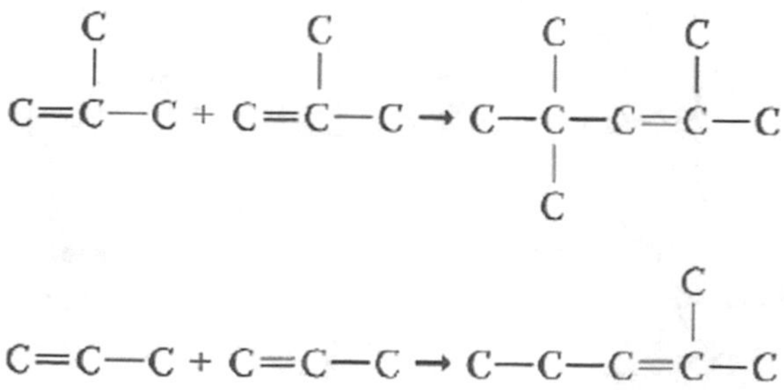

Fig. 10.17 Typical polymerization reactions

Olefin gases can be polymerized to liquid products that are suitable for gasoline in the carbon number range between 5 and 12. The term "polymerization" has a unique meaning in the gasoline industry. It is not the same as making high polymers, such as polyethylene and polypropylene. A better name for gasoline "polymerization" is "oligomerization" because of the low number of repeating olefin units in the reaction. Polymer gasoline (poly gasoline or polymerate) exhibits high octane values. A typical reaction is shown in Fig. 10.17. The usual feedstock is propylene and butylene from cracking processes. Catalytic polymerization came into use in the 1930s and was one of the first catalytic processes to be used in the petroleum industry.

Figure 10.18 shows a simplified diagram for the polymerization of olefin unit. The feed is propylene and butylenes from thermal and catalytic cracking units. Sulfur- and oxygen-containing compounds are removed from the feed. Amine treating removes H_2S and caustic washing removes mercaptans. The gas is then passed through a scrubber in which H_2O to removes caustic or amines. A small amount of H_2O is added to compensate for water loss during the polymerization, thus maintaining catalyst activity. The catalysts can be sulfuric acid, copper pyrophosphate, solid phosphoric acid (SPA) on pellets of kieselguhr (a porous sedimentary rock), or some similar material. The reaction conditions are temperature at 150–220 °C (300–425 °F) and pressure at 150–1200 psi (10–80 atm). The reaction is exothermic. Cold recycled propane is used as a coolant to quench the reaction. The product mixture is sent to a flasher to separate propylene and butylene for recycle, while the bottom enters a stabilizer to separate C_3/C_4 from polymer gasoline product. Polymer gasolines derived from propylene and butylene have octane numbers above 90.

Figure 10.19 shows a flow diagram for the UOP SPA polymerization process. The SPA catalyst is a combination of Kieselguhr and phosphoric acid [26]. The conditions can be varied to produce maximum polymer yield at almost complete conversion of the olefins. In the polymerization of butylenes the process can be carefully controlled to yield largely isooctenes, which on hydrogenation will yield isooctanes having an

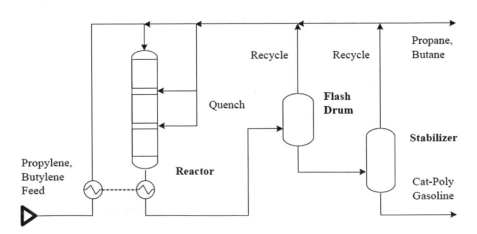

Fig. 10.18 Catalytic polymerization of olefins

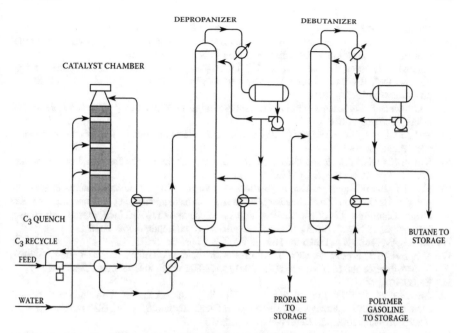

Fig. 10.19 UOP solid phosphoric acid polymerization unit

octane rating of 90–96. Pretreatment of feed for removal of sulfur compounds is accomplished by countercurrent contact with caustic solution. Removal of nitrogen compounds is accomplished by contact with a stream of water in a countercurrent packed tower. Water must be present in the feed to maintain the required equilibrium, i.e., to prevent the catalyst from becoming dehydrated.

References

1. Hsu CS, Robinson PR (2017) Gasoline production and blending. In: Hsu CS, Robinson PR (eds) Springer handbook of petroleum technology, New York, Springer
2. Dean LE, Harris HR, Belden DH, Haensel V (1959) The Penex process for pentane isomerization. Platimum Metal Rev. 3(1):9–11
3. The petroleum handbook: edition 6 by shell, Elsevier (1986)
4. Holcombe TC (1980) Total Isomerization Process, US Patent 4210771A, Jul 1, 1980
5. Zak T, Behkish A, Shum W, Wang S, Candela L, Ruszkay J (2009) "Isomerization of Butenes: LyondellBasell's isomplus technology developments", presented at Production and Use of Light Olefins, DGMK Conference 28–30 Sept, Dresden, Germany
6. Antos GJ, Aitani AM (eds) (2007) Catalytic naphtha reforming. 2nd edn. New York, Marcel Dekker
7. Hsu CS, Robinson PR (2017) Gasoline production and blending. In: Hsu CS, Robinson PR (eds) Springer handbook of petroleum technology, New York, Springer
8. LeGoff P-Y, Kostka W, Ross J (2017) Catalytic reforming for fuels and aromatics, In: Hsu CS, Robinson PR (eds) Springer handbook of petroleum technology, Heidelberg, Springer

9. Emms NJ (1958) Catalytic Reforming. Can J Chem Eng 36(6):267–270
10. (a)   Gasoline,   https://www.uop.com/processing-solutions/refining/gasoline/#naphtha-reforming; (b) http://documents.mx/documents/ccr-platforming.html
11. Lapinski M, Baird L, James R (2016) UOP platforming process: http://library.certh.gr/libfiles/PDF/GEN-PAPYR-5580-UOP-by-LAPINSKI-in-CH-4-1-BK-PP-4-3-4-32-Y-2004. pdf. Accessed 15 Sept 2016
12. The powerforming process in Big Chemical Encyclopedia: http://chempedia.info/info/170935/. Accessed 20 Sept 2016
13. Stefanidakis G, Gwyn JE (1993) Alkylation. In: McKetta JJ Chemical processing handbook, CRC Press. pp 80–138
14. Krantz K (2003) Alkylation chemistry: mechanisms, operating variables, and olefin interactions. Leawood, Kansas, STRATCO
15. Erwin J (1994) Vapor pressure interactions of ethanol with butane and pentane in gasoline preprint. In: Division of fuel chemistry, 207th American chemical society national meeting, San Diego, California, 13–17 March 1994. https://web.anl.gov/PCS/acsfuel/preprint%20archive/Files/39_2_SAN%20DIEGO_03-94_0310.pdf, accessed September 9, 2008
16. Mrtick AV, Smith KA, Pinkerton RD (1951) Adv Chem Ser 5:97
17. Simpson MB, Kester M (2007) "Hydrofluoric acid alkylation", ABB Review, 3/2007. https://library.e.abb.com/public/1b9c3c80511554ef8325734b004198cf/22-26%203M774_ENG72dpi.pdf
18. Branzaru J (2001) Introduction to sulfuric acid alkylation unit process design, Stratco, 11/2001: http://www2.dupont.com/Clean_Technologies/en_US/assets/downloads/AlkyUnitDesign2001.pdf. Retrieved 29 Sept 2018
19. H_2SO_4 vs. HF: http://www.dupont.com/content/dam/dupont/products-and-services/consulting-services-and-process-technologies/consulting-services-and-process-technologies-landing/documents/H2SO4_vs._HF.pdf
20. Liu ZC, Zhang R, Xu CM, Xia RG (2006) Ionic liquid alkylation process produces high-quality gasoline. Oil Gas J 104(40):52–56
21. PetroChina Lanzhou Greenchem ILs (Center for Greenchemistry and Catalysis), Ionilkylation process: (a) http://www.ionicliquid.org/en/application/2014-04-24/40.html. Accessed 15 Sept 2016. (b) https://www.uop.com/?press_release=honeywell-uop-introduces-ionic-liquids. Accessed on 25 Sept 2016
22. Honeywell UOP introduces ionic liquids alkylation technology: https://www.uop.com/?press_release=honeywell-uop-introduces-ionic-liquids. Accessed 26 Sept 2016
23. Timken HK "Isoalkyl Technology" Chevron presented at SCAQMD PR 1410 Working Group, August 2, 2017. http://www.aqmd.gov/docs/default-source/rule-book/Proposed-Rules/1410/chevron-presentation.pdf. Retrieved 29 Sept 2018
24. http://www.cbi.com/getattachment/61818074-13d9-4b08-9c5f-1261ccdefad2/AlkyClean-Solid-Acid-Catalyst-Alkylation-Technolo.aspx. Accessed 26 Sept 2016
25. http://investors.albemarle.com/phoenix.zhtml?c=117031&p=irol-newsArticle&ID=2123306. Accessed 25 Sept 2016
26. Ipatieff VN, Corson BB, Egloff G (1935) Polymerization, a new source of gasoline, Ind Eng Chem 27(9):1077–1081
27. Weinert E (1951) Polymerization with solid phosphoric acid catalyst In: 3rd World petroleum congress, 6 May 28–June 1951, The Hague, The Netherlands. https://www.onepetro.org/conference-paper/WPC-4319. Accessed 23 Sept 2016

Chapter 11
Cracking

As mentioned previously, cracking larger molecules produces gasoline, kerosene, jet fuel, diesel, heating oils and fuel oils. It also generates light gases (including olefins), LPG, and butanes. Cracking can be carried out thermally through free radical chemistry or by catalysis (catalytic cracking) through carbocation chemistry. Modern thermal cracking processes include steam cracking, coking and visbreaking. Catalytic processes include fluid catalytic cracking and hydrocracking. Cracking requires heat to break C–C bonds. In hydrocracking, hydrogenation of cracked fragments generates heat, so on a net basis hydrocracking is exothermic.

11.1 Thermal Cracking

Thermal cracking was first invented and patented by a Russian engineer, Valdimir Shukhov, in 1891. However, its development was not pursued beyond the laboratory. American engineers, William Merriam Burton and Robert E. Humphreys independently developed a similar process in 1908, but Burton filed for a patent in 1912; the patent, which is of great significance in the history petroleum refining, was granted in 1913 (U.S. 1,049,667). The first battery of twelve stills used in thermal cracking went into operations at the Witting Refinery of Standard Oil of Indiana (now BP) in 1913. Early versions were batch processes. However continuous processes eventually took over in the 1920s. Once catalytic cracking was invented in the early 1940s, it became much more popular than thermal cracking. More sophisticated forms of thermal cracking have been developed for various purposes, including steam cracking, visbreaking and coking. Modern high-pressure thermal crackers, which operate at absolute pressures of about 7000 kPa, are the basis for economically important production of olefins for polymers.

Thermal cracking is similar to what occurs below the surface of the Earth when kerogen is broken down into lighter components. The actual reaction mechanism behind cracking is very complex, and computers model hundreds or thousands of reactions for the process, but the main reactions are all related to free radical chem-

© Springer Nature Switzerland AG 2019
C. S. Hsu and P. R. Robinson, *Petroleum Science and Technology*,
https://doi.org/10.1007/978-3-030-16275-7_11

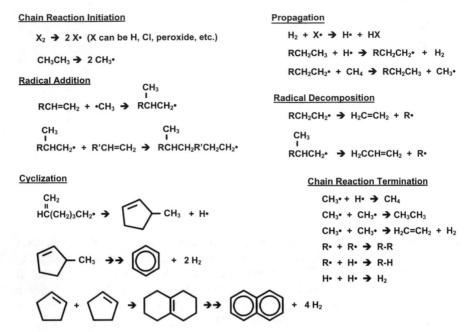

Fig. 11.1 Thermal cracking mechanism: radical chain reaction

istry. At the high temperatures, free radicals are more likely to form. A free radical is a molecule with an unshared electron attached to it. This unshared electron can break other bonds leading to a series of chain reactions. Each subsequent reaction will either create another free radical, which can react further, or combine with another free radical to end the chain.

Figure 11.1 outlines mechanisms postulated by Greensfelder, et al [1]. for the thermal cracking of hydrocarbons. This chain-reaction mechanism includes the following steps:

(1) Chain Initiation: Radicals—neutral atoms or compounds with a free electron and no charge—are formed due to direct thermal rupture of a chemical bond. Common initiators (other than heat) are organic sulfur and oxygen compounds, which are present in delayed coker feeds. H_2O_2 and Cl_2 are potent initiators. In steam cracker feed, heteroatom initiators are absent, so the chain reaction is initiated by the rupture of a C–C bond at very high temperature.
(2) Propagation: A radical abstracts hydrogen from another compound, turning it into a different radical.
(3) Radical addition: A radical attacks an olefin group, forming a C–C bond and a larger radical.
(4) Isomerization: A primary radical isomerizes into a more stable secondary radical.

(5) Radical decomposition (cracking): A radical decomposes into an olefin and a smaller radical.
(6) Cyclization: radicals with 5 or more carbons and an olefin group react within the molecules to form ring compounds. At the high temperatures at which thermal cracking occurs, aromatics and polyaromatics are thermodynamically favored, so saturated or partially saturated ring compounds readily undergo dehydrogenation and form aromatics.
(7) Chain reaction termination: Two radicals react to produce a saturated product or an olefin and hydrogen.

The radical mechanism is consistent with the fact that thermal cracking produces significant amounts of hydrogen, methane, ethane, ethylene and higher olefins. Feeds to delayed coking units contain organic sulfur, nitrogen, and oxygen compounds, which form free radicals more easily than hydrocarbons. These heteroatoms end up in two types of molecules—light ones (H_2S, ammonia and water) and heavy ones, from which they are hard to remove.

Figure 11.2 represents a thermal cracking unit for gas oils and residues. The feed is heated to 700–1100 °F in a furnace before entering a reactor under 50–1500 psig pressure. For residues, 2-stage thermal crackers are often used; the 1st stage produces gas oils for 2nd stage cracking. The reactor product is introduced into a flasher to separate residue from the product mixture, which is fractionated into olefin-containing gas, naphtha, distillates, and gas oil products. The fractionator bottom product is mainly residue, which is mixed with fresh feed and charged to the reactor. Linear alpha olefins can be produced from wax by thermal cracking. The cracking is accompanied with simultaneous removal of distillates as a semi-continuous process.

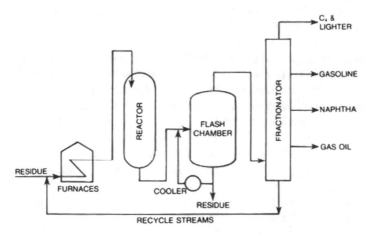

Fig. 11.2 Residue thermal cracking unit

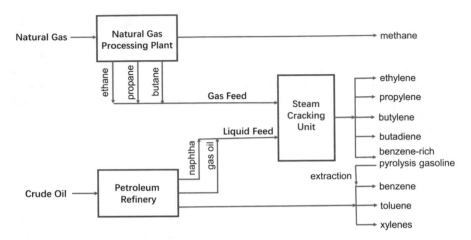

Fig. 11.3 Links between a refinery and a steam cracking unit

11.2   Steam Cracking

Steam cracking, shown in Fig. 11.3, is a process of breaking down saturated hydro-carbons into smaller, often unsaturated, hydrocarbons by reactions with steam in a bank of pyrolysis furnaces. It is the principal industrial method employed at olefin plants for producing ethylene and propylene, important feedstocks for polyolefins (polyethylenes, polypropylenes, etc.), which account for 50–60% of all commercial organic chemicals. Propylene is also produced by other methods, such as propane dehydrogenation [2]. Fluid catalytic cracking (FCC) is an important commercial source of propylene and butylene. Steam cracking in olefin plants co-produces hydrogen, which is often used in refineries.

The feed can be gaseous or liquid hydrocarbons, such as ethane, LPG (propane and butane), naphtha, and unconverted oil from hydrocracking units. C_4 olefins, including butadiene, and benzene-rich pyrolysis gasoline are produced when heavy liquid feeds, such as naphtha and hydrocracker gas oils, are used. The feeds are diluted with steam and briefly heated to 1050 °C in a furnace without the presence of oxygen and fed to Cr–Ni reactor tubes. The hydrocarbon chains preferentially crack at the center of molecules at 400 °C. Cracking shifts toward the end of the molecule with increasing temperature, leading to larger quantities of preferred low molecular weight olefin products. Thus, the reaction temperature is very high, with a reactor outlet temperature around 850 °C. The residence time is very short to improve yield and to avoid coke formation and oligomerization. In modern furnaces, reaction time can be reduced to milliseconds, with velocities faster than the speed of sound. The reaction is stopped by rapid quenching to 300 °C. The product mixture is scrubbed to remove H_2S and CO_2 and then dried before it is sent to a series of separation columns to separate and recover methane, ethane, ethane, propane, propylene and C_4 hydrocarbons. Figure 11.4 shows a steam cracking unit outside London with tall

1. A debutaniser which separates the C_4 hydrocarbons from the C_1 - C_3 hydrocarbons
2. A depropaniser which separates out the C_3 hydrocarbons
3. A deethaniser which separates out the C_2 hydrocarbons
4. A demethaniser which separates out the methane
5. A C_3 splitter which separates propene from propane
6. A C_2 splitter which separates ethene from ethane

Fig. 11.4 A steam cracking unit outside London [3]

separation columns in series for the production and separation of light hydrocarbon gases.

Steam cracking product distribution depends on the feed composition, hydrocarbon to steam ratio, cracking temperature and residence time in the furnace. Light hydrocarbon gas feeds, such as natural gas, ethane or LPG yield product streams rich in the light olefins, essentially ethylene and propylene. In addition to light olefins, heavier liquid feedstocks can also produce butylenes, butadiene and products rich in aromatic hydrocarbons suitable for blending into gasoline and fuel oil, or routed through an extraction process to recover BTX aromatics (benzene, toluene and xylenes).

11.3 Catalytic Cracking [4]

Thermal cracking was used prior to 1925 for producing naphtha through thermal decomposition of larger molecules into smaller molecules. In the late 1920s, Eugene Houdry demonstrated that a catalytic cracking process yielded more gasoline of higher octane-number, with less heavy fuel oils and light gases. Thermally cracked naphtha is quite olefinic, while catalytically cracked naphtha contains fewer olefins and large amounts of aromatics and branched compounds. The first commercial fixed bed cat cracking unit began production in 1937. The catalysts are covered by a deactivating layer of coke in a short time during the process. The catalysts can be regenerated by burning off the coke, but the time is relatively slow compared to reaction time. It is more efficient to move the catalyst from one reactor for cracking hydrocarbons to another reactor for catalyst regeneration.

In refining, catalytic cracking falls into two categories: catalytic cracking in the absence of external hydrogen—primarily fluid catalytic cracking (FCC)—and catalytic cracking in the presence of external hydrogen—hydrocracking.

Catalysts for both kinds of catalytic cracking contain strong acid sites. The mechanism involves carbocations, also known as carbenium ions or carbonium ions. Acidity is provided by amorphous silica/alumina or a crystalline zeolite. Commercial catalysts for these and other processes are discussed in a subsequent section.

FCC units produce aromatics and other heavy products via cyclization, alkylation and polymerization of intermediate olefins. Polyaromatics can grow into larger polyaromatics, which eventually can form coke.

Key features of catalytic cracking chemistry include a preponderance of branched-chain paraffin products and low yields of methane and ethane. If there is a deficiency of hydrogen, such as in the FCC process, hydrogen is produced and significant amounts of small olefins are formed.

11.3.1 FCC Feed Pretreating

These days, most refiners pretreat FCC feeds in a fixed-bed hydrotreater. The hydrotreater removes trace metal contaminants such as nickel and vanadium. Otherwise, nickel would increase coke formation and decrease liquid yields. Vanadium reduces conversion, decreases liquid yields, and destroys the catalyst. In addition to removing Ni and V, the pretreater decreases concentrations of sulfur, nitrogen, and aromatics. In the FCC regenerator, sulfur on the coked catalyst is converted to sulfur oxides (SOx) in the flue gas. Clean air regulations restrict SOx emissions, which cause acid rain. Therefore, removing sulfur from the FCC feed—thereby reducing SOx formation—is highly beneficial. Removing nitrogen is beneficial, too, because basic feed nitrogen suppresses the activity of highly acidic FCC catalysts. Pretreating also saturates aromatics. As we have seen, saturating aromatics makes them more crackable, so pretreating increases FCC conversion, often by more than 10 vol.%.

11.3.2 Catalytic Cracking Mechanism [4, 5]

The underlying mechanism for catalytic cracking is essentially the same for both fluid catalytic cracking and hydrocracking. Process differences are due to differences in

Table 11.1 Zeolitic catalytic cracking versus cracking catalyzed by an amorphous silica-alumina [6]

	Silica-Alumina	Zeolite
Coke, wt%	4	4
Conversion, vol.%	55	65
C_5 + gasoline, vol.%	38	51
C_3 gas, vol.%	7	6
C_4's, vol.%	17	16

catalysts, equipment, and operating conditions. FCC produces a high yield of naphtha (gasoline) and LPG. It is considered as the heart and workhorse of a fuels refinery. FCC produces more than half the world's gasoline. It generates middle distillate streams (cycle oils) for further refining or blending. It also produces, via heat exchange, a large quantity of high-quality steam.

FCC chemistry is a complex mixture of many reactions. The strong acid sites, both Lewis and Brønsted, are supplied by zeolites, which are key components of modern FCC catalysts. Influenced by their unique pore geometry, zeolites (usually H–Y) are far more efficient at generating gasoline and middle distillate products than either thermal cracking or cracking catalyzed by amorphous silica-alumina. A comparison of zeolite and amorphous Al–Si cracking is shown in Table 11.1.

11.3.2.1 Reaction Initiation

There are several possible initiation steps, shown in Fig. 11.5 [7]. One type of initiation step (Reaction 11.1) is mild thermal cracking to generate free radicals, followed by radical recombination to generate olefins. FCC reactions are mainly catalytic as evidenced by the fact that they produce relatively small amounts of methane, the generation of which is significant in thermal cracking. Methane can be formed by hydrogen abstraction by a methyl radical or by combining a methyl radical with hydrogen radical; the latter is a termination step of free-radical reactions. The main initial steps involve Brønsted acid protonation of olefins to generate secondary (2°) carbocations, shown in Reaction 2, and hydride abstraction from alkanes at Lewis base catalyst sites, shown in Reaction 3, also to generate (2°) carbocations. Of minor importance is the reversible dehydrogenation of alkanes at metal-containing catalyst sites.

11.3.2.2 Carbocation Ion Rearrangement

Figure 11.6 shows how 2° carbocations rearrange. There can be 2°–2° rearrangement and 2°–3° (tertiary) rearrangement; 3° carbocations are more stable.

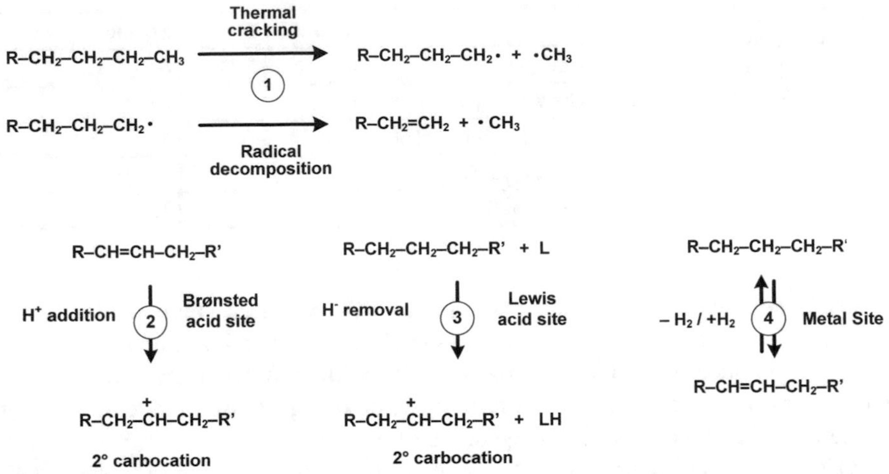
Thermal cracking
R–CH2–CH2–CH2–CH3
R–CH2–CH2–CH2• + •CH3
1
R–CH2–CH2–CH2•
R–CH2=CH2 + • CH3
Radical decomposition
R–CH=CH–CH2–R'
H+ addition
2
Brønsted acid site
R–CH2–CH–CH2–R'
2° carbocation
R–CH2–CH2–CH2–R' + L
H- removal
3
Lewis acid site
R–CH2–CH–CH2–R' + LH
2° carbocation
R–CH2–CH2–CH2–R'
– H2 / +H2
4
Metal Site
R–CH=CH–CH2–R'

Fig. 11.5 Catalytic cracking mechanism: initiation steps

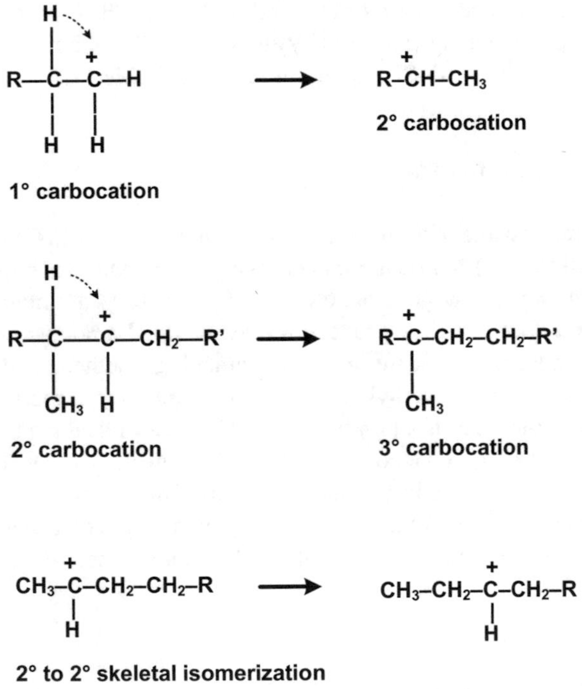
1° carbocation
R–CH–CH3
2° carbocation
2° carbocation
R–C–CH2–CH2–R'
CH3
3° carbocation
CH3–C–CH2–CH2–R
CH3–CH2–C–CH2–R
2° to 2° skeletal isomerization

Fig. 11.6 Catalytic cracking mechanism: carbocation rearrangement

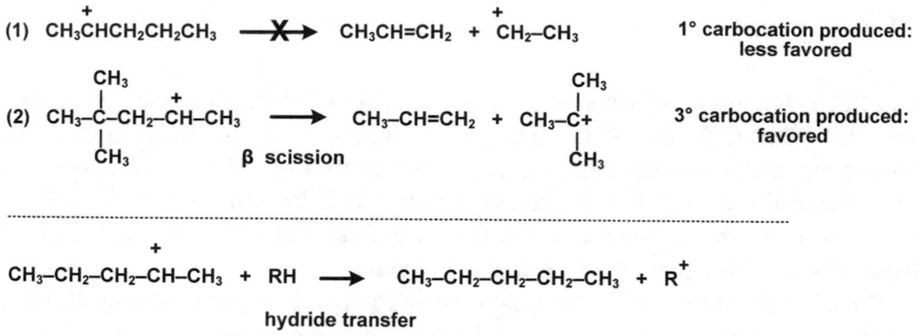

(1) $CH_3\overset{+}{C}HCH_2CH_2CH_3 \quad \xrightarrow{\quad X \quad} \quad CH_3CH=CH_2 \; + \; \overset{+}{C}H_2-CH_3$ — 1° carbocation produced: less favored

(2) $CH_3-\underset{\underset{CH_3}{|}}{\overset{\overset{CH_3}{|}}{C}}-CH_2-\overset{+}{C}H-CH_3 \quad \xrightarrow{\quad\quad} \quad CH_3-CH=CH_2 \; + \; CH_3-\underset{\underset{CH_3}{|}}{\overset{\overset{CH_3}{|}}{C}}+$ — 3° carbocation produced: favored

β scission

$CH_3-CH_2-CH_2-\overset{+}{C}H-CH_3 \; + \; RH \quad \longrightarrow \quad CH_3-CH_2-CH_2-CH_2-CH_3 \; + \; R\overset{+}{}$

hydride transfer

Fig. 11.7 Catalytic cracking mechanism: beta scission and hydride transfer

11.3.2.3 Cracking of C–C bonds via beta scission

Beta scission involves cleavage of a C–C bond in the position beta to the carbon atom that carries the positive charge. It is the primary cracking reaction in FCC. It is endothermic and favored at higher temperatures; however, if the temperature is too high, thermal cracking can become significant. Beta scission, shown in Fig. 11.7, is characteristic of mechanisms with carbocation intermediates.

11.3.2.4 Hydride Transfer (Propagation)

Hydride transfer also is shown in Fig. 11.6. If hydride transfer is substantial, paraffins (normal and branched) will be the predominant products: FCC catalysts can be designed for greater or lesser hydrogen transfer.

11.3.2.5 Termination Reactions

Termination reactions involve proton transfer to the catalyst surface or desorbing an olefin while regenerating a Brønsted site. Proton transfer to a carbocation ion generates a paraffin and regenerates a Lewis acid site. Radical termination reactions involve radical-radical reactions, such as:

$$H \bullet + H \bullet \to H_2$$
$$H \bullet + R \bullet \to RH$$
$$R \bullet + R' \bullet \to R - R'$$

11.3.3   Fluid Catalytic Cracking Processes

In FCC, a full range of smaller molecules is formed from the breakup of large molecules. Due to the overall deficiency of hydrogen, significant amounts of olefins are formed. Since the feed contains large aromatic and naphthenic molecules with long side chains attached, side chain cleavage, which can also initiate the chain of reactions, is common. The molecules from which side chains have been removed have higher specific gravity, i.e., lower API gravity.

Cracking also generates coke via aromatic ring growth by successive cyclization, polymerization and dehydrogenation. The coke coats the catalyst, rendering it inactive. Hence, the catalyst needs to be regenerated by burning off the coke in order to restore activity.

To allow onstream catalyst regeneration, a moving-bed unit, Thermofor catalytic cracking (TCC), was developed in 1941 with catalyst cycled between the reactor and regenerator. It has become obsolete and replaced by more advanced fluidized bed units with greatly improved catalysts.

A simplified diagram of the TCC process is shown in Fig. 11.8 [8]. Its essential elements are a reactor for continuously contacting hydrocarbon feed with a moving bed of granular catalyst for conversion of the hydrocarbons, and a kiln (regenerator) for removing carbon deposit from the catalyst during the cracking operation by controlled combustion with air. The catalyst flows by gravity through both vessels, which stand side by side, and is transferred from the bottom of one to the top of the other by means of bucket elevators. All required process heat is supplied by the highly exothermic combustion of coke in the regenerator. Catalyst temperatures reach 1200–1500 °F. In contrast, the cracking reaction is endothermic. The feed is preheated by heat exchange to 500–800 °F before entering the reactor. There, additional heat is supplied by mixing with the hot recycled catalyst to reach a temperature of 900–1050 °F. The reactor outlet vapor is sent to a fractionator, where it is separated into different fractions—gas (offgas), cracked naphtha, fuel oil, and slurry oil, which can be mixed with fresh feed and recycled back to the reactor for further processing.

A typical FCC unit comprises three major sections—riser/reactor, regenerator, and disengaging vessel. In the riser/reaction section, preheated oil is mixed with hot, regenerated catalyst. The mixture acts as a fluid because the catalyst particles are about the size of sifted flour. The hot catalyst vaporizes the oil, and the vaporized oil carries the catalyst up the riser/reactor. The cracking reaction is very fast, achieving completion in just a few seconds or even less. It produces light gases, high-octane gasoline, and heavier products called light cycle oil (LCO), heavy cycle oil (HCO), slurry oil, and decant oil. It also leaves a layer of coke on the catalyst particles, rendering them inactive.

There are two basic types of FCC units: "side-by-side," where the reactor and the regenerator adjacent to each other, shown in Fig. 11.9, and the stacked type, in which the reactor is mounted on top of the regenerator, shown in Fig. 11.10.

At the top of the riser, the riser outlet temperature (ROT) can reach 900–1020 °F (482–549 °C). The ROT determines conversion and affects product selectivity, so

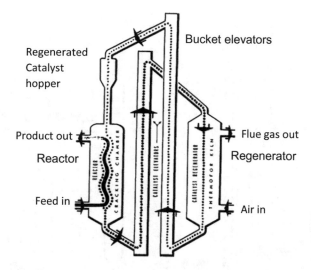

Fig. 11.8 Simplified diagram of Thermofor catalytic cracking system

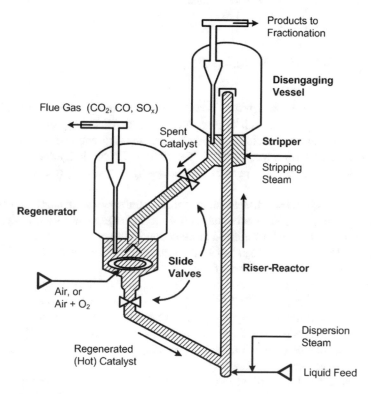

Fig. 11.9 FCC process flow

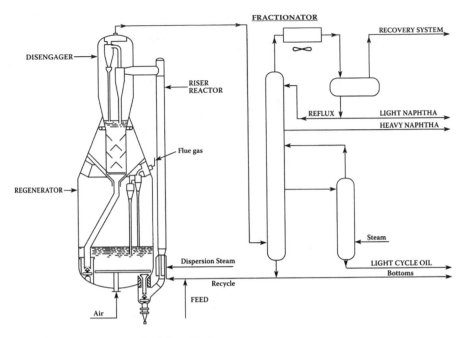

Fig. 11.10 FCC Unit—M. W. Kellogg Design

FCC operators control it as tightly as possible. Higher temperatures favor production of olefin-rich light gases, especially propylene, at the expense of gasoline; many FCC units maximize refinery-grade propylene for purification and use in olefin plants. Moderate temperatures favor gasoline production. Lower temperatures decrease gasoline yields and increase heavier products—light cycle oil (LCO) and heavy cycle oil (HCO).

In the disengaging section, steam helps separate the now-deactivated catalyst from the reaction products. The spent catalyst goes to the regenerator by gravity, where the coke is burned away by fluidized combustion in the presence of air or oxygen-enriched air. The regenerated catalyst is hot, with temperatures up to 1350 °F (732 °C). It returns to the riser/reactor, where the cycle begins again.

In a 60,000 barrels-per-day unit processing a typical mixture of vacuum gas oils, the total catalyst in the unit (the "inventory") is 400–500 tons. To maintain activity, about 0.5–1 wt% of the inventory is replaced each day. If the feed to the unit contains significant amounts of residue, the replacement rate is higher. The discharged catalyst is cooled and shipped either to a land fill for disposal or to another refiner, which might need "conditioned" FCC catalyst.

The stacked type FCC unit is represented by a Kellogg design, shown in Fig. 11.8. Again, it constitutes a riser/reactor, a regenerator and a disengager. The spent catalysts from the reactor drop from the disengager to the regenerator by gravity, with less hardware, space and maintenance required than the side-by-side type. In the figure, a fractionator to separate product components is also shown. The top effluent contains light gases that are sent to a recovery system to remove sour gases for recovery of flue gas, LPG, light olefins, etc. The bottom stream can be recycled by mixing with feed for further processing.

11.3.4 FCC Heat Balance

FCC units must be heat-balanced, or they won't run. The burning of coke deposited on the catalyst in the regenerator provides all of the heat required by the process. In fact, FCC units are significant sources of high-quality steam for other refinery units. Table 11.2 gives a representative breakdown of FCC heat requirements.

11.3.5 Residue FCC

Residue FCC (RFCC) units, also known as heavy oil crackers, can process significant amounts of atmospheric residue (650 °F+) and vacuum residue to produce gasoline and lighter components [9, 10]. It generates substantially more coke than conventional FCC feeds. Excess heat is generated when the extra coke is burned in the regenerator, and residues contain high amounts of trace metals, particularly nickel and vanadium. Those metals destroy FCC catalysts. Residue FCC units must handle both challenges.

The metals are removed in upstream hydrodemetalation (HDM) units [11]. Catalyst coolers and supplemental regenerators recover the excess heat. The catalyst coolers (e.g., steam coils) are installed usually on the second-stage regenerator. The UOP catalyst cooler is an external vertical shell-and-tube heat exchanger [12]. Catalyst flows across the tube bundle in the dense phase. UOP's air lance distribution

Table 11.2 Representative FCC heat consumption

Factor	Portion of total
Heat up and vaporize fresh feed	40–50%
Heat recycled oil	0–10%
Heat of reaction (endothermic)	15–30%
Heat steam	2–8%
Heat losses	2–5%
Heat air to regenerator temperature	15–25%
Heat coke from the reactor to regenerator temperature	1–2%
Total heat duty	500–1000 Btu/lb 1160–2325 kJ/kg

system ensures uniform air distribution within the tube bundle and a uniform heat transfer coefficient. The generation of steam (up to 850 psig) from the circulating water is used to remove heat from the regenerated catalyst.

Three different styles of catalyst coolers—flow-through, back-mix and hybrid—have been designed and commercialized to accommodate a wide range of heat removal duties as well as physical and plot-space constraints [13].

11.3.6 FCC Gasoline Post-treating

Conventional hydrotreating does a good job of removing sulfur from FCC feed, which leads to lower sulfur in FCC products. Unfortunately, despite pretreatment, FCC gasoline can still contain up to 150 ppmw sulfur—far more than the present specification of 10 ppmw. Hydrotreating removes sulfur from the gasoline, but it also reduces octane by saturating C_6–C_{10} olefins.

In processes such as Prime-G+ [14], offered for license by Axens, full-range naphtha is split into light and heavy fractions. The light fraction contains most of the high-octane olefins but not much of the sulfur. After diolefins are removed via selective hydrogenation, the light cut is ready for gasoline blending. The heavy fraction contains most of the sulfur but not much of the olefins. It is hydrotreated conventionally.

The S-Zorb process [15], invented by ConocoPhillips, uses selective adsorption to remove sulfur from FCC gasoline. The feed is combined with a small amount of hydrogen, heated, and injected into an expanded fluid-bed reactor, where a proprietary sorbent removes sulfur from the feed. A disengaging zone in the reactor removes suspended sorbent from the vapor, which exits the reactor as a low-sulfur stock suitable for gasoline blending. The sorbent is withdrawn continuously from the reactor and sent to the regenerator section, where the sulfur is removed as SO_2 and sent to a sulfur recovery unit. The clean sorbent is reconditioned and returned to the reactor. The rate of sorbent circulation is controlled to help maintain the desired sulfur concentration in the product.

11.4 Petroleum Refining Catalysts

Many refining processes require catalysts. Catalysts do not change thermodynamics, which govern whether or not a reaction can proceed, but an appropriate catalyst increases the rate of the desired reaction. A catalyst facilitates chemical reactions, either by increasing the rate at a given set of process conditions, or by lowering required temperatures and/or pressures and/or residence times, thereby reducing the size of process equipment. Catalysts do this by decreasing activation energy. The reaction heat might vary with temperature and pressure, but ultimately, reaction heat is governed by thermochemistry.

In practice, catalysts do change as time goes by and more materials are processed. They degrade due to coking, attrition, feed contamination and/or process upsets. Some catalysts last for years before they have to be replaced. In other processes, such as FCC and CCR catalytic reforming, they are regenerated and reused inside the process during normal operation. In the FCC process, degraded catalyst is removed and replaced with fresh catalyst as needed. The chemistry of coke formation on catalysts is similar to the coke formation mechanism presented in Chap. 9, Sect. 4.

Catalysts facilitate reactions by decreasing activation energies. Consider ammonia synthesis:

$$N_2 + 3H_2 \rightarrow 3NH_3$$

The reaction is favored by thermodynamics, but molecular nitrogen is exceedingly stable due to the strength of the N≡N triple bond. To break the N≡N bond with heat alone requires 3000 °C (5400 °F). Before the catalytic Haber-Bosch process, ammonia was synthesized from nitrogen and hydrogen in electric arc reactors, which in essence generate mini-lightning bolts while consuming immense amounts of energy. In catalytic ammonia synthesis, the N≡N bond is weakened when it "dissolves" in the catalyst, allowing it to react with hydrogen at <750 °F (<400 °C).

Most petroleum refinery catalysts are solids, with active metals supported on high-strength supports such as γ-alumina, silica or aluminosilicates. The major exceptions are alkylation catalysts, which employ liquid-phase acids—HF or H_2SO_4 or ionic liquids. The separation of liquid and vapor products from solids is straightforward, so it is easier to recover the catalysts and regenerate or dispose of them.

As mentioned in the FCC chemistry discussion, FCC and hydrocracking (discussed below) employ solid-acid catalysts, primarily based on synthetic zeolites. Hydrocracking employs both zeolites and amorphous silica/alumina catalysts. Zeolites possess microporosity, which gives them high surface area and contributes to their high activity. They also possess a mesopore structure, which affects feed flexibility and product selectivity.

Related materials are used in other processes. For example, metal-promoted silica alumina phosphates such as SAPO-11 are used to isomerize normal paraffins in catalytic dewaxing units. Figure 11.11 shows six zeolite structures [16]. In cracking catalysts, HY zeolite is the most common, but beta and ZSM-5 are used as well. The building blocks of A, X, and other zeolites are tetrahedral units of Si and Al oxides. In the ultra-stable Y (USY) zeolites employed in cracking catalysts, the Si/Al ratio is >10.

The acidity of zeolites (and amorphous silica/aluminas) comes from their structure [17]. Fig. 11.12 shows how one can visualize these materials as a silica (SiO_2) superstructure, in which every so often an aluminum atom replaces a silicon atom. The silicon atoms have a valence of +4, and each one binds to four oxygen atoms.

Replacing Si (+4) with Al (+3) creates electron-accepting Lewis acid sites. Carbocations can be formed from olefins by hydride transfer to the empty d-orbital on aluminum. The oxygen atoms in associated hydroxyl groups are Brønsted acids (pro-

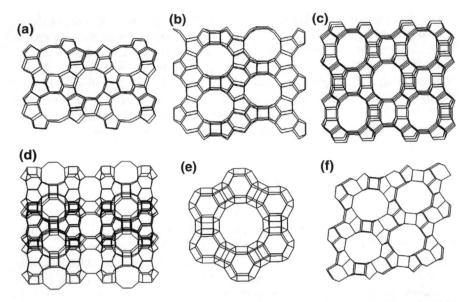

Fig. 11.11 Structures of zeolites: ZSM-5 (**a**), mordenite (**b**), beta (**c**), MCM-22 (**d**), zeolite Y (**e**), and zeolite L (**f**)

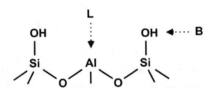

Fig. 11.12 Brønsted (B) and Lewis acid sites (L) in zeolites and amorphous silica/aluminas

ton donors). The H⁺ can be replaced by other positive counter-ions such as Na⁺, K⁺, and NH₄⁺. The counter ions can be swapped via ion exchange. For example, when Na-Y zeolite is exchanged with an ammonium salt, the Na⁺ ion is replaced by NH₄⁺. When NH₄-Y is heated to the right temperature, the ammonium ion decomposes, releasing NH₃ (gas) and leaving behind highly acid H-Y zeolite. A typical Si/Al ratio for a zeolite with high cracking activity is >5. In hydrocracking catalysts with less activity but greater selectivity for production of middle distillates, the Si/Al might be about 30.

ZSM-5 is a shape-selective zeolite made by including a soluble organic template in the mix of raw materials. Templates for this kind of synthesis include quarternary ammonium salts. ZSM-5 enhances distillate yields in FCC units and catalytic dewaxing in hydroprocessing units, where due to its unique pore structure, it selectively cracks waxy *n*-paraffins into lighter molecules.

The amorphous silica/alumina (ASA) catalysts used for hydrocracking are less active, but they do a better job of converting straight-run VGO into diesel with

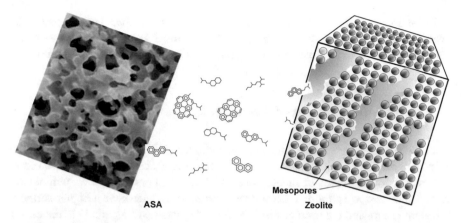

Fig. 11.13 Schematic comparison of amorphous alumina silica (ASA) and zeolite structures

minimal co-production of naphtha and lighter products. Figure 11.13 provides a conceptual comparison of amorphous ASA catalysts with crystalline zeolite catalysts. The pore diameters of H-Y zeolite catalysts are uniform and relatively small, around 7.5 Å. Mesopores have larger diameters, which admit mid-sized molecules such as those found in heavy vacuum gas oils. ASA catalysts include small, medium and large pores. The width of ASA pores can exceed 100 Å. The larger pores can accommodate larger molecules, which explains why ASA catalysts do a better job of cracking feeds with endpoints >1000 °F (538 °C). The pore diameter for ZSM-5 is 6.3 Å.

11.4.1 FCC Catalysts and Additives

Finished FCC catalysts are powders with the consistency of sifted flour. They are given their final form by spray drying in which a slurry of catalyst components is converted into a dry powder by spraying into hot air. The method gives a consistent particle size distribution. The particles are roughly spherical, with bulk densities of 0.85–0.95 g/cm^3 and average diameters of 60–100 μm.

The catalysts include four major components—one or more *zeolites* (up to 50%), a *matrix* based on non-crystalline alumina, a *binder* such as silica sol, and a clay-based *filler* such as kaolin. The zeolite is ultra-stable (US) H-Y (faujasite) or very ultra-stable (VUS) H-Y. The extra stability is provided by de-alumination with hydrother-mal or chemical treatment. The H-Y zeolite is sometimes augmented with ZSM-5 to increase propylene yield. To provide thermal stability and optimize the relative amounts of different active sites, a mixture of rare earths (RE), such as lanthanum-rich mixture of rare earth (RE) oxides, is incorporated into the zeolite structure by ion exchange. The RE mixture can contain up to 8 wt% ceria, up to 80% La$_2$O$_3$, up to 15 wt% Nd$_2$O$_3$, with traces of other rare earths.

Other components can be part of the core catalyst or introduced as external additives. The extra components can provide NOx and/or SOx reduction, Noble metal combustion promoters enhance the conversion of CO to CO_2 in the regenerator.

11.4.2   Catalysts for Fixed-Bed Reactors

Fixed-bed catalysts are shaped into spheres, cylinders, hollow cylinders, lobed extrudates, pellets. Some look like small wagon wheels. Figure 11.14 shows some of these. A cross-section of a lobed extrudate can look like a 3-leaf or 4-leaf clover without the stem. Compared to cylindrical extrudates, shaped extrudates have a higher surface-to-volume ratio, and the average distance from the outside of a particle to the center is shorter. This increases activity by decreasing the average distances traveled by molecules to reach active catalyst sites. To make extrudates, a paste of support material forced through a die. The resulting spaghetti-like strands are dried and broken into short pieces with a length/diameter ratio of 2–4; for main-bed hydrotreating catalysts, diameters range from 1.3 to 4.8 mm. The particles are calcined, which hardens them and removes additional water and volatile molecules such as ammonia.

Spherical catalysts are made by (a) spray-drying slurries of catalyst precursors, (b) spraying liquid onto powders in a tilted rotating pan, or (c) dripping a silica-alumina slurry into hot oil. Pellets are made by compressing powders in a dye. FCC catalysts are made by spray drying.

Impregnation distributes active metals within the pores of a catalyst support. Like sponges, calcined supports are especially porous. Far more than 99% of the surface area is inside the pores. When the pores are exposed to aqueous solutions containing active metals, capillary action pulls the aqueous phase into them. After drying, the catalyst might be soaked in another solution to increase the loading of the same (or a different) active metal. Catalysts can also be made by co-mulling active metal oxides with the support. Co-mulling tends to cost less because it requires fewer steps. It also produces materials with different activities—sometimes higher, sometimes lower—than impregnation.

Eventually, refinery catalysts deactivate and must be replaced. The major causes of deactivation are feed contaminants (trace metals, particulates, etc.) and catalyst coking; the latter is discussed above in some detail. In fluid catalytic cracking (FCC), continuous catalyst replacement (CCR) processes, and ebullated-bed hydrocracking, aged catalysts are continuously removed and replaced with fresh. But in fixed-bed units, catalyst replacement requires a shutdown. For a 40,000 barrels-per-day hydrocracker with an upgrade value of $15-20 per barrel, every day of down time for a cost $600,000 to $800,000. Lost production during a 4-week catalyst changeout can amount to $18 to $24 million.

Fig. 11.14 Catalyst loading scheme showing size/shape grading. Photo used with kind permission from Haldor Topsoe Inc

11.5 Comparison of Catalytic and Thermal Cracking

Table 11.3 compares thermal cracking and catalytic cracking. The fundamental differences are: thermal cracking does not use a catalyst. It operates at higher temperature; the pressure can be higher, as in steam cracking, or lower as in delayed coking. The mechanism involves free radical chemistry. Catalytic cracking uses a catalyst at lower temperature and pressure, and the mechanism involves ionic (carbocation) reaction chemistry.

Table 11.4 compares the yields on similar topped crude feed by thermal and catalytic cracking. Catalytic cracking produces more gasoline and olefins than paraffins of the same carbon number. Thermal cracking produces more residual oil than catalytic cracking. Catalytic cracking gives volume swell—the volume of liquid products is greater than the volume of feed—but thermal cracking decreases liquid

Table 11.3 Comparison between thermal cracking and catalytic cracking

Thermal cracking of VGO	Catalytic cracking of VGO with FCC
No catalyst	Use a catalyst
Higher temperature	Lower temperature
Higher Pressure	Lower pressure
No regeneration of catalysts needed	More flexible in term of product slates
Free radical reaction mechanism	Ionic reaction mechanism
Moderate thermal efficiency	Higher thermal efficiency
Gas yield is relatively high and feedstock dependent	Good integration of cracking and regeneration
Moderate yields of gasoline and other distillates	Higher yields of gasoline and other distillates
Alkanes produced but yields are feedstock dependent	Low gas yield
Low to moderate C_5+ product selectivity	High $C_5 +$ product selectivity
Low octane number gasoline	High octane number gasoline
Some chain branching in alkanes	Low n-alkane yield
Low to moderate yield of C_4 olefins	Chain branching and high yield of C_3 and C_4 olefins
Low to moderate yield of aromatics	High yield of aromatics

volume. Also, catalytic cracking produces coke that can deactivate the catalyst, but thermal cracking of VGO does not.

11.6 Hydrocracking [18]

From the crude distillation complex, straight-run gas oils and especially VGO can be sent to hydrocracking units. Hydrocrackers produce LPG, light gasoline, heavy naphtha, middle distillate fuels (jet and diesel), FCC feed, lube basestock, olefin plant feed, and isobutane, which is an important feed component for alky plants.

Hydrocrackers also process cracked stocks, such as coker gas oils and FCC cycle oils. Typically, if a refinery has both a hydrocracker and an FCC unit, straight-run VGO goes to the FCC while cracked stocks go to the hydrocracker. In North America, most hydrocrackers process cracked stocks.

Refineries which own recycle hydrocrackers and switch from maximum production of middle distillate fuels in the winter to maximum production of naphtha in the summer to meet heavier gasoline demands. Kerosene is a swing product, which can be minimized or maximized by adjusting cut points. Hydrocracker middle distillate are excellent blendstocks for diesel and jet. Hydrocracker naphtha has low octane, but due to high naphthene content, it is a superb feed for catalytic reformers.

Table 11.4 Thermal versus Catalytic Cracking Yields on Similar Topped Crude Feed

	Thermal cracking		Catalytic cracking	
	wt%	vol.%	wt%	vol.%
Fresh feed	100.0	100.0	100.0	100.0
Gas	6.6		4.3	
Propane	2.1	3.7	1.3	2.2
Propylene	1.0	1.8	6.8	10.4
Isobutane	0.8	1.3	2.6	4.0
n-Butane	1.9	2.9	0.9	1.4
Butylene	1.8	2.6	6.5	10.4
C_5 + gasoline	26.9	32.1	48.9	59.0
Light cycle oil	1.9	1.9	15.7	15.0
Decant oil			8.0	7.0
Residual oil	57.0	50.2		
Coke	0		5.0	
Total	100.0	96.5	100.0	109.9

Friedrich Bergius developed hydrocracking technology for converting lignite coal into synthetic fuels with hydrogen at high pressure and temperature [19]. His lab-scale unit operated at about 450 °C (840 °F) and up to 130 bars. Bergius received a patent for this work in 1913. In 1931, he and Karl Bosch shared the Nobel Prize in Chemistry for their respective contributions in high-pressure chemistry. Together with Matthias Pier, Bergius designed a demonstration plant, which was constructed by Goldschmidt AG near Essen. Due to World War I, production didn't begin until 1919. Subsequently, Bergius sold his patent to BASF.

Commercial plants based on improved versions of the Bergius-Pier process were operated between 1927 and 1944 to supply Germany with fuels and lubricants from coal. Meanwhile, between 1925 and 1930, Standard Oil of New Jersey (Esso) collaborated with I.G. Farben to extend Bergius-Pier technology into a process for heavy oil conversion.

In 1977, with support from the German government, Ruhr Kohle and Veba Oel, existing coal liquefaction technology evolved into the Veba COMBIcracking process, which was demonstrated commercially in a 200 tons/day unit in Bottrop, Germany. The process is now known as VCC™ and is licensed jointly by BP Refining Technology and KBR Technology. The integrated process includes a slurry-phase 1st stage hydrocracking section directly linked to a fixed-bed 2nd stage hydroprocessing section. The first commercial integrated unit, which was designed by KBR, started operation in Yulin, China in January 2015.

Modern fixed-bed hydrocracking was commercialized by Chevron in 1958. The UOP Unicracking™ process, based on zeolite-containing hydrocracking catalysts, was co-invented by Union Oil, Esso, and Union Carbide. The first Unicracker started operation in 1964 at the Unocal refinery in Wilmington, California. Acquired from Unocal by UOP in 1995, Unicracking is now world's leading fixed-bed hydrocracking process. Hydrocracking is also licensed by Shell Global Solutions, Haldor Topse, and Axens.

Ebullated bed (e-bed) hydrocracking was commercialized in 1968. The leading e-bed processes are H-Oil RC, now offered by Axens Technology Licensing, and LC-Fining®, now licensed by CB&I.

In 1993, Isodewaxing technology, invented by Chevron, revolutionized lube base-stock production. Together with Isofinishing, it replaced aromatics extraction and solvent dewaxing with hydroprocessing, thereby increasing overall base stock yields.

Catalytic hydrocracking uses catalytic cracking in the presence of hydrogen to convert high boiling and heavy hydrocarbons into lower-boiling and lower molecular weight hydrocarbons. Breaking C–C bonds in paraffins and naphthenes (with ring opening) is endothermic; aromatic C–C bonds are not broken. Hydrogenation of reaction intermediates (olefins) is exothermic. Hence, the heat released by hydrogenation more than compensates for all the heat required for cracking. In addition, hydrogenation minimizes coke formation. As with FCC, isoparaffins are produced preferentially.

Table 11.5 lists length and energy of selected bonds, and Table 11.6 compares heat of reaction calculated from bond energies with that from heat of formation in hydrocracking of n-hexane (Reaction 11.1) and hydrodesulfurization of ethyl sulfide (Reaction 11.2).

$$C_6H_{14} + H_2 \rightarrow 2C_3H_8 \tag{11.1}$$

$$C_2H_5SH + H_2 \rightarrow C_2H_6 + H_2S \tag{11.2}$$

Table 11.5 Average bond lengths and energies of selected bonds

Bond	Length (pm)	Energy (kJ/mol)	Bond	Length (pm)	Energy (kJ/mol)
C–H	109	413	H–H	74	436
C–C	154	348	H–N	101	391
C=C	134	614	H–O	96	366
C≡C	120	839	H–S	–	347
C–N	147	308	N–N	145	170
C–O	143	360	N≡N	110	945
C=O	–	745	O–O	148	145
C=O	in CO$_2$	803	O=O	121	498
C–S	182	272	S–S	–	266

Fixed-bed catalytic hydrocracking employs high hydrogen partial pressures (1000–3000 psi) and, relative to FCC, moderate temperatures (600–800 °F). Other than hydrogenation, the role of hydrogen is to prevent buildup of coke on the catalyst.

In addition to fixed-bed hydrocracker, there are slurry-phase thermal hydrocrackers such as the VCC units in China. Since catalysts cannot economically recovered from slurry, a slurry hydrocracker uses low cost catalysts that elute off with bottoms streams or non-catalytic solid additives. Process conditions for slurry-phase and fixed-bed hydrocracking are as follows:

Slurry-phase Thermal Hydrocracking

- Feedstocks: coal, vacuum residue, FCC slurry oil
- Normal operating temperature: 460–490 °C (860–920 °F)
- Normal operating pressure: 170–200 barg (2500–3000 psig)
- Conversion: up to 95 wt% conversion of vacuum residue
- Non-catalytic additive used in the VCC process. Low-metal catalysts used in Uniflex and other
- No need to shut down for catalyst or additive replacement.
- Products must be stabilized by subsequent hydrotreating.

Fixed-bed Catalytic Hydrocracking

- Feedstocks: gas oil, FCC light cycle oil, coker gas oil, vacuum gas oil, FCC heavy cycle oil
- Normal operating temperature: 315–440 °C (600–825 °F)
- Normal operating pressure: 68–200 barg (1000–3000 psig)
- Liquid hourly space velocity (LHSV): 0.5–2.0
- Conversion: up to 99 wt% conversion of VGO
- Catalyst cycle life: 1–5 years, typically 2–3 years.

Ebullated-bed Catalytic Hydrocracking

- Feedstocks: atmospheric and vacuum residue
- Normal operating temperature: 420–450 °C (730–840 °F)
- Normal operating pressure: 136–200 barg (2000–3000 psig)
- Liquid hourly space velocity (LHSV): 0.1–1.5
- Conversion: up to 70 wt% conversion of vacuum residue
- No need to shut down for catalyst or additive replacement.

Table 11.6 Comparison of heats of reaction (kJ/mol) calculated from bond energies with that heats of formation for Reactions 11.1 and 11.2

	Bonds broken	Bonds formed	Net bond energy	Calc. from ΔH_f
Reaction 11.1	C–C, H–H	2 C–H	−42	−42.3
Reaction 11.2	C–S, H–H	C–H, S–H	−52	−58

1. $C_6H_{14} + H_2 \rightarrow 2\,C_3H_8$
2. $C_2H_5SH + H_2 \rightarrow C_2H_6 + H_2S$

Hydrocracking can use hydrogen from a variety of sources: catalytic reformers, steam/hydrocarbon reformers (including steam/methane reformers (SMR)), olefin manufacturing plants, partial oxidation facilities, and electrolysis plants.

Compared to the FCC process, hydrocracking has several advantages: greater feedstock flexibility, greater product naphtha boiling flexibility, relatively high ratio of iso- to normal butane, and vastly superior middle distillate products. FCC works best with straight-run paraffinic atmospheric and vacuum gas oils. Due to the lack of external hydrogen, it makes substantial quantities of olefins. Fixed-bed hydrocrackers can be designed for a greater variety of feedstocks. Most units in Asia, and a large percentage of the units in Europe, process straight-run vacuum gas oil. Most units in North America also process aromatic cycle oils and coker vacuum distillates, which are relatively unreactive in catalytic cracking. Instead of making olefins, hydrocracking produces light paraffins with iso/normal ratios that are higher than predicted by thermodynamics. For example, the isobutane:n-butane ratio is >2.0.

In hydrocracking units, abundant high-pressure hydrogen quickly saturates olefin intermediates, suppressing the growth of polyaromatic rings. The reaction mechanism is similar to catalytic cracking, but with concurrent hydrogenation. For the hydrocracking of fluorene ($C_{13}H_8$), Lapinas, et al [20]. proposed a mechanism that might apply generally to the hydrocracking of polyaromatics.

Figure 11.15 outlines the reaction sequence for hydrocracking a heptyl ethyl naphthalene isomer into lighter compounds. The reactions include:

- Removal of part of a side chain by dealkylation
- Saturation of an aromatic ring
- Isomerization of the saturated ring
- Opening the saturated ring
- Paraffin hydrocracking.

The first dealkylation reaction removes hexanes from the heptyl group through β-cleavage, not heptanes. The ethyl group is not removed, because the production of methane by the β-scission of a two-carbon C–C bond isn't possible thermodynamically. Due to resonance stabilization, it is difficult to break C–C bonds in an aromatic ring. Saturation is required before rings can be opened.

It is important to note that hydrocracking is a bifunctional process, combining acid-catalyzed catalytic cracking and metals-catalyzed hydrogenation. Hydrogenation is exothermic and cracking is endothermic. However, the heat released from hydrogenation is much greater than the heat consumed by cracking. Hence, cold hydrogen is added in quench decks to lower temperature, thereby controlling reaction rates. Heat control is the primary concern during the design of hydrocracking units, because loss of control can lead to loss of containment, causing fires and even explosions [21]. This issue is discussed in detail later.

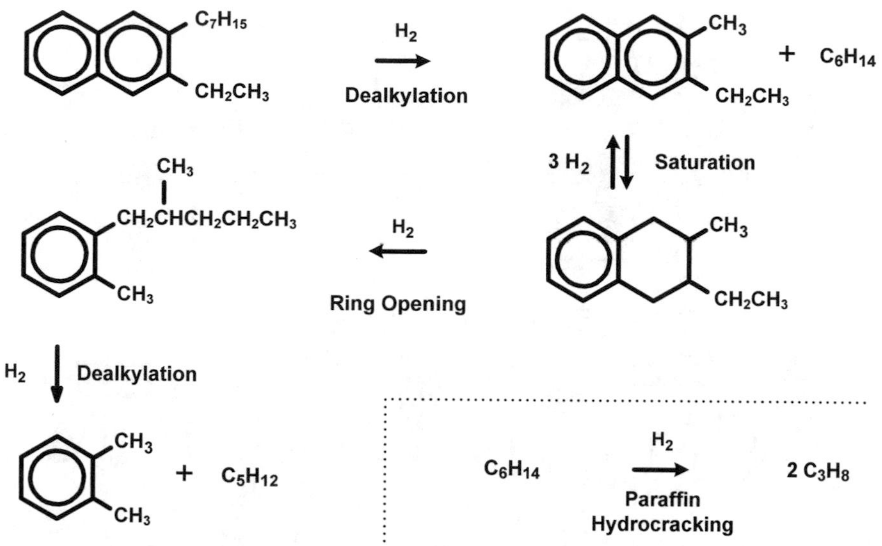

Fig. 11.15 Reaction sequence for the hydrocracking of polyaromatics

11.6.1 Fixed-Bed Catalytic Hydrocracking

The chemical mechanism for catalytic hydrocracking is essentially the same as that for FCC catalytic cracking. The processes themselves are considerably different due to different catalysts, equipment, and operating conditions. An integral part of hydrocracking is hydrotreating, which in fixed-bed units removes almost all oxygen, sulfur, nitrogen and trace elements from the feed before it reaches the cracking catalyst. Organic nitrogen poisons acidic cracking sites, so its removal is essential. Along the way, hydrocracking also isomerizes n-paraffins and saturates olefins and aromatics. With respect to equipment and process flow, fixed-bed hydrocrackers are similar to fixed-bed hydrotreaters.

As shown in Fig. 11.16, hydrocrackers can have many different configurations; for simplicity, pumps, heaters and heat exchangers are not shown. Most commercial units have two reactors, but at least one unit (at Chevron El Segundo) has 6 reactors. Some reactors have two beds. Other reactors have seven beds. Some units are once-through; the unconverted oil goes off-plot. Others recycle unconverted oil to near-extinction. Some have two flash drums, others have four. Some have amine treaters, others do not. In some, only the gas is heated—hot gas is mixed with warm feed just before the 1st reactor. In others, gas and oil are mixed before heating.

- **Sketch 1** shows a once-through unit in which the oil flows in series through the pretreat catalysts directly to the cracking catalysts.
- **Sketch 2** shows two options for recycle of unconverted oil (UCO). The UCO can go either to R1 (first reactor) or R2 (2nd reactor).

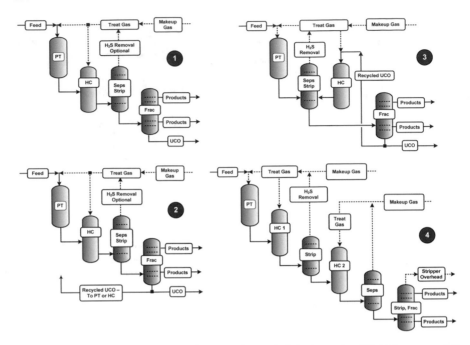

Fig. 11.16 Hydrocracker Configurations (PT: pretreater, HC: hydrocracker, HP: high pressure, LP: low pressure, UCO: unconverted oil). The makeup gas is hydrogen-rich with H_2 ranging from 80 to 99.9%

- **Sketch 3** shows a unit in which the pretreated oil is stripped or fractionated before moving on to a hydrocracking reactor.
- **Sketch 4** shows a unit with two independent recycle gas loops, one for PT and HC-1 and another for HC-2. This configuration has two major process advantages: in R3, the pressure can be lower and the environment is nearly sweet—sweet because H_2S and ammonia are stripped, and almost all of the nitrogen and sulfur are removed in PT and HC-1. Temperatures required by uninhibited sweet hydrocracking are lower. This leads to increased saturation of aromatics, which improves the quality of middle distillate products. Some refiners still use noble-metal catalysts, due to the higher activity and lower gas production over such catalysts. The Sketch 4 option seems like it would be more expensive, because it contains additional equipment. But in one recent head-to-head comparison, due to the lower pressure and less expensive metallurgy, the estimated installation cost of a Sketch 4 unit was nearly the same as that for a Sketch 2 unit.

Figure 11.17 shows a two-reactor once-through hydrocracker with four catalyst beds, along with some typical oil properties at various points in the process. The liquid feed in this case is a typical straight-run VGO.

The makeup hydrogen can come from a steam-methane reformer (SMR), a catalytic reformer, purified refinery purge gas, an olefins plant, or even an electrolysis

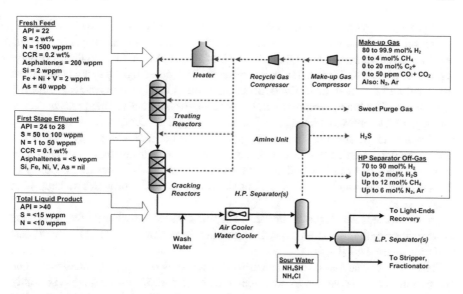

Fig. 11.17 Once-through single-stage (series flow) hydrocracking flow sketch with selected stream properties

plant. The highest purity makeup (99.9 + % H$_2$) comes from an SMR in which purification is achieved with a pressure swing adsorption (PSA) unit. Makeup gas from an SMR equipped with Benfield purification can contain 4 vol.% methane. Olefin plant hydrogen has up to 5 vol.% methane, and electrolysis hydrogen can contain troublesome traces of HCl. Purified refinery purge gas can have purities ranging from 90 to 95% H$_2$ for a membrane unit to 99.9 + % H$_2$ for a PSA. FCC and thermal cracking units produce a lot of low-purity hydrogen containing olefins; it is uneconomical to purify such gases.

Hydrotreating catalysts are described above. They remove metals, organic sulfur, and organic nitrogen. Sulfur doesn't harm the hydrocracking catalysts in R2, but nitrogen inhibits acid-induced cracking.

Hydrocracking catalysts are bifunctional, containing metals for saturation and a solid acid for cracking. The active metals can be either palladium, which is an expensive noble metal, or a base-metal sulfide (MoS$_2$ or WS$_2$) promoted by NiS. The acid is either an amorphous aluminosilicate or a crystalline zeolite. These were discussed in the refining catalyst section.

A mixture of liquid feed and hydrogen-rich gas enter the first reactor at 550 °F (start-of-run) to 750 °F (end-of-run); these temperatures correspond to 290 and 400 °C, respectively. Temperatures increase by as much as 100 °F (56 °C) but in no case are they allowed to exceed the metallurgical limit (usually 825 °F, 440 °C). Between beds, relatively cold quench gas is added to reduce the temperature prior to the subsequent bed. The effluent from the last first-stage reactor is sent through a heat exchanger train, to a series of two to four separators (flash drums). Before the cold high-pressure separator (CHPS), wash water is added to dissolve NH_4SH and NH_4Cl, which otherwise would foul tubes in the reactor effluent air cooler (REAC). The resulting sour water is drained from the boot of the CHPS. In the cold high-pressure separator, hydrogen laden with H_2S, methane, and other light gases is recovered overhead and recycled. Sometimes, an amine unit is used to remove H_2S from the recycle gas. The high-pressure separator bottom stream goes to low-pressure separator(s), which remove residual gases before the liquid product proceeds to stripping and fractionation. Some of the recycle gas is purged to maintain purity. The purge stream, which contains 80–85% hydrogen, can go to hydrotreaters. As mentioned above, purge gases can be purified with a membrane unit which boosts the purity of diffused (low-pressure) gas to >95%. Purification with a PSA provides makeup with $99.9 + \% H_2$.

From the separators, the liquids go to a fractionator, which cuts the full-range product into individual streams—light naphtha, heavy naphtha, middle distillates, and unconverted oil (UCO).

There can be one middle distillate draw (diesel) or two (kerosene and diesel). The fractionator can be operated to give sales-quality diesel. Kerosene quality is highly feed dependent. It often meets sales specifications, but toward the end of a catalyst cycle, high aromatics might keep it from meeting smoke-point specifications.

As mentioned, the UCO can go off-plot to an FCC unit or olefins plant. It can also be used as lube base stock. The UCO from hydrocracking is a premium product. Compared to the VGO from the vacuum distillation column, it has higher isoparaffin concentration, high viscosity index, less sulfur, less nitrogen, lower aromatics, and higher hydrogen content. When recycled to the reactors for additional conversion, overall conversions can exceed 98 wt%.

The hydrocracker is a pivot point in the refinery—the "process in between." It takes straight-run VGO from the crude distillation complex, but it also takes feed from other conversion units. It generates finished products—light naphtha and middle distillates—but it also provides feeds to other units.

Feeds   Straight-run VGO, coker gas oil, FCC cycle and slurry oils, DAO, extract.
Intermediates   Isobutane => alkylation
                Light naphtha => gasoline blending
                Heavy naphtha => catalytic reforming
                Middle distillates => jet, diesel
                Unconverted oil (UCO) => recycle, FCC, olefins plant, lube basestock

As shown in Table 11.7, a fixed-bed recycle hydrocracking unit can have significant product flexibility, producing either large amounts of C_4-plus naphtha or large amounts of middle distillates. In petroleum refining, this kind of process flexibility is unique.

Catalyst cycles last from 1 to 4 years, typically for two years. Units run to achieve specified targets, such as "this much" conversion or "that much" production of FCC

Table 11.7 Hydrocracking product flexibility

Feed	Straight-run vacuum gas oil		
Boiling range, °C	340–550		
Boiling range, °F	644–1020		
API gravity	22.0		
Specific gravity	0.9218		
Nitrogen, wppm	950		
Sulfur, wt%	2.5		
Primary product objective	Naphtha	Kerosene	Gas oil
Weighted average reactor temp, °C	base	−6	−12
Weighted average reactor temp, °F	base	−11	−22
Product yields, vol.%			
Butanes	11	8	7
Light naphtha	25	18	16
Heavy naphtha	90	29	21
Kerosene or gas oil	–	69	77
Total C_4-plus	126	124	121
Chemical H_2 consumption			
Nm^3/m^3	345	315	292
Scf/bbl	2050	1870	1730
Product qualities			
Light naphtha (C_5-82 °C)			
RON Clear	79	79	80
Heavy naphtha			
P/N/A	45/50/5	44/52/4	–
RON clear	41	63	67
End point, °C (°F)	216 (421)	121 (250)	118 (244)
Kerosene			
Flash point, °C (°F)	–	38 (100)	–
Freeze point, °C (°F)	–	−48 (−54)	–
Smoke point, mm	–	30	–
FIA aromatics, vol.%	–	7	–
End point, °C (°F)	–	282 (540)	–
Gas oil			
Cloud point, °C (°F)	–	–	−15 (5)
API gravity	–	–	44
Cetane number	–	–	55
Flash point, °C (°F)	–	–	52 (126)
End point, °C (°F)	–	–	349 (660)

or olefin-plant feed. As a catalyst cycle progresses, it's necessary to raise the average temperature about 1–3 °F per month to compensate for loss of catalyst activity.

Higher hydrogen partial pressure improves almost every aspect of hydrocracker operation. The maximum operating pressure of an existing unit is fixed, but hydrogen partial pressure can be increased by purging recycle gas or improving makeup gas purity.

Conversion is a function of residence time, i.e., feed rate. In an existing unit, increasing the feed rate decreases conversion. To compensate, it's necessary to increase temperature.

An increase in feed nitrogen content might decrease conversion, because organic nitrogen inhibits catalyst activity. However, pretreat reactors are operated to maintain constant nitrogen in the feed to the hydrocracking catalyst, typically 10–30 ppmw, so the impact of feed nitrogen on the cracking catalyst is dampened.

If a unit is not designed for high levels of feed sulfur, corrosion of equipment could impair unit performance. Feed sulfur affects the pretreat catalyst more than it does the hydrocracking catalyst. For both catalysts, removing hydrogen sulfide from the recycle gas with an amine treater diminishes the impact of feed sulfur content.

The buildup of heavy polynuclear aromatics from recycled fractionator bottoms can cause fouling of heat exchanger. Reducing feed end point or removing 2% of the UCO in a drag stream may be necessary.

11.6.2 Ebullated Bed Hydrocracking

Heavy feeds, such as residual fuel oils and reduced crudes contain high concentrations of asphaltenes and ash, and they contain more sulfur, nitrogen, and metal-containing components (such as metalloporphyrins) than gas oils. Typically, catalytic residue upgrading processes are applied to atmospheric residues (AR). The type of catalysts and operation conditions in AR hydrocracking are different than those used for gas oils. Vacuum residues (VR) have low hydrogen/carbon ratio and high metal content, which deactivates catalysts rapidly. They typically are processed in non-catalytic processes, such as solvent extraction, delayed coking, Flexicoking, and thermal hydrocracking. (However, certain thermal hydrocracking additives can have catalytic activity.)

In contrast to fixed-bed hydrocrackers, ebullated bed (e-bed) units can process large amounts of atmospheric and vacuum residues (AR and VR) with high metals, sulfur, nitrogen, asphaltenes and solid contents. They employ catalysts with both hydrotreating and hydrocracking activity; in such units, it is impossible to segregate catalysts, so the catalyst accomplishes both hydrotreating and hydrocracking. As discussed, hydrotreating removes sulfur and nitrogen and hydrogenates aromatic rings. Hydrocracking entails catalytic breaking of C–C bonds. E-bed processes convert residue-containing feeds into distillates and upgraded bottoms for FCC and other conversion units.

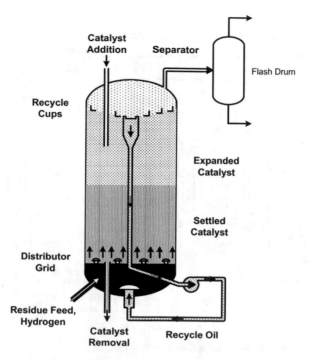

Fig. 11.18 H-Oil Reactor as designed by IFP

Process-wise, catalyst life does not limit these units, because fresh catalyst is continually added as spent catalyst is removed. Economics determine whether or not the benefits of residue conversion offset the cost of catalyst replacement.

In ebullated bed reactors, hydrogen-rich recycle gas is bubbled up through a mixture of oil and catalyst particles. This provides three-phase turbulent mixing, which is needed to ensure a uniform temperature distribution. The process can tolerate significant differences in feed quality, because in addition to manipulating temperature, operators can change catalyst addition rates. Catalyst consumption is determined by the concentrations of trace metals—particularly Fe, Ni and V—in the feed.

Figure 11.18 shows an H-Oil reactor that uses ebullated-bed hydrocracking technology to process heavy feedstock residues such as vacuum gasoils (VGO), deasphalted oils (DAO), and Coal derived oils. A fresh catalyst is continuously added and the spent catalyst withdrawn to control the level of catalyst activity in the reactor, enabling constant yields and product quality over time.

At the top of the reactor, catalyst is disengaged from the process fluids, which are separated in downstream flash drums. Most of the catalyst is returned to the reactor. Some is withdrawn and replaced with fresh catalyst. When compared to fixed-bed processes e-bed technology offers the following advantages:

• The ability to achieve more than 70 wt% conversion of atmospheric residue.

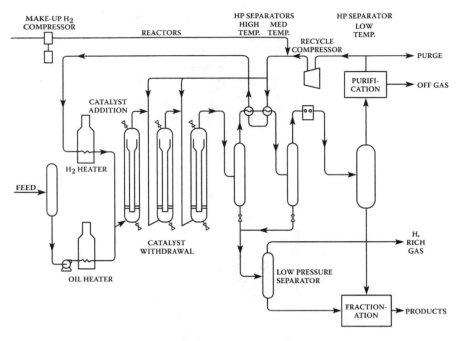

Fig. 11.19 LC-fining expanded-bed hydroprocessing unit [22]

- Ample free space between catalyst particles, which allows entrained solids to pass through the reactor without accumulation, plugging, or build-up of pressure drop.
- Better liquid-product quality than delayed coking.

Disadvantages versus fixed-bed processes include high catalyst attrition, which leads to high rates of catalyst consumption; higher installation costs due to larger reactor volume and higher operating temperatures; and sediment formation. Recent improvements include second-generation catalysts with lower attrition; catalyst rejuvenation, which allows the reuse of spent catalysts; improved reactor design leading to higher single-train feed rates; and two-reactor layouts with inter-stage separation.

Another version of fluidized (or extended) bed hydrocracking is LC-fining developed by Lummus (now CB&I), shown in Fig. 11.19.

11.6.3 Slurry Phase Hydrocracking

Slurry-phase hydrocracking (Fig. 11.20) achieves up to 95 wt% conversion of vacuum residue, FCC slurry oil, coal or coal tar. Reaction temperatures are about 840 °F (450 °C) and pressures range from 2000 to 3000 psig (14,000–20,800 kPa). Slurry-phase hydrocracking employs finely divided solid additives, which are proprietary and may or may not be infused with catalytic metals.

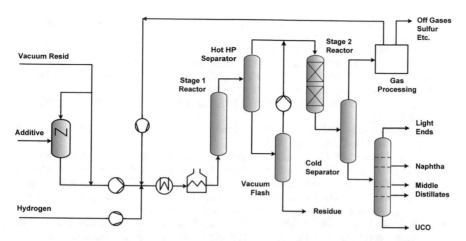

Fig. 11.20 Two-stage slurry-phase hydrocracking process flow. Based on drawings supplied by KBR Technology. Used with kind permission from KBR, Inc. [23]

Inside the reactor, the liquid/additive mixture behaves as a single phase due to the small size of the additive particles. The additives prevent bulk coking by providing highly dispersed nucleation sites for "micro coking." The additive isn't recovered. Instead, it ends up in a pitch fraction, which comprises <5 wt% of the feed.

Slurry-phase hydrocracking has several advantages:

- It can achieve >95 wt% conversion of vacuum residue and high conversion of coal.
- In two-stage designs, which incorporate fixed-bed hydrotreating and hydrocracking, product quality is excellent.
- Feeds can include vacuum residue, FCC slurry oil, and even coal.
- The additive is low-cost and disposable.
- For a given volume of residue feed, total slurry phase reactor volume is lower than the reactor volume of e-bed processes.

The main disadvantage of the KBR slurry-phase hydrocracking is the unconverted pitch. The pitch quality is so low that it is exceptionally difficult to dispose.

A better process is LC-MAX, offered by CB&I. LC-MAX is a combination of LC FINING and solvent deasphalting. The conversion to fuels is somewhat lower, but all products, including the asphalt, can be sold conventionally.

References

1. Greensfelder BS, Voge HH, Good GM (1949) Catalytic and thermal cracking of pure hydrocarbons. Ind Eng Chem 41(11):2573–2584
2. Zhu G, Xie C, Li Z, Wang X (2017) Catalytic processes for light olefin production (Chapter 36). In: Hsu CS, Robinson PR (eds) Springer handbook of petroleum technology. Springer, New York

3. https://www.quora.com/What-does-a-steam-cracker-look-like-and-what-are-its-essential-components. Accessed 6 Aug 2014
4. Hsu CS, Robinson PR (2017) Gasoline production and blending (Chapter 17). In: Hsu CS, Robinson PR (eds) Springer handbook of petroleum technology. Springer, New York
5. Speight J (2017) Fluid-bed catalytic cracking (Chapter 19). In: Hsu CS, Robinson PR (eds) Springer handbook of petroleum technology. Springer, Heidelberg
6. Avidan A, Edwards M, Owen H (1980) Experiments performed at constant coke production. Oil Gas J 88(1):52
7. von Ballmoos R, Harris DH, Magee JS (1995) Catalytic cracking (Section 3.10). Encyclopedia of Catalysis. Wiley, London, pp 1955–1985
8. Van Antwerpen FJ (1944) Thermofor catalytic cracking. Ind Eng Chem 36(8):694–698
9. AIChE (2009) Chemical engineers and energy: In: Chemical engineers in action—innovation
10. Palmas P (2009) Traces of the history of RFCC and provides guidelines for choosing the appropriate regenerator style: https://www.uop.com/?document=uop-25-years-of-rfcc-innovation-tech-paper&download=1. A reprint from hydrocarbon engineering
11. Park JI, Mochida M, Marah AMJ, Al-Mutairi A (2017) Modern approaches to hydrotreating catalysis (Chapter 21). In: Hsu CS, Robinson PR (eds) Springer handbook of petroleum technology. Springer, New York
12. Resid Upgrading, By Honey UOP: https://www.uop.com/processing-solutions/refining/residue-upgrading/#resid-fcc. Accessed 31 Aug 2016
13. Honeywell UOP, "Catalyst Cooler," https://www.uop.com/equipment/fcc/catalyst-cooler/. Accessed 1 Nov 2018
14. Prime-G +: The benchmark technology for ultra-low sulfur gasoline by Axens: https://zh.scribd.com/doc/209103920/Prime-G. Accessed 20 Sep 2106
15. Laan JV ConocoPhillips S Zorb Gasoline Sulfur Removal Technology: http://www.icheh.com/Files/Posts/Portal1/S-Zorb.pdf. Accessed 20 Sep 2016
16. Schroder M (2010) Functional metal organic frameworks: gas storage. Sep Catal 293:175–205
17. Olah GA (1994) My search for carbocations and their role in chemistry. In: Nobel Lecture
18. Robinson PR, Dolbear GE (2017) Hydrocracking (Chapter 22). In: Hsu CS, Robinson PR (eds) Springer handbook of petroleum technology. Springer, New York
19. Bergius Process, Project Gutenberg, http://self.gutenberg.org/articles/Bergius_process. Retrieved 4 Aug 2015
20. Lapinas AT, Klein MT, Gates BC, Macris A, Lyons JE (1991) Catalytic hydrogenation and hydrocracking of fluorene: reaction pathways, kinetics, and mechanisms. Ind Eng Chem Res 30(1):42–50
21. https://archive.epa.gov/emergencies/docs/chem/web/pdf/tosco.pdf. Retrieved 9 Oct 2018
22. Chevron Technology Marketing: LC Fining: http://www.chevrontechnologymarketing.com/CLGtech/lc_finishing.aspx Retrieved 4 Aug 2015
23. Motaghi M, Ulrich B, Subramanian A (2011) Slurry-phase hydrocracking: possible solution to refining margins. Hydrocarbon Proces 90(2):37

Chapter 12
Coking and Visbreaking

12.1 Coking

Coking is a process of thermal cracking long-chain hydrocarbon molecules in the residual oil feed from the atmospheric or vacuum distillation column into shorter-chain lower-boiling molecules leaving behind the excess carbon in the form of petroleum coke—a coal-like material. Other than the straight-run residua, the feedstock can also be cracked residua from other refinery units, oil sand bitumen and whole heavy and extraheavy crude oils. In fact, coking is a primary method of refining extraheavy oils and bitumen to produce synthetic crude oils (syncrudes). There are three major types of processes: delayed coking, fluid coking and Flexicoking. Only cokes from delayed coking can be sold as products. Flexicoking does not yield coke at all. All of its coke is gasified with steam/air at very high temperature to yield synthesis gas (syn gas). The purged coke from fluid coking can only be used in the refinery as fuel.

12.1.1 Delayed Coking [1]

Coke was an undesirable product in early refineries during production of kerosene from crude oils through thermal cracking. For example, the in Dubbs process, invented in 1919, the oil was heated to 900 °F at 150 psia at a low per-pass yield to prevent serious coke formation in the heating zone. The first delayed coker was built by Standard Oil at the Whiting Refinery in Indiana in 1929. In the Standard process, the heater and the coke drum were separate. A sharp increase in coker units occurred in World War II, when it was essential to convert heavy feed into useful fuel products. The production of heavier crudes further increased the demand for coking.

Coke formation was discussed in Chap. 9, Sect. 4 starting with aromatics. It can also start with aliphatic paraffins that form cycloparaffins through cyclization. Aromatics can then be formed by dehydrogentation. The formation of aromatic free radi-

© Springer Nature Switzerland AG 2019
C. S. Hsu and P. R. Robinson, *Petroleum Science and Technology*,
https://doi.org/10.1007/978-3-030-16275-7_12

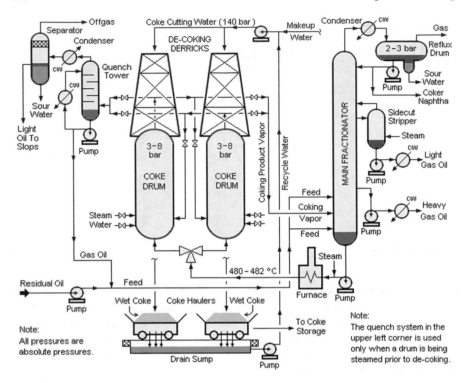

Fig. 12.1   Delayed coking unit [2]

cal intermediates leads to polymerization and rearrangement. Consequent alkylation, cyclization, ring condensation, polymerization, rearrangement and dehydrogenation will eventually form coke. These reactions occur in the vapor phase or on solid surfaces before the reactant are quenched into liquid. The reaction sequence between compound types can be simplified as:

$$\text{Saturates} \rightarrow \text{Aromatics} \rightarrow \text{Resins} \rightarrow \text{Asphaltenes} \rightarrow \text{Coke}$$

Delayed coking is a cyclic process that employs more than two coke drums—typically at least four, as a semi-continuous or semi-batch process. The drums operate on staggered 18–24 h cycles, during which a drum is full for about 12–18 h. Each cycle includes preheating the drum, filling it with hot oil, allowing coke and liquid products to form, cooling the drum, and decoking. A typical coking unit with two coke drums is shown in Fig. 12.1.

The feedstock of the coking unit (coker) is typically residue from the vacuum distillation unit. It goes to the coke furnace to be heated to about 900–970 °F (487–520 °C) at 90 psi pressure. At this stage, steam is injected to minimize cracking of heavy liquid into smaller molecules as gas and liquid products. The products of coker furnace continue to be pumped into a large drum (soaker) . Thermal cracking begins immediately, generating coke and cracked products. Coke accumulates in the drum at 845–900 °F while the vapors rise to the top then go to a product fractionator. Meanwhile, hot feed keeps flowing into the drum until it is filled with solid coke. When gas formation stops, the soaking step is complete. At this point, the top and bottom heads of the drum are removed. A rotating cutting tool uses multiple high-pressure jets of water (500–3500 psi) to drill a hole through the center of the coke from top to bottom as a pilot nozzle for decoking. In addition to cutting the hole, the water also cools the coke, forming steam as it does so. Next, the high-pressure water is switched from vertical pilot nozzle to horizontal reamer. The cutter (pilot nozzle and reamer) is then raised, step by step, (or moved up and down vertically) cutting the coke into lumps, which fall out the bottom of the drum. The wet coke goes through dewatering process to final dry coke product.

Light products include gases, coker naphtha, light coker gas oil (LCGO), and heavy coker gas oil (HCGO). All of these require further processing due to their high content of sulfur, nitrogen, and olefins, which makes them unstable and poorly suited for direct blending into finished products. The gases are treated with amine to remove hydrogen sulfide. The coker naphtha and LCGO are hydrotreated. The HCGO can go either to an FCC unit or a hydrocracker.

Coke accounts for up to one-third of the product. The raw coke from the coker is referred to as green (unprocessed) coke. It can be upgraded to fuel grade (relatively high in sulfur and metals) and anode grade (low in sulfur and metals). Desirable forms include sponge coke and needle coke. Shot coke is dangerous and undesirable.

12.1.2 Fluid Coking

Fluid coking, also called continuous coking, is a moving-bed process for which the operating temperature is higher than the temperatures used for delayed coking. The process burns only enough of the coke to satisfy the heat requirement of the reactor and feed preheat. It typically consumes 20–25% of the coke produced in the reactor. The rest of the coke is withdrawn from the burner vessel, which can be used as furnace fuel in the refinery. The first commercial fluid coker was built at the Billings, Montana refinery of the Carter Oil Company.

In continuous coking, hot recycled coke particles are combined with liquid feed in a radial reactor at about 50 psig (446 kPa) as shown in Fig. 12.2. The feed, such as vacuum residue is preheated to 500–700 °F before entering the coker, which is operated at 900–1050 °F at 10 psi pressure with recycled hot coke from a burner at 1100–1200 °F.

Vapors are taken from the top of the reactor, quenched to stop any further reaction and fractionated. Metals and Conradson carbon residue in the feed are rejected with the coke, but the liquid products still contain sulfur and nitrogen. The VGO goes to a hydrocracker or an FCC feed pretreater. The hydrotreated naphtha goes to blending or a catalytic reformer, and the hydrotreated middle distillates go to blending. Operating conditions are selected for maximum yield of distillate products. Less coke is made

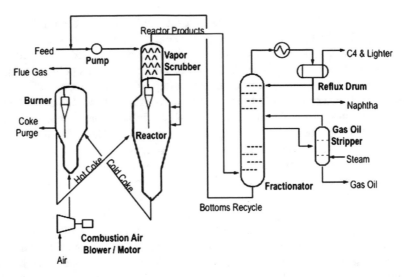

Fig. 12.2 Fluid coking (fluidized-bed coking) [3]

than in delayed coking. The coke goes to a burner, where coke is partially burned to provide sufficient heat for cracking: the heat is transferred when the hot coke particles return to the coker. The remaining unburned coke (up to 75%) cannot be sold as product, but is purged out of the coker as fuel for other parts of refinery, such as furnaces. The yield of distillates from coking can be improved by reducing residence time. However, fluid cokers can be designed and operated to produce CO and H_2 from the coke. Installation costs for fluid coking are considerably higher than for delayed coking, but feeds can be heavier, heat losses are lower, and conversion is higher. Unfortunately, much of the incremental conversion is to C_3^- material. Fluid coking makes more fuel gas than delayed coking.

12.1.3 Flexicoking [4, 5]

Another version of fluid coking was developed and commercialized by ExxonMobil in 1976. It integrates fluid coking and air gasification to eliminate petroleum coke production. Instead of withdrawing unburned coke as in fluid coking, coke is gasified at 1500–1800 °F in an integrated steam/air gasifier to produce synthesis gas, a mixture of CO and H_2. Hence, Flexicoking should not be considered simply as fluid coking although both are based on the same principle of recycling regenerated hot coke.

As shown in Fig. 12.3, the fresh feed is mixed with a recycle stream containing some of the higher-boiling point hydrocarbons (~975 °F+ or 525 °C+). Light overhead product vapors in the scrubber go to conventional fractionation. The blended feed is thermally cracked in the reactor fluidized bed to a full range of gas and liquid

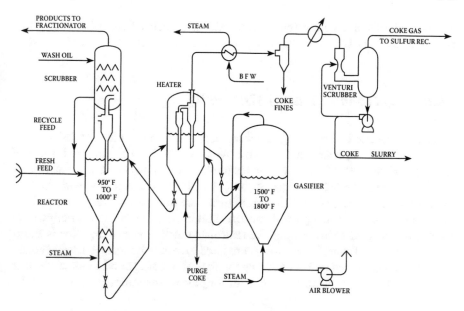

Fig. 12.3 Simplified diagram for a flexicoking [5]

products and coke. The coke circulates from the reactor to the heater via the cold coke transfer line. In the heater, the coke is heated by the gasifier gas (a low BTU mixture of $CO/H_2/CO_2$) and circulated back to the reactor via the hot coke transfer line to supply the heat that sustains the thermal cracking reaction. The excess coke in the heater is transferred to the gasifier, where it reacts with air and steam to produce a gas. The gasifier products, consisting of a gas and coke mixture, return to the heater to heat up the coke. The gas exits the heater overhead and goes to steam generation and to the integrated FLEXSORB™ desulfurization process to reduce H_2S in the flue gas. The low BTU gas is typically fed to a CO boiler for heat recovery but can also be used in modified furnaces/boilers; atmospheric or vacuum distillation tower furnaces; reboilers; waste heat boilers; power plants and steel mills; or as hydrogen plant fuel, which can significantly reduce or eliminate purchases of expensive natural gas. The small amount of residual coke produced can be sold as boiler fuel for generating electricity and steam or as burner fuel for cement plants.

12.2 Visbreaking

Visbreaking is a mild form of thermal cracking that achieves about 15% conversion of atmospheric residue into gas oils and naphtha. At the same time, a low-viscosity residual fuel is produced. The primary purpose of visbreaking is not for conversion, but to reduce the viscosity and pour point of fuel oil, such as No. 6 fuel oil, to acceptable levels for specifications. The reaction is quenched before the residual fuel can form coke. The two main types of visbreaking are "short-contact" and "soaker."

12.2.1 Coil (Short Contact) Visbreaking

In short-contact coil visbreaking, cracking occurs in furnace tubes (coils) and the product quenched by a stream of cold oil. As shown in Fig. 12.4, the feed is heated to about 900 °F (480 °C) in furnace tubes (coils) and sent to a soaking zone reactor at 140–300 psig (1067–2170 kPa). Steam (~1 wt%) is injected along with the feed to generate turbulence, control the liquid residence time and prevent coking along the tube wall. The elevated pressure allows cracking to occur while restricting coke formation. To avoid over-cracking, the residence time in the soaking zone is short—several minutes (typically 1–3 min) compared to several hours in a delayed coker. The hot oil is quenched by heat exchange with feed or cold gas oil to inhibit further cracking and sent to a vacuum fractionator for product separation.

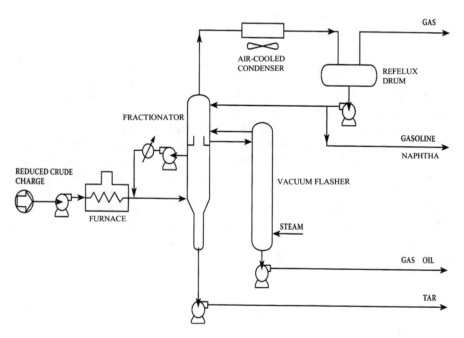

Fig. 12.4 Coil visbreaker

12.2.2 Soaker Visbreaking

Soaker visbreaking, shown in Fig. 12.5, keeps the hot oil at relatively lower temperature, about 800 °F (430 °C), and at 50–200 psi in a soaker drum after the furnace for a longer time (15–25 min) to allow more cracking to occur before quenching, This increases the yield of middle distillates. Low-viscosity visbreaker gas oil can be sent to an FCC unit or hydrocracker for further processing, or used as heavy fuel oil.

12.3 Comparison of Cracking/Coking/Visbreaking Processes

Table 12.1 compares the product yield for catalytic and thermal cracking processes. The catalytic cracking is represented by FCC and hydrocracking, and thermal cracking by coking, visbreaking and steam cracking. Thermal cracking to produce gasoline has been replaced by catalytic cracking. However, steam cracking has become a primary method for producing ethylene and propylene for the polymer production, while catalytic cracking is the key method for producing fuels from heavy distillates.

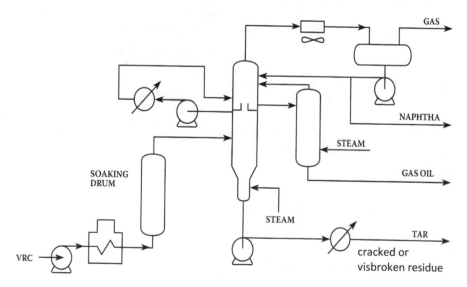

Fig. 12.5 Soaker visbreaker

Table 12.1 Comparison of product yields for catalytic and thermal cracking processes [6]

Process	Type	Product characteristics
FCC	Catalytic cracking in the absence of external H_2	C_1 and C_2: low to moderate Iso/normal paraffin ratio: high C_3 and C_4 olefin yields: can be significant H_2 production: moderate Aromatics: higher than feed Coke formation: high Alkyl aromatics: β-scission next to the ring
Catalytic hydrocracking	Catalytic cracking in the presence of external H_2	C_1 and C_2: low Iso/normal paraffin ratio: high Olefins: removed by pretreating Aromatics: significantly lower than feed Coke formation: minimal Alkyl aromatics: β-scission next to the ring
Slurry-phase hydrocracking	Thermal cracking in the presence of external H_2	C_1 and C_2: high Iso/normal paraffin ratio: similar to feed Olefin production: low Aromatics: depends on feed Coke formation: low Alkyl aromatics: scission within the side chain
Coking/visbreaking	Thermal cracking	C_1 and C_2: high Iso/normal paraffin ratio: similar to feed Olefin production: high Aromatics: moderate Alkyl aromatics: scission within the side chain
Steam cracking	Thermal cracking in the presence of steam	Olefin production: high H_2 production: high Aromatics production: high Coke formation: low

References

1. Wisecarver K (2017) Delayed coking. Chapter 30 in Springer handbook of petroleum technology, Hsu CS, Robinson PR (eds) Springer, New York
2. https://en.wikipedia.org/wiki/Coker_unit. Retrieved 9 Oct 2017
3. https://www.google.com/search?q=fluid+coking+process&biw=1137&bih=516&tbm=isch&tbo=u&source=univ&sa=X&ved=0CDoQsARqFQoTCPi1mO2MlsgCFQVcHgodEfEOXw&dpr=1.25#imgrc=XrVvzvMpdIK30M%3A. Retrieved 9 Oct 2017
4. Flexicoking: The flexible resid upgrading technology: http://cdn.exxonmobil.com/~/media/global/files/catalyst-and-licensing/em_flexicoking_brochure13.pdf. Retrieved 9 Oct 2017
5. Flexicoking: http://abarrelfull.wikidot.com/flexicoking. http://cdn.exxonmobil.com/~/media/global/files/catalyst-and-licensing/em_flexicoking_brochure13.pdf. Retrieved 9 Oct 2017
6. Greensfelder BS, Voge HH, Good GM (1949) Catalytic and thermal cracking of pure hydrocarbons. Ind Eng Chem 41(11):2573–2584

Chapter 13
Lubricant Processes and Synthetic Lubricants

Lubricant oil production exists in a relatively small number of refineries. Although it represents a small fraction of refinery production, making lubricant oil is highly profitable compared to making commodity fuels. Lubricants contain two major components: base oil (base stock) and additives. Additives can comprise 0.1–30% of the finished product.

Making base stocks requires recovery of desired molecules. Traditional lubricant manufacturing is based on solvent refining for separating different types of hydrocarbons, e.g., removing aromatics and waxes. Many newer technologies based on hydroprocessing, such as hydrocracking and hydroisomerization, have been developed to increase feedstock flexibility and meet higher demands in various applications. There are many synthetic lubricants made from petrochemicals for specific specifications and applications.

The world demand for lubricant base stock is expected to remain steady or rise only slightly, but the quality of the base stocks is expected to change dramatically due to the tougher specifications for lubricants used in automobiles. The anticipated benefits are better fuel economy, lower emissions and longer lubricant and engine life. Hence, the base stocks are expected to have lower viscosity, lower volatility, higher saturate content, and higher viscosity index for wider operating temperature ranges. Lower viscosity improves cold start performance and fuel economy by reducing friction in the engine. Lower volatility directly lowers emissions and minimizes oil thickening for extending lubricant life, leading to less oil change frequency and loss. Higher saturate content, with less aromatics and heteroatom-containing hydrocarbons, extends oil life by decreasing susceptibility to oxidation.

The lubricant properties and performance of base oils are improved by the additives, which give formulated lubricants the properties they need to become finished products.

© Springer Nature Switzerland AG 2019
C. S. Hsu and P. R. Robinson, *Petroleum Science and Technology*,
https://doi.org/10.1007/978-3-030-16275-7_13

13.1 Use of Lubricants and Specialty Oils

The use of petroleum heavy fractions for lubrication started in early civilization. For example, Egyptians use pitch for lubricating chariot wheels more than 4000 years ago. Today, there are many uses for lubricants and specialty oils, with some examples listed below:

Automotive   Engine oils, automatic transmission fluids (ATF's), gear oils.
Industrial   Machine oils, greases, electrical oils, gas turbine oils.
Medicinal   Food grade oils for ingestion, lining of food containers, baby oils.
Specialty   Food grade waxes, waxes for candles, fire logs, cardboard.

13.2 Key Properties of Lubricant (Lube) Base Stocks

The key properties of lubricant base stocks include the following:

Viscosity is a measure of the fluidity (internal resistance to flow) of the oil. There are two types of viscosity measurements: *kinematic* viscosity, which is flow due to gravity, ranges from ~3 to 20 cSt (centistokes) for solvent neutrals and about 30–34 cSt at 100 °C for bright stock; and *dynamic* viscosity, which is flow due to applied mechanical stress. Cold-cranking simulator (CCS) viscosity at −25 °C for engine oils and Brookfield viscosity at −40 °C for automatic transmission fluids (ATF) are examples of dynamic viscosity measurements. The SI unit of dynamic viscosity is the poise (P).

Kinematic viscosity is the ratio of dynamic viscosity to density, with the SI unit as mm^2/s (or centistoke, cSt). In the industry, the commonly used units are cSt at 40–100 °C and Saybolt Universal Second (SUS) at 100–210 °F. ASTM D445 and ASTM D2161 are commonly used methods for measurements [1].

For conversion from centipoise (cP) to centistokes (cSt), simply multiply cP by density. The density of lubricant oil is about 0.85–0.9, so the cSt value is about 10–15% higher than the cP value.

The names of neutral lubricant base stocks are based on viscosity grade, such as 100 neutral (100 N) and 600 N, which come from the viscosity measured at 100 °F. For the very heavy grades, such as 150 Bright Stock (BS150), the viscosity is measured at 210 °F.

Viscosity index or *VI* is one of the most important properties of lube base stocks and base oils. It is the measure of the change in viscosity as a function of temperature on an arbitrary scale. The higher the VI, the smaller is the change in viscosity for a given change of temperature. In other words, the higher the VI, the less an oil will thicken as it gets cold and the less it will thin out at higher temperatures—providing better lubricant performance at both temperature extremes. This characteristic is essential to good performance in multi-grade engine oils, such as SAE 5W-30 or 10W-40. Naphthenic crudes can have negative VI values, while paraffinic crudes can have VI values of about 100. Specialty products can have VI's > 130.

Normal paraffins have the higher VI, but also have the highest pour point, discussed below. Hence, they must be removed by solvent dewaxing or catalytic dewaxing processes.

Pour point is the temperature at which the fluid ceases to pour and is nearly a solid. Low pour point ensures ease of starting and proper start-up lubrication on cold days. *Viscosity* pour point is measured by lowering temperature until flow stops. *Wax* pour point occurs abruptly as paraffinic wax crystals precipitate from solution and the oil solidifies.

Cloud point is the temperature at which wax crystals first appear.

Boiling range is a factor along with viscosity for the selection of cut points for lubes.

Volatility is a measure of oil loss due to evaporation. The volatility is generally lower for higher viscosity and higher VI base stocks.

ASTM D5800 (Noack method) [1] is the commonly accepted method for measuring volatility of automotive lubricants. The test simulates the evaporation loss in operating internal combustion engine for high-temperature services.

Simulated distillation (SimDist) provides a convenient means of measuring the front end of boiling point curves for volatility measurements. ASTM D2887, ASTM D6352 and ASTM D7213 are commonly used [1].

Flash Point is a measure of the temperature at which there is sufficient vapor in the oil to ignite, which also reflects the front of the boiling point curve. Flash point is a required safety specification for many base stocks. ASTM D92 can be applied for the measurements.

Saturates, naphthenes and aromatics are the content of these molecular types present in the base stock. They are determined by chromatography and mass spectrometry at the molecular level [2].

Oxidative resistance is the resistance to rapid oxidation at high temperature, especially when oil comes in contact with the piston head. Oxidation causes the formation of coke and varnish-like asphaltic materials from paraffinic base oils and sludge from naphthenic base oils. Phenolic additives and zinc dithiophosphates (ZDDP's) are added to suppress oxidation and its effects.

Acidity causes corrosion of bearing metals. Acids are formed from oxidation of lube oil hydrocarbons and introduced into the crankcase by piston blow-by of engine combustion by-products. Paraffinic base oils exhibit lower acidities than naphthenic base oils due to excellent thermal and oxidation stabilities.

Color is commonly measured by ASTM D1500 [1]. Most premium base stocks have a D1500 reading less than 0.5 when they are fresh. An increase in the D1500 reading is an indication of formation of oxidation products. However, many manufacturers have their own proprietary methods for measuring oxidation and storage stability of base oils.

Conradson carbon (CCR) or *Micro-Carbon Residue (MCR)* is a measure of the ash left after flame burning.

13.3   Terms, Names and Categories

Lubes made from virgin distillates and solvent extracts are usually referred to neutrals, such as 100 N (or S100 N), 150 N, 600 N, etc. Bright stock comes from Deasphalted Oil (DAO) from vacuum resid. An example is BS150. The grades shown in the finished oils are expressed as SAE 5, 10, 30, or ISO 22, 32, etc.

The Society of Automotive Engineers (SAE) developed a scale for both engine oils and transmission oils. The multi-grade notation as "xWy" is commonly used for the motor oils on shelves today. "x" refers to viscosity when cold where "W" stands for winder grade, which rates the oil's flow at 0 °F (−17.8 °C). The lower the number, the less the oil thickens in the cold. A low number before "W", such as 0W20 or 5W30, mean that engines start more easily and are better protected in cold weather. "y", on the other hand, refers to viscosity measured at 212 °F (100 °C), which rates the oil's resistance to thinning at high temperatures. The higher the number, the more readily the oil thins. Hence 20W-30 will thin more than 20W-40 at higher temperatures. The difference between the x and y values reflects viscosity index (VI). Low differences indicate higher VI. Monograde designations, such as SAE 20, 30, or 40, are no longer used.

The American Petroleum Institute (API) organizes lube base stocks into five categories, listed in Table 13.1. In general, Groups I–III base stocks are from crude oils and are made through solvent refining or hydroprocessing (hydrocracking or hydroisomerization), while Groups IV and V are synthetic.

13.4   Lube Plant Feedstocks

Paraffinic crudes, such as West Texas and Arab Light, have higher wax content and require n-paraffin removal, via isomerization or dewaxing, to remove the wax. Naphthenic crudes, such as Venezuelan and many California crudes, have low wax.

Figure 13.1 shows how lubricant feedstocks are derived from crude oil fractions [3]. The reduced or topped crude from the atmospheric distillation tower is sent to a

Table 13.1 Five API categories of lube base stocks

Group	Saturates (%)	Sulfur (%)	VI	Typical manufacturing process
I	<90	>0.03	80–120	Solvent processing
II	>90	<0.03	80–120	Hydroprocessing
III	>90	<0.03	120+	Gas-to-liquid (GTL); wax isomerization, severe hydroprocessing
IV				Polyalphaolefins (PAO)
V				All other base stocks

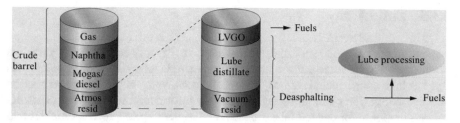

Fig. 13.1 Lube plant feedstocks [3]

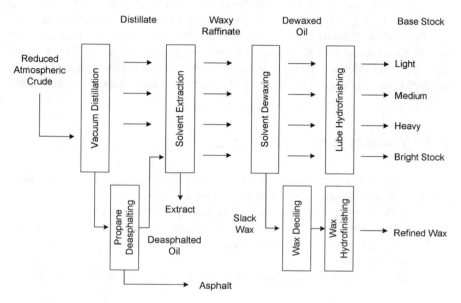

Fig. 13.2 Typical lube vacuum tower [3]

vacuum distillation tower to fractionate the feed into light vacuum gas oil (LVGO) which can be used as distillate fuels after further upgrading, heavy vacuum gas oil (HVGO) for lube processing, and vacuum resid (residue) which can also be used as lube feedstock after deasphalting.

The lube plant can also have its own vacuum tower, taking reduced crude from the refinery, shown in Fig. 13.2. It produces several lube feedstock fractions for different grades of lube oils—light, medium and heavy. Bright stock comes from vacuum resid.

13.5   Lube Assay

Among all the crude oils available, the first step is to identify economically attractive crudes through characterization and assays that include lube yields and product quality. The lube assays include atmospheric distillation, vacuum distillation, extraction and dewaxing. Computer modeling using assay information is employed to predict process response for desired lube products and to investigate process variables and operating optimization for distillation, extraction, and dewaxing. An important goal is to assess manufacturing flexibility. Even the best models have limitations, so a plant test is performed, including blending into a formulated lube for further testing. Periodic re-evaluation is needed to assess lube quality variation over time due to feed changes and process changes caused by equipment fouling. During ongoing economic evaluation, lube yields and product qualities are evaluated. Important parameters include viscosity, sulfur, density, refractive index, oxidative stability, etc.

13.6   Viscosity Index Improvement

The viscosity index of hydrocarbon types is listed in Table 13.2. Normal paraffins have the highest viscosity index, but those boiling in the lube range form wax deposits, especially in cold weather. Hence, dewaxing is an important process to remove normal paraffins while retaining isoparaffins and mononaphthenes that also have high VI. Aromatics have the lowest VI. They are removed by extraction from lube base stock, along with polynaphthenes.

The viscosity index can be calculated using the following formula:

$$VI = [(L - U)/(L - H)] \times 100$$

where U is the oil's kinematic viscosity at 40 °C (104 °F), and L and H are values based on the oil's kinematic viscosity at 100 °C (212 °F). L and H are the values of viscosity at 40 °C for oils of VI = 0 and 100 respectively, having the same viscosity at 100 °C as the test oil. L and H values can be found in ASTM D2270 [2]. If the kinematic viscosity of the oil is less than or equal to 70 mm^2/s (cSt), VI can be obtained by linear interpolation. If the kinematic viscosity is above 70 mm^2/s (cSt), L and H are calculated by the equation below:

Table 13.2   Viscosity index of hydrocarbon types

Hydrocarbon type	Viscosity index
n-Paraffins	175
i-Paraffins	155
Mononaphthenes	142
Dinapththenes	70
Aromatics	50

$$L = 0.8353\,Y^2 + 14.67\,Y - 216$$
$$H = 0.1684\,Y^2 + 11.85\,Y - 97$$

where:

L   Kinematic viscosity at 40 °C of an oil of VI = 0 having the same kinematic viscosity at 100 °C as the oil whose VI is to be calculated,
Y   Kinematic viscosity at 100 °C of the oil whose VI is to be calculated,
H   Kinematic viscosity at 40 °C of an oil of VI = 100 having the same kinematic viscosity at 100 °C as the oil whose VI is to be calculated.

There are several processes that can be used for viscosity index improvements, including:

- Hydrogenation via hydrotreating to convert polyaromatic hydrocarbons to poly-naphthenic ring compounds.
- Hydrocracking of VGO to increase paraffin concentration and viscosity index.
- Higher severity hydrocracking to convert polyaromatics into mononaphthenes and isoparaffins.
- Hydroisomerization (Isodewaxing) to catalytically convert $C_{20}+$ normal paraffins to $C_{20}+$ isoparaffins with minimum cracking.
- In both hydrogenation and hydroisomerization, dehydrogenation of naphthenes to aromatics should be avoid or minimized.
- Separation of aromatic hydrocarbons from paraffins and naphthenes through solvent extraction.

13.7   Lubricants Through Solvent Refining [3]

Figure 13.3 shows a flow diagram for a conventional solvent refining lube base stock production plant. Solvent refining processes include solvent deasphalting, solvent extraction, and deoiling/dewaxing processes. Vacuum distillation of atmospheric residue yields vacuum residue, which undergoes propane deasphalting to remove asphalt. Solvent deasphalting reduces coke and sludge formation in the bright stock. The deasphalted oil goes to a solvent extraction tower, which removes most aromatics. Lighter lube fractions from different levels of the vacuum tower also undergo solvent extraction. The low-aromatics raffinates undergo solvent dewaxing. Solvent extraction improves viscosity index by removing aromatics. Solvent dewaxing and lowers cloud and pour points of the base stocks. Hydrotreating (hydrofinishing) or clay treating improves color and oxidation stability and lowers organic acidity. The wax is a useful and marketable by-product. It is refined through wax deoiling, wax hydrotreating, or hydrofinishing.

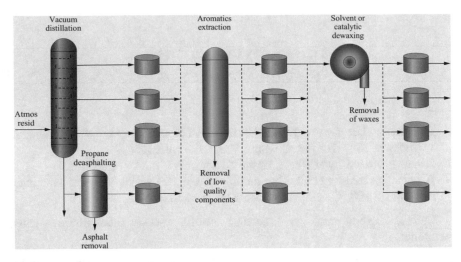

Fig. 13.3 Lube plant process flow [3]

13.8 Solvent Deasphalting

Lubricant feedstocks come from heavy distillation fractions, including heavy gas oil, atmospheric resid (topped crude), vacuum gas oils and vacuum resid, which may contain significant amounts of asphaltene. Deasphalting becomes the first step of lubricant oil manufacturing. Also, solvent deasphalting takes advantage of the fact that undesirable aromatic compounds are insoluble in paraffins.

Propane or *n*-pentane are commonly used in industry to precipitate asphaltenes from residual oils. The deasphalted oil (DAO) is sent to hydrotreaters, FCC units, hydrocrackers, or fuel-oil blending. In FCC units, DAO is easier to process than the corresponding straight-run residue. This is because, by definition, the asphaltene content of DAO is very low. The asphaltenes in straight-run residue easily form coke and often contain catalyst poisons such as nickel and vanadium. In hydrocrackers, DAO can be harder to process than straight-run VGO and FCC cycle oils, because although DAO no longer contains asphaltenes, it is rich in resins and hence still has a very high endpoint and high CCR.

The earliest commercial applications of solvent deasphalting used propane as the solvent to extract high-quality lubricating oil bright stock from vacuum residue, shown in Fig. 13.4. In propane deasphalting, the residual oil and propane are pumped to an extraction tower at 150–250 °F (65–120 °C) and 350–600 psig (2514–4240 kPa). Separation occurs in a tower, which may include a rotating disc contactor. The raffinate flowing off at the top of extraction tower contains DAO, which goes to a high-pressure flash tower to recover propane, which is recycled. The DAO then goes to a stripper to get rid of other dissolved impurities. The final DAO product can become bright stock in a lube plant, or sent as a feed for resid upgrading. With substantially lower levels of metals and carbon contaminants, DAO is far easier to

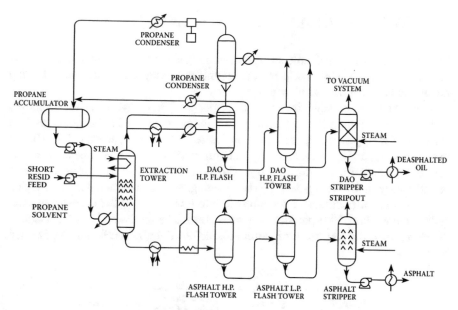

Fig. 13.4 Propane deasphalting unit 1 [3]

process than conventional resid. The bottoms of the extractor go to a high pressure then a low-pressure flash tower to recover additional dissolved propane for recycle. Asphalt is obtained from an asphalt stripper. In this process, DAO is considered as an extract while asphalt is a raffinate.

An advanced version of solvent deasphalting is "residuum oil supercritical extraction," or ROSE™. The ROSE Process was developed by the Kerr-McGee Corporation and now is offered for license by KBR. In this process, residue and solvent are mixed and heated to above the critical temperature of the solvent. Liquid yields are higher under supercritical conditions, because the lighter part of the oil becomes more soluble. In addition to giving higher yields, the process is more energy efficient and has lower operating costs due to improved solvent recovery. The ROSE process can employ three different solvents, the choice of which depends upon process objectives:

Propane Preparation of lube base stocks
Butane   Asphalt production
Pentane Maximum recovery of liquid.

The ROSE process can make poor quality asphalt or provide feeds for FCC, hydrocracking, or coking units. It can also be a unit in a lube complex. Unfortunately, due to its high residue content, ROSE DAO is especially difficult to process in conventional hydrotreaters and hydrocrackers.

13.9   Molecular Aspects of Lube Processing [3]

Figure 13.5 shows typical molecules processed in the lube processing plants. Extraction eliminates the undesirable molecules and dewaxing eliminates wax, leaving molecules between the two sloped lines.

Lube hydroprocessing includes three steps. First, hydrotreating removes heteroatom-containing compounds and saturates aromatics and poly-aromatics. The removal of nitrogen is required prior to the second step: hydrocracking. During hydrocracking, additional aromatics are hydrogenated, mono-naphthenes are converted into isoparaffins by ring opening, and multi-ring naphthenes are converted into naphthenes with fewer rings, also by ring opening. With conventional hydrocracking catalysts, some n-paraffins are transformed into i-paraffins. With isodewaxing catalysts, large amounts of n-paraffins are isomerized. The third step is hydrofinishing, which removes the last traces of sulfur and olefins, thereby stabilizing the finished base stock.

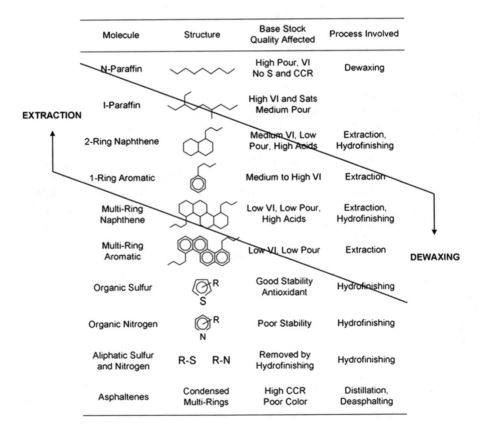

Fig. 13.5   Typical molecules in lube oil [3]

13.10 Solvent Extraction

Solvent extraction is used to remove aromatics and other impurities from the paraffinic and naphthenic portion of lube base stocks and grease stocks. This improves viscosity index, color and oxidation resistance of the base stock, and reduces carbon and sludge formation. As shown in Fig. 13.6, the feedstock is dried, then contacted with the solvent in a counter-current or rotating disk extraction unit. The undesirable polyaromatics are accumulated in solvent-rich extract phase. The desirable paraffinic and naphthenic components are accumulated in light raffinate phase, which can be hydrotreated prior to dewaxing. The solvent is separated from the product stream by heating, evaporation, or fractionation. Remaining traces of solvent are removed from the raffinate by steam stripping or flashing. Electrostatic precipitators can enhance the separation of inorganic compounds. The solvent is then regenerated and recycled.

To provide intimate contact between the solvent and the oil, packed or trayed towers are used, shown in Fig. 13.7. Newer extraction units are rotating disk contactor (RDC) or centrifugal extractors for smaller volumes needed for extraction.

Today, phenol, furfural, N-methyl-2-pyrtolidone (NMP) and cresylic acid are widely used as solvents. Liquid sulfur dioxide, chlorinated ethers, and nitrobenzene also have been used. Table 13.3 compares the properties of N-methyl-2-pyrrolidone (NMP), furfural and phenol.

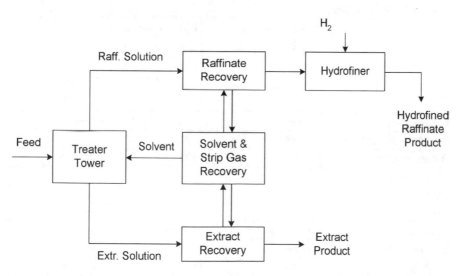

Fig. 13.6 Extraction process flow [3]

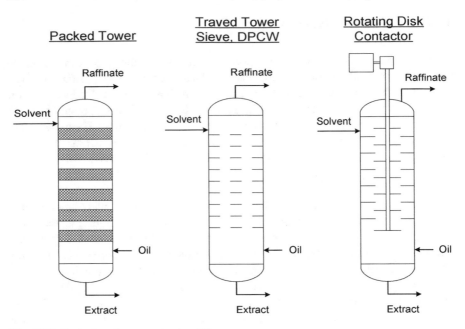

Fig. 13.7 Typical continuous extractors [3]

Table 13.3 Typical solvents for extraction

Characteristics	NMP	Furfural	Phenol
Specific gravity at 20 °C	1.16	1.03	1.07
Boiling point (°C)	162	202	182
Melting point (°C)	−37	−24	41
Flash point (°C)	58	86	79
Viscosity at 60 °C	0.95	1.02	2.58
Selectivity	Excellent	Very good	Good
Solvent power	Good	Excellent	Very good
Stability	Good	Excellent	Very good
Biodegradability	Good	Good	Good
Toxicity	Moderate	Low	High

13.10.1 Furfural Extraction

Furfural is the most widely used solvent for lubricant oil manufacture. In furfural extraction, the feed is introduced into a continuous countercurrent extractor at a temperature that is a function of viscosity. The temperature at the top of tower is below the miscibility temperature between furfural and oil, usually in the range of 220–300 °F (104–149 °C). The gradient from top to bottom of the tower is between 60–90 °F (33–50 °C), hot at the top. The furfural-dispersed phase flows downward through the continuous oil phase. Extract is recycled at a ratio of 0.5:1. Furfural-to-oil ratios range from 2:1 for light stocks to 4.5:1 for heavy stocks. Furfural is easily oxidized and polymerized, so it must be protected with inert gas and carefully controlled heat exchange. A complex solvent recovery system is needed due to the formation of an azeotrope between furfural and water. The most important operation variables are the furfural-to-oil ratio, extraction temperature and extract cycle ratio.

13.10.2 Phenol Extraction

The phenol process was first commercialized in 1930 and has been extensively used ever since, being the second only to the furfural process. Phenol is much easier to recover than furfural. Although phenol has high solvent power and low cost, its toxicity necessitates special handling facilities.

In phenol extraction, phenol is introduced at the top of tower at higher temperature than the oil coming up from the bottom. The oil coming out of phenol-rich phase at lower portion of the tower reverses direction and rises to the top as reflux. The temperature at the top is below the miscible temperature of the mixture, while the temperature at the bottom is maintained at 20 °F (11 °C) lower than the top. The most important operating parameters are phenol-to-oil ratio (treat rate), extraction temperature, and percentage of water in phenol. Treat rates range from 1:1 to 2.5:1 depending on the quality and viscosity of the feed and the quality of product desired.

13.10.3 NMP Extraction

NMP is a weak base and is an aprotic solvent with low volatility and toxicity. In the late 1970s, NMP was offered as a selective solvent for lube oil refining by Texaco and Exxon. It was developed to replace phenol for safety, health, and environmental concerns. Compared to phenol NMP boils at 40 °F (22 °C) higher, has a 115 °F (64 °C) lower melting point, and a 69% lower viscosity at 122 °F (50 °C). NMP is completely miscible and forms no azeotrope with water. As shown in Fig. 13.8, a portion of distillate or deasphalted oil feed is used as the lean oil in an absorption tower to remove the NMP from exiting stripping steam. The rich oil is combined

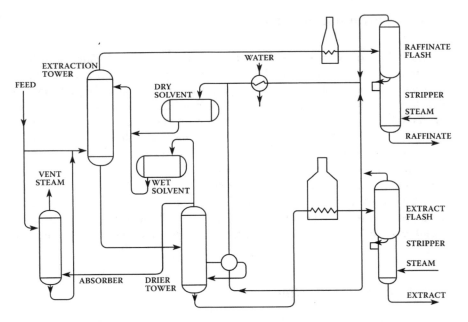

Fig. 13.8 NMP extraction unit

with the remainder of the feed and heated to a desired temperature before going into the treat tower. The raffinate is the desired product, while the extract is enriched with aromatics.

NMP has higher recovery rate than phenol, with losses less than half of that of phenol. It also allows greater throughput for a given size tower. The treat rate and product quality are the same as for phenol extraction. NMP is better than furfural and phenol in terms of VI improvement, whereas furfural performs better than NMP and phenol in terms of raffinate yield.

13.11   Dewaxing

Heavy *n*-paraffins are waxy. They solidify in cold weather, so during lube base stock production, they must be removed, either by extraction, conversion, or isomerization. The dewaxing of lubrication oil represents the largest use of scraped-surface continuous crystallizers.

The dewaxing technologies in commercial units today are solvent dewaxing and catalytic dewaxing. Figure 13.9 shows solvent and catalytic dewaxing processes in lubricant base stock manufacturing [3]. Note that solvent dewaxing produces slack wax as a co-product, but slack wax is not produced in catalytic dewaxing. Solvent dewaxing processes are based on a sequence of distillation and solvent extraction,

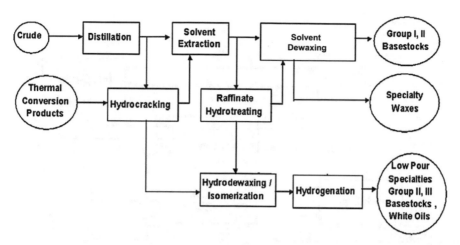

Fig. 13.9 Solvent and catalytic dewaxing processes for lubricant base stock manufacturing [4]

and catalytic dewaxing of raffinates includes cracking or isomerization, typically under hydrogen. Hydrocracking converts the heavier and higher viscosity molecules into lighter and lower viscosity base stocks while also producing fuels; on the other hand, hydroisomerization does not change the carbon numbers of molecules, it just changes them into isomers with lower viscosities. Some refiners use combinations of solvent and catalytic processes to attain specialty products.

13.11.1   Solvent Dewaxing

Solvent dewaxing removes wax (*n*-paraffins > C_{20}). The main process steps include mixing the raffinate with the solvent, chilling the mixture to crystallize wax, and recovering the solvent, shown in Fig. 13.10. The solvent reduces viscosity for filtration. The polarity of the solvent-oil mixture increases, decreasing the solubility of wax in oil and promoting the formation of more-compact wax crystals. Commonly used solvents include methyl ethyl ketone (MEK), methyl isobutyl ketone (MIBK), MEK/MIBK, toluene, MEK/toluene or propane. MEK/MIBK refrigeration requirements are lower than for MEK/toluene due to lower wax solubility.

Both dewaxed oil (DWO) and wax are valuable products. For use as lube base stock, important DWO properties include pour point, cloud point and low-temperature fluidity. For wax, the properties of interest are oil content, melting point and needle penetration depth.

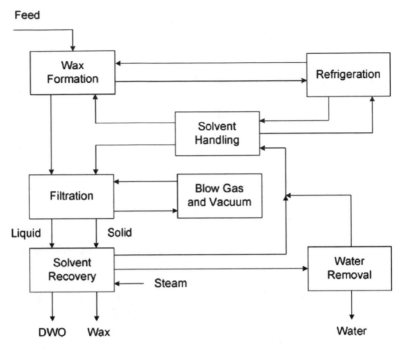

Fig. 13.10 Schematic diagram of solvent dewaxing [3]

13.11.1.1   Propane Dewaxing

Propane dewaxing uses liquid propane as the solvent. The vessels in the unit must be able to operate at elevated pressure. Figure 13.11 shows a propane dewaxing plant. The feed is mixed with a dewaxing aid and passes through a pre-chiller (heat exchanger) before entering the chillers. Propane is used both as a refrigerant and a diluent for direct chilling. The process requires careful pressure control. The chilled solution is sent to a filter to separate wax from oil. Both the filtrate and wax go through separate sets of heat exchangers to a propane recovery unit.

13.11.1.2   Ketone Dewaxing

Figure 13.12 shows an incremental ketone dewaxing plant. It uses a combination of a scraped surface exchanger (SSE) that uses cold filtrate for cooling, followed by a scraped surface chiller (SSC) that uses a refrigerant, such as propane, for cooling. The raffinate/feed slurry passes through a heat exchanger to melt any wax crystals that may have formed in tankage. The slurry then passes through a pre-cooler before entering the tube side of the SSE. The cold filtrate from the rotary filter is used to cool the feed below the cloud point to initiate crystallization. Solvent may be added

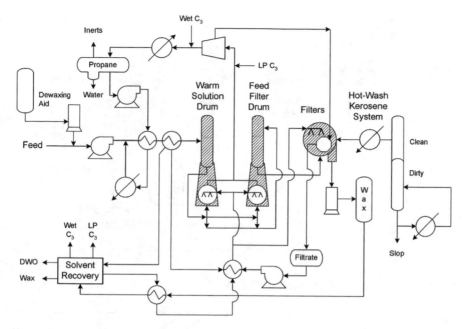

Fig. 13.11 Propane dewaxing plant [3]

before and after the SSE to reduce the slurry viscosity and enhance heat transfer. After exiting the SSC, the cold slurry enters the continuously rotating filter feed drum, where the wax crystals are filtered under vacuum. Wax is scraped off the filter surface and sent to the wax recovery section. The solvent/filtrate is sent to the shell side of the SSE on its way to the dewaxed oil (DWO) recovery section. The filtrate may also be recycled back to the slurry to adjust the final dilution ratio before filtration.

13.11.1.3 Comparison Between Propane and Ketone for Dewaxing

Table 13.4 compares the advantages and disadvantages of propane and ketones for dewaxing. Propane is readily available, less expensive and easier to recover. Propane dewaxing uses direct chilling to reduce capital and maintenance costs. It has high filtration rates and is accompanied by resin and asphaltene rejection. There is a large difference between the filtration temperature and pour point of the dewaxed oils (25–45 °F). Propane dewaxing requires the use of dewaxing aides to get good filtration rates, and it requires higher pressure equipment.

For ketone dewaxing, there is a small difference between the filtration temperature and pour point of dewaxed oil (9–18 °F). Hence, it has lower pour point reduction capability. But the process has greater heat efficiency and lower refrigeration require-

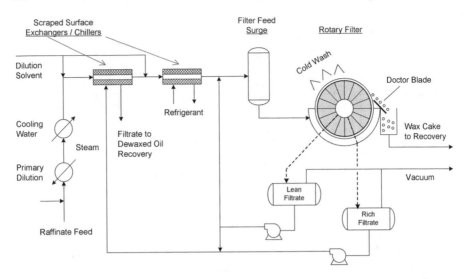

Fig. 13.12 Incremental ketone dewaxing plant [3]

Table 13.4 Propane versus ketones dewaxing [3]

	Propane	Ketone
Filter rate		
150 N	30–50	7–9
600 N	18–30	4–5
Bright stock	10–15	2–3
Dilution rate		
150 N	1.2–1.6	2
600 N	1.4–2	0.3
Bright stock	2–2.2	4
Economics		
Investment	−40%	Base
Operating costs	More	Base
Utilities	Less	Base
Pour point	Limited to −15 °C	
2-stage	Yes	Yes
Deoiling	Yes	Yes

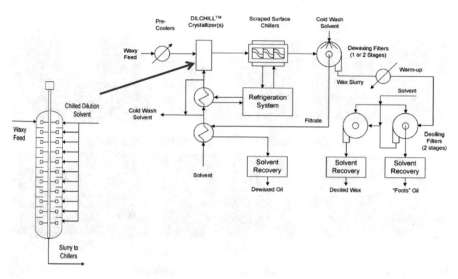

Fig. 13.13 Dilchill dewaxing [3]

ments. The chilling rate is fast, and the filtration rate is good (but lower than with propane).

13.11.1.4   Dichill Dewaxing

In ExxonMobil's Dilchill dewaxing, shown in Fig. 13.13, a multistage crystallizer (shown in the left lower corner) replaces the SSE for dilution and chilling. Cold solvent is added at each stage and is mixed with the slurry with an impeller. Due to rigorous mixing, less oil is occluded in the wax crystals, resulting in spherical crystals as shown in Fig. 13.14. The remaining sections of the unit are similar to those described above.

13.11.2   Catalytic Dewaxing

Until recently, the improvements in lubricant oils for passenger and commercial vehicles were largely achieved by the use of better additives, such as antioxidants, antiwear agents, and viscosity improvers. However, additives alone have not been able to meet the fuel economy and performance requirements for newer, improved vehicles. Other than synthetic lubricants, which will be discussed later, several new technologies, including catalytic dewaxing, have been developed to meet such demands.

Most catalytic dewaxing processes involve hydroprocessing or hydroconversion [4, 5]. Two main processes are (1) hydrocracking to break down large wax

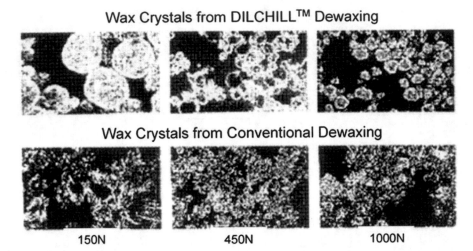

Fig. 13.14 Comparison of wax crystals by Dilchill and conventional dewaxing [3]

molecules and (2) hydroisomerization to isomerize linear or semi-linear molecules into branched molecules with reduced pour point. An all-hydroprocessing lubricant manufacturing train is shown in Fig. 13.9.

13.11.2.1   Wax Hydrocracking

The Mobil Lube Dewaxing (MLDW) process was developed in the mid-1970s as an alternative to solvent dewaxing. The MLDW catalyst uses medium pore ZSM-5, a catalytic shape-selective catalyst discovered by Mobil in the late 1960s. This zeolite selectively cracks the wax molecules on the acid sites in the pores while preserving the valuable base stock molecules. The ZSM-5 is shape selective. The constrictive, intersecting two-dimensional pore geometry on the order of 5 Å allows linear paraffins and linear portion of slightly branched isoparaffins to enter the pores to be cracked into smaller hydrocarbons over acid sites, while rejecting highly branched isoparaffins, naphthenes (cycloparaffins), and aromatics. Large sulfur and nitrogen species cannot get deep inside the ZSM-5 pores to deactivate the zeolite. Therefore, it is robust for a full range of raffinates. The cracking products include propane, butane, naphtha, and middle distillates with lower carbon numbers.

The uncracked material has a much lower pour point and oxidative stability. The desired reaction conditions are determined by the feed and product objectives. Temperatures range from 560 to 700 °F (293–371 °C) under a wide range of hydrogen partial pressures: from 300 to 2000 psig (20–180 bar). Gas rates can range from 500–5000 scf/bbl.

Compared to solvent dewaxing, ZSM-5 removes more wax when achieving the same pour point on a common feed, with lower yield. The cracking produces a small amount of olefins that must be removed by hydrofinishing.

13.11.2.2 Wax Hydroisomerization

Wax hydroisomerization technology was invented by Chevron in the early 1990s. Instead of cracking wax into relatively low value LPG and fuels, this process, named ISODEWAXING® [5, 6], isomerizes wax into low-pour-point high value base stocks while minimizing production of lighter products.

In isodewaxing units, waxy normal paraffins are catalytically isomerized into isoparaffins over silica-alumina-phosphate (SAPO) based catalysts that achieve wax isomerization with minimal cracking. [5, 6] Isodewaxed products have considerably lower cloud and pour points, which are indications of fluidity at low temperatures, and excellent thermal and oxidation stability. The impact of heavy paraffin isomerization on melting point is illustrated below for $C18H38$. Going from the n-isomer to the 3-methyl isomer lowers the melting point by 34 °C. The melting points of compounds in mixtures differ from pure-compound melting points, and melting point is not the same as cloud point or pour point. Even so, the magnitude of the changes in melting point is consistent with the observed changes in cloud and pour points.

Compound	Melting Point, °C
n-octadecane	28
2-methylheptadecane	6
3-methylheptadecane	−6

Data from ChemSpider.com

Isodewaxing with SAPO catalysts under hydrogen gives minimal cracking, so it provides higher base stock yields than dewaxing with ZSM-5 catalysts, which cracks C20+ n-paraffins into naphtha and gasoil. While preserving a feedstock's paraffinicity, isodewaxing also produces higher product VI than other dewaxing processes. A broader range of feeds can be processed – from feeds with low wax content to those with close to 100% wax – to produce a broad range of base oils with VI ranging from 95 to 140 and higher. Pour points may range from low (-9 to -15°C) to ultra-low (< -40°C).

ExxonMobil developed a version of hydroisomerization dewaxing technology, MSDW (Mobil Selective Dewaxing) [4, 5], and implemented it at the Singapore-Jurong lube plant in 1997, as described in Fig. 13.15. Shell hydroisomerization/dewaxing technology is used in the GTL Pearl Plant in Qatar.

The hydroisomerization catalysts are composed of zeolitic molecular sieve materials and at least one noble metal. The zeolite provides isomerization activity and the metal provides hydrogenation activity. The pore opening is narrow, on the order

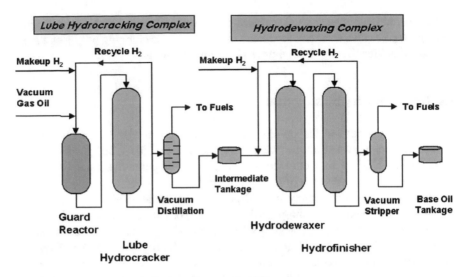

Fig. 13.15 Hydrocracking-hydrodewaxing complex [4]

of 0.6 nm. The pore geometry allows long-chain wax molecules to enter the catalyst pores while excluding isoparaffins, which have a larger effective dimeter due to branching. The acidity of the catalytic sites is less than in ZSM-5. This results in less cracking and greater preservation of molecules in the base oil range. At the noble metal sites, intermediate iso-olefins are rapidly saturated. In addition to preventing cracking, this prevents coke formation and maintains catalyst activity and stability. Final stabilization is achieved in the hydrofinishing unit.

13.12 DWO Quality Improvement

Several processes improve the color, odor, thermal and oxidative stability, and demulsibility of dewaxed oil (DWO) for use as lube base stock, including selective hydrocracking, hydrofining and clay contacting.

Hydrofinishing is a mild hydrotreating process using Co-Mo catalysts to remove sulfur, nitrogen and oxygen compounds; such compounds increase the color and decrease the color stability of lube oils. In conventional hydrofinishing, reaction temperatures range from 400 to 650 °F (204–343 °C) under 500–800 psig (34–55 bar) pressure. The LHSV is 0.5–2.0 v/h/v, and hydrogen gas rate is 3–4 times consumption. In certain hydrofinishing processes, operating at higher pressures (up to 3000 psig, 200 bar) and lower temperatures provides essentially complete removal of aromatics in addition to hetero-atom removal.

Clay Contacting contacts DWO with activated clay, which adsorbs aromatic, sulfur and nitrogen compounds to improve stability of the oil.

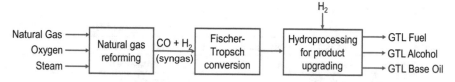

Fig. 13.16 Gas-to-liquid lubricant base stock through Fischer-Tropsch process [4]

13.13 GTL (Fischer-Tropsch) Process

The Fischer-Tropsch (F-T) process is a collection of chemical reactions that converts a mixture of carbon monoxide and hydrogen into liquid hydrocarbons. The ideal ratio of H_2:CO ratio is 2:1. Fe, Co and Ru are most widely used catalysts.

$$(2n + 1)\, H_2 + n\, CO \;\rightarrow\; C_n H_{(2n+2)} + n\, H_2O$$

where n is an integer.

Recent advances in catalysts and process design have enabled the commercialization of gas-to-liquid (GTL) processes through F-T chemistry. As shown in Fig. 13.16, natural gas or methane is converted into a mixture of carbon monoxide and hydrogen, often called synthesis gas or syngas [4]. Then the syngas is converted into a mixture of linear paraffins and some oxygenates. In most cases, GTL process conditions are adjusted to maximize diesel. However, by-products with higher (and lower) carbon numbers also are generated. $C_{20}+$ GTL products require considerable hydrodeoxygenation and selective isomerization of linear paraffins into isoparaffins to make them suitable as lubricant base stocks.

13.14 Wax

The raw wax from dewaxing units, called slack wax, is soft because it still contains some oil. The oil must be removed to convert slack wax into hard wax the meets "food grade" specifications for oil content, melting point, and needle penetration. Methyl isobutyl ketone (MIBK) is used in a *wax deoiling* process to prepare food-grade wax, Food-grade wax serves as a coating for milk cartons and other packaging. It is a component of certain medicines, and it is blended directly into foods, such as "construction chocolates," to provide hardness and/or modify melting properties. In recrystallization deoiling, the wax crystals from the dewaxing plant are melted in a heat exchanger in the presence of a lean solvent. The wax is then recrystallized in SSEs and SSCs to get rid of excess oil. Wax can be fractionated by consecutive slow cooling and filtration at various temperatures, with collection of wax crystals at each specific temperature.

Table 13.5 Comparison of waxes from different sources

	Wax	Microcrystalline wax
Source	Diesel or vacuum distillates	Vacuum resids
Crystal shape	Large crystal, plate	Microcrystal
Molecular weight	300–450	450–800
Carbon number	C_{17}–C_{35}	C_{35}–C_{60}
Density	0.88–0.94	–
Melting point	30–70 °C	70–95 °C
Composition	Mostly n-paraffins, less amounts of isoparaffins, naphthenes and minute amounts of aromatics	Mostly naphthenes with n-alkyl or isoalkyl side chains

Petroleum wax is of two general types: paraffin wax from petroleum distillates and microcrystalline wax from residua. Table 13.5 compares waxes from these different sources.

13.15   Finished Lube Oils and Specialty Products

There are many forms of finished lubricant oil products and specialty products for a wide variety of applications. Finished Lube Oils include engine oils, transmission fluids, gear oils, turbine oils, hydraulic oils, metal working (cutting) oils, greases, paper machine oils, etc. Specialty products include white oils for foods, pharmaceuticals and cosmetics; agricultural oils such as those used as orchard sprays; electrical oils for electrical transformers (heat transfer media), etc.

13.16   Synthetic Lubricant Oils (Groups IV and V Oils)

In the early 1950s, extreme demands for high- and low-temperature lubricity in jet engines could not met by mineral lube oils. Polyol esters provide superior thermal oxidative stability, lubricity and volatility. During oil drilling in Alaska in the mid-1960s, mineral lube oils solidified under severe Alaskan cold weather conditions and could not function. It was found that poly alpha olefins (PAO), initially studied by Socony-Mobil in the early 1950s, provide excellent low temperature flow properties. In 1973, Mobil introduced PAO-based SHC in Europe, followed by fuel saving SAE 5W-20 Mobil 1 in the U.S.

Synthetic lube base stocks are more expensive, but they are preferred: (1) when conventional lubes cannot meet specific performance demands, (2) when a synthetic lube can offer economic benefits for overall operations, and (3) for modern machines and equipment that are operated under increasingly severe conditions, where it is especially desirable to reduce maintenance and to improve energy efficiency. Even for ordinary automobiles, for which conventional lubricants are considered good enough, a significant percentage of drivers are willing to pay more for synthetic products.

The base stocks synthesized from chosen chemicals are designed for improving lube performance with optimized properties:

- Viscosity Index (VI)—The desired VI range for 5 cSt oil is 85–100. The VI of most synthetic lube is >120.
- Pour point—synthetic lubes have superior cold temperature lubrication pour points at −30 to −70 °C, compared to conventional mineral oils in the range of 0 to − 20 °C.
- Thermal and oxidation stability—synthetic lubes have improved thermal and oxidative stability. Hence, they respond well to antioxidants and resist aging processes.
- Volatility—The volatility is minimized with narrow molecular weight distribution.
- Others—Other properties, such as biodegradability, friction coefficient, tracking coefficient, etc., can be optimized for intended uses.

Figure 13.17 summaries the synthetic lube base stocks derived from petrochemicals, mainly ethylene, propylene, butenes, benzene and xylene [7].

13.16.1   Polyalphaolefins (PAO)—Group IV

Polyalphaolefins (PAO) comprise a class of molecularly engineered base stocks with optimized viscosity index, pour point, volatility, oxidative stability and other important properties [7]. Low viscosity PAO (4–6 cSt) accounts for 80% of the total volume.

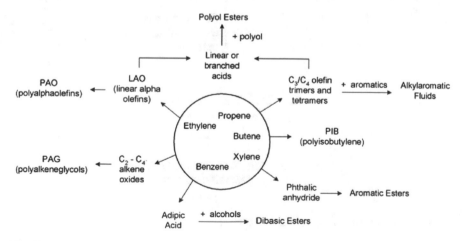

Fig. 13.17   Synthetic lube base stocks derived from petrochemicals [7]

The remaining PAOs have viscosities of 10–100 cSt. Among the PAO of different carbon numbers synthesized from ethylene, 1-decene is the most commonly used starting material. It can produce higher oligomers, shown in Fig. 13.18, depending on the catalyst, BF_3 or $AlCl_3$, and reaction conditions in a multistage, continuously stirred tank reactor (CSTR). Table 13.6 lists some properties—kinematic viscosity, viscosity index and pour point—of various PAO oligomers synthesized from different linear 1-olefins. The combination of C_{30} and C_{40} provides a very narrow molecular weight distribution for a 4 cSt oil, with less lower carbon number components. This is important because the smaller components can lower flash point by degrading oil volatility.

PAO with a given carbon number is a mixture of many isomers with different types of branching. This is because the use of BF3 or AlCl3 to catalyze oligomerization of LAO via carbonium ion intermediates can have the charge center moved along the hydrocarbon chain, resulting in non-uniform connections between the monomers, shown in Figure 13-19. The irregular branching is beneficial for very low poor point properties of PAOs. Using metallocene as a catalyst, the oligomerization of LAO produces oligomers with a uniform comb-like structure shown in Fig. 13.20. Thus, it can produce high molecular weight PAOs with narrow molecular weight distributions, with higher VI, wider viscosity range, and lower pour points.

PAOs do not contain ring hydrocarbons, naphthenes and aromatics, which gives them intrinsic oxidative stability compared to conventional Groups I, II and III min-

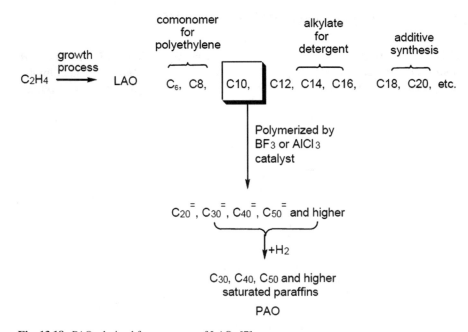

Fig. 13.18 PAOs derived from a range of LAOs [7]

Table 13.6 Properties of PAO derived from different linear 1-olefins [7]

Name	Carbon number	Kinematic viscosity cSt at			Viscosity index	Pour point (°C)
		100 °C	40 °C	−40 °C		
Propylene decamers	C_{30}	7.3	62.3	>99,000	70	–
Hexane pentamers	C_{30}	3.8	18.1	7850	96	–
Octene tetramers	C_{32}	4.1	20.0	4750	196	–
Decene trimers	C_{30}	3.7	15.6	2070	122	<−55
Undecene trimers	C_{33}	4.4	20.2	3350	131	<−55
Dodecene trimers	C_{36}	5.1	24.3	13,300	144	−45
Decene tetramers	C_{40}	5.7	29.0	7475	141	<−55
Octene pentamers	C_{40}	5.6	30.9	10,225	124	–
Tetradecene trimers	C_{42}	6.7	33.8	Solid	157	−20

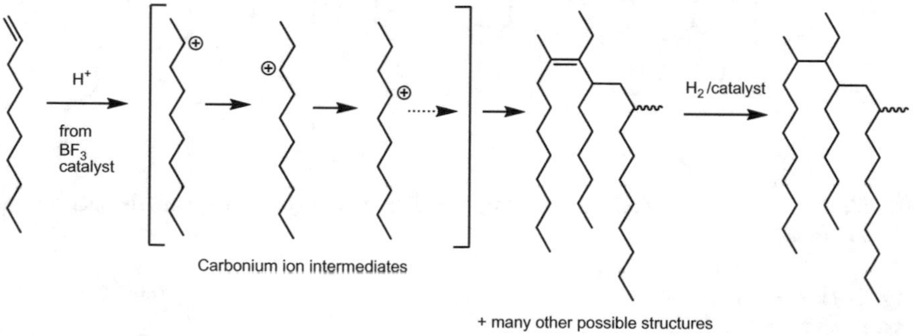

Fig. 13.19 Formation of PAO via carbonium ion intermediates using BF_3 or $AlCl_3$ as a catalyst

eral oil base stocks. PAOs also have superior viscometric properties compared to the mineral oil base stocks, with lower volatility due to narrow molecular weight ranges.

The largest volume of PAO is made for premium automotive engine oils, which provide improved engine protection, extended oil drain interval with reduce oil consumption, improved fuel economy, excellent low-temperature fluidity and pumpability, and high temperature oxidative resistance. Figure 13.19 compares viscosity changes in engine tests of PAO with Group I/II and Group III oils [7]. After over 200-h operations, the mineral oils become too thick for viscosity measurement, while PAO is still in a suitable operation range. The high viscosity PAOs are used as industrial oils and greases, with longer fatigue life, a wider operating range due to higher

VI, and better thermal-oxidative stability. In compressor oil applications, PAO-based lubricant oils are resistant to chemical attack due to their better chemical inertness.

13.16.2 Ester Base Stocks

Ester based lubricants of natural sources, such as lard and vegetable oil, have been known throughout human history. However, demands for synthetic ester lubricants started after World War II to meet jet engine requirements. Esters are made from the reaction of acids with alcohols. Among many types of ester lubricants, dibasic esters, polyol esters and aromatic esters are the most commonly used. A comparison of their properties is shown in Table 13.7.

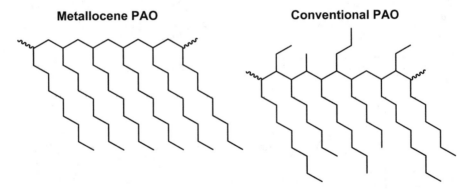

Fig. 13.20 Comparison of PAO structures by metallocene catalyst versus conventional BF$_3$ or AlCl$_3$ as catalyst

Fig. 13.21 Comparison of engine test results of PAO and mineral base oils in viscosity changes

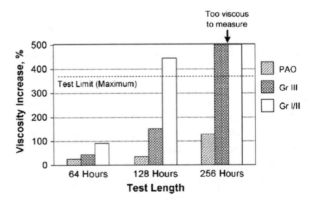

Ester base stocks have lower volatility than PAO and mineral oil base stocks of comparable viscosities because of their higher polarity. The general rank of volatility is PE ester < TMP ester < dibasic ester < PAO ≪ Group I or II mineral oils. Dibasic and polyol esters have excellent biodegradability. Aromatic esters in general are less biodegradable than aliphatic dibasic and polyol esters.

Many commonly used additives are more soluble in ester base stocks than in mineral oil base stocks. The high solubility of organic acids and sludge during services makes low viscosity ester fluids ideal for co-base stocks of PAO for solvency and dispersancy. A major issue for ester fluids is the formation of corrosive acids when contaminated with water. Esters made from aromatic and other sterically hindered acids improve hydrolytic stability. Ester base stocks provide mild film protection at lower temperatures. Higher molecular weight acids, upon decomposition, are bound to metal to provide some degree of wear protection and friction reduction.

Table 13.7   Basic properties of ester base stocks [7]

Acid	Alcohol	Viscosity (cSt)		VI	Pour point (°C)	Volatility (wt%)	Biodegrada bility
		100 °C	40 °C				
Dibasic esters							
Adipate	Iso- $C_{13}H_{27}$	5.4	27	139	−51	4.8	92
Sebacate	Iso- $C_{13}H_{27}$	6.7	36.7	141	−52	3.7	80
Polyol esters							
n-C_8/n-C_{10}	PE	5.9	30	145	−4	0.9	100
n-C_5/n- C_7/iso-C_9	PE	5.9	33.7	110	−46	2.2	69
n-C_8/n-C_{10}	TMP	4.5	20.4	137	−43	2.9	96
Iso-C_9	TMP	7.2	51.7	98	−32	6.7	7
n-C_9	NPG	2.6	8.6	145	−55	31.2	97
Aromatic esters							
Phthalate	Iso- $C_{13}H_{27}$	8.2	80.5	56	−43	2.6	46
Phthalate	Iso-C_9	5.3	38.5	50	−44	11.7	53
Trimellitate	Iso- $C_{13}H_{27}$	20.4	305	76	−9	1.6	9
Trimellitate	n-C_7/n- C_9	7.3	46.8	108	−45	0.9	69

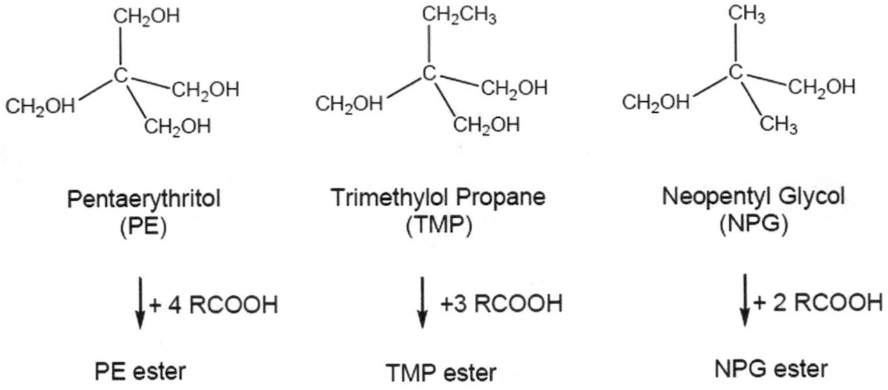

CH₂OH   CH₂CH₃   CH₃

Pentaerythritol   Trimethylol Propane   Neopentyl Glycol
(PE)   (TMP)   (NPG)

↓ + 4 RCOOH   ↓ +3 RCOOH   ↓ + 2 RCOOH

PE ester   TMP ester   NPG ester

Fig. 13.22 Polyol Esters

13.16.2.1   Dibasic Esters

Dibasic esters are made from the reaction of carboxylic diacids with alcohols. The most commonly used diacid is adipic (hexanedioic acid), usually combined with ethylhexanol or isotridecanols to give balanced high VI and low temperature properties (see Table 13.7). Dibasic esters are often used with PAO as a co-base stock to improve solvency with additives and swell properties of final formulated lubricant products.

13.16.2.2   Polyol Esters

Polyol esters are made from reaction of monoacids with polyols. The most commonly used polyols are pentaerythriol (PE), trimethylolpropane (TMP) and neopenttylglycol (NPG), shown in Fig. 13.22. High VI and low pour point base stocks can be made by careful choice of acids with branching.

A polyol ester molecule lacks β-H adjacent to carbonyl oxygen, so it cannot undergo low energy β-H transfer to decompose into olefin and acid as dibasic esters, shown in Fig. 13.23. Hence, it can only be decomposed by C-O or C-C cleavage into radicals at extremely high temperatures. Therefore, polyol esters are thermally stable up to 250 °C.7 Polyol esters exhibit thermal stability in the following order: PE esters > TMP esters > NPG esters.

13.16.2.3   Aromatic Esters

The reaction between phthalic anhydride or trimellitic anhydride with alcohols results in aromatic esters. They have relatively low VI's, as shown in Tables 13.4, 13.5 and 13.6, and only are used in special industrial oil applications.

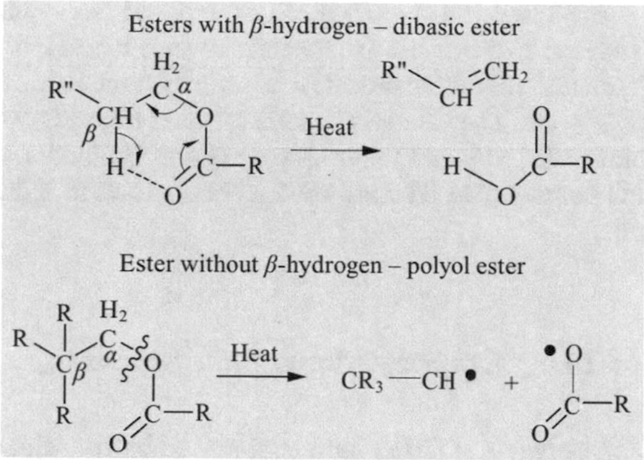

Fig. 13.23 Ester cracking mechanisms

Esters with β-hydrogen – dibasic ester

Ester without β-hydrogen – polyol ester

13.16.3 Polyalkylene Glycols (PAG)

Polyalkylene glycol (PAG) is mainly used to make polyurethanes. About 20% of PAGs is used for lubrication.7 PAG was first developed as fire-resistant, water-based hydraulic lubricants for military use during World War II.

PAGs are produced via oligomerization of alkylene oxides over a base catalyst (such as NaOH) with water or alcohol as an initiator. Ethylene oxide (EO), propylene oxide (PO) and butylene oxide (BO) are used as starting materials. Longer chain alkylene oxides can be added for improved compatibility with hydrocarbons. PAGs have a wide range of molecular weights, viscosities, VIs, pour points, water solubilities, and oil solubilities, shown in Table 13.8.

Table 13.8 Lubricating properties of selected PAGs [7]

Fluid Type	Mol. Wt.	00 °CcSt	V40 °C cSt	VI	Pour point, °C	Four ball wear scar, mm[a]	Four ball seizure load, kg[a]	Friction coeff.[b]	Soluble in[c]
EO/PO	4.6	19	161	Fluid Type	−46	0.53	120–140	0.15	water
EO/PO	15	76	218	Fluid Type	−42	0.44	180–200	0.11	water
PO	6	27	179	Fluid Type	−44	0.53	160–180	0.19	oil
PO	14	73	193	Fluid Type	−35	0.57	120–140	0.12	none

(a) by DIN 51350 method
(b) determined by oscillation of a steel ball on a steel disc at 30 °C under a load of 200 N
(c) determined by mixing equal proportions of water and PAG or oil and PAG.
(d) partially soluble in oil

EO-based PAGs are typically waxy and have poor low-temperature properties. They are typically used to formulate water-based lubricants, especially fire-resistant hydraulic oils. PO-based PAGs are excellent lubricant oils with high VI and low pour point. They are less soluble to water than EO-based PAGs, but are not oil miscible. EO/PO-based PAGs have a better combination of VI and low pour points than PO-based PAGs. The are used as base stock in industrial circulation/bearing/gear oils.

13.16.4   Other Synthetic Base Stocks

Polyisobutylene (PIB) fluid is produced by the oligomerization of isobutylene in a mixed C_4 stream over a BF_3 or $AlCl_3$ catalyst. PIBs are seldom used by themselves. They are typically used as additives to increase lubricant viscosity.

Alkylbenzenes and alkylnaphthalenes are obtained from Friedel-Craft alkylation of benzene and naphthalene over BF_3 or $AlCl_3$. They have very low pour points as components for refrigeration compressor oil. Alkyl benzene was used in the early 1960s for engine lubricants in Alaska oil production because of its superior flow at extremely low temperatures where mineral-based lubricant oils were frozen solid. However, it is rarely used today in lubricant formulation.

Alkylnaphthalenes have superior thermal oxidative and hydrolytic stability. They are used to replace esters for fully formulated synthetic motor and industrial lubricant oils. For example, an industrial turbine lubricant containing 10% alkylnaphthalenes in PAO with additives showed no change in total acid number for a stability test of 3000 h, compared to the significant degradation of 10% polyol esters in PAO with the same additives only after 1000 h of the test. Alkylnaphthalenes are used in automotive engine oil, rotary compressor oil, and other industrial oils.

Phosphate esters are produced from phosphorus oxychloride with various alcohols or phenols. They generally have good thermal and oxidative stabilities and fire resistance, but poor VI-pour point balance. They have very limited uses except for fire-resistant hydrolytic oils.

References

1. 2016 Annual Book of ASTM Standards (2016) Section 5: petroleum products, liquid fuels, and lubricants, vols 05.01 and 05.02. American Society for Testing and Materials (ASTM) International, West Conshohocken, PA
2. Hsu CS (ed) (2003) Analytical advances for hydrocarbon research. Kluwer Academic/Plenum Publishers, New York
3. (a) Beasley BE (2006) Conventional lube base stock manufacturing. In: Hsu CS, Robinson PR (eds) Practical advances in petroleum processing. Springer, New York (Chapter 15). (b) Beasley BE (2017) Conventional lube base stock. In: Hsu CS, Robinson PR (eds) Springer handbook of petroleum technology. Springer, New York (Chapter 33)
4. Cody IA (2006) Selective hydroprocessing for new lubricant standards. In: Hsu CS, Robinson PR (eds) Practical advances in petroleum processing. Springer, New York (Chapter 16)

5. Lee SK, Rosenbaum JM, Hao Y, Lei GD (2017) Premium lubricant base stocks by hydroprocessing. In: Hsu CS, Robinson PR (eds) Springer handbook of petroleum technology. Springer, New York (Chapter 34)
6. Lei G-D, Dahlberg A, Krishna K (2015) All hydroprocessing route to high quality lubricant base oil manufacturing using Chevron ISODEWAXING Technology. http://www.nt.ntnu.no/users/skoge/prost/proceedings/aiche-2008/data/papers/P138184.pdf. Accessed 15 Aug 2015
7. (a) Wu MM, Ho SC, Forbus TR (2006) Synthetic lubricant base stock processes and products. In: Hsu CS, Robinson PR (eds) Practical advances in petroleum processing. Springer, New York (Chapter 17). (b) Wu MM, Ho SC, Luo S (2017) Synthetic lubricant base stock. In: Hsu CS, Robinson PR (eds) Springer handbook of petroleum technology. New York: Springer, 2017 (Chapter 35)

Chapter 14
Other Refining Processes

14.1 Hydrogen Production

As mentioned earlier, catalytic reforming is a major source of high-grade hydrogen for use elsewhere in the refinery. Major consumers of hydrogen include hydrotreating, hydrocracking, and isodewaxing. A relatively small amount of hydrogen is consumed by isomerization units, where it to prevent catalyst coking, and in hydrofinishing units, where it improves color and product stability.

As discussed in previous chapters, hydrotreating accomplishes hydrodesulfurization (HDS), hydrodenitrogenation (HDN), hydrodeoxygenation (HDO), and hydrodemetallation (HDM). Hydrocracking achieves conversion of heavy molecules into lighter molecules with higher value. Catalytic dewaxing and hydroisomerization convert waxy normal paraffins into isoparaffins, thereby increasing yields of high-quality lube base stock.

When additional hydrogen is needed, the process of choice is steam hydrocarbon reforming. This process is the heart of refinery hydrogen plants. High-grade hydrogen and synthesis gas also are generated by partial oxidation of natural gas, coke, and residua.

Low-grade hydrogen is generated by other processes, including coking, visbreaking, and catalytic cracking. Due to low purity and contamination by olefins, the main use for low-grade hydrogen is fuel gas.

14.1.1 Steam-Hydrocarbon Reforming/Water Gas Shift Reaction/Partial Oxidation

Steam-hydrocarbon reforming is commonly called steam-methane reforming (SMR) because methane is the most common feedstock. Another common feed is naphtha, which also has high hydrogen content. SMR produces hydrogen through the reaction of steam with light hydrocarbons at very high temperatures—around 1500 °F

© Springer Nature Switzerland AG 2019
C. S. Hsu and P. R. Robinson, *Petroleum Science and Technology*,
https://doi.org/10.1007/978-3-030-16275-7_14

(816 °C). The product of the initial reaction is a mixture of H_2, CO, CO_2, residual methane, and in some cases traces of other hydrocarbons. The initial product goes to one or more water shift reactors, where the shift reaction between H_2O and CO yields CO_2 and additional hydrogen.

$$\text{Primary reaction:} \quad CH_4 + H_2O \rightarrow CO + 3H_2$$
$$\text{Water Shift reaction: } CO + H_2O \rightarrow CO_2 + H_2$$

At high temperature in the presence of steam, the reaction can be:

$$CH_4 \rightarrow 2H_2 + C$$
$$C + H_2O \rightarrow CO + H_2$$

Another common hydrogen source is partial oxidation (POX). With methane as an example:

$$2CH_4 + O_2 \rightarrow CO + 4H_2$$

14.1.2 Hydrogen Purification Using Methanation

In older units, residual CO was removed by methanation over a nickel-based catalyst, and residual CO_2 was removed by adsorption with activated molten potassium carbonate (Benfield process) or monoethanolamine (MEA). The product hydrogen (95–97% pure) contained 3–5% methane. Figure 14.1 shows a flow diagram of a hydrogen plant with a methanator (or methanizer). A steam methane reformer (SMR) first converts hydrogen to a synthesis gas (syngas) mixture (CO and H_2) and then employs a high-temperature shift converter (HTSC) and low-temperature shift converter (LTSC) to shift most of the CO to H_2 and CO_2. A methanator is used after CO_2 removal to convert the remaining CO and CO_2 to methane and water.

$$\text{Methanation:} \quad CO + H_2 \rightarrow CH_4$$

14.1.3 Pressure-Swing Adsorption (PSA)

In newer units, pressure-swing adsorption (PSA) removes nearly all contaminates, yielding a product containing 99.99% hydrogen. A PSA unit employs a bed of solid adsorbents—molecular sieves, activated carbons, silica gels, and activated aluminas—to separate impurities from the hydrogen stream. For carbon, silica, and alumina, separation is due to their differential tendency to adsorb different gases.

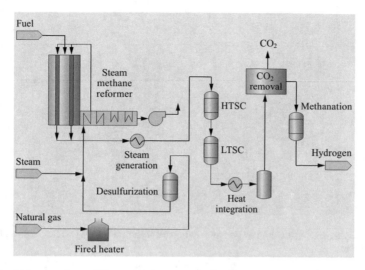

Fig. 14.1 Hydrogen plant with a methanizer [1] HTSC: high-temperature shift converter; LTSC: low-temperature shift converter

PSA cycles include pressurization, depressurization, sweeping, and purging. In other words, the hydrogen is purified by swing between high and low pressures. At high pressure, hydrogen is separated from tail gas because it is the most difficult gas to be condensed upon pressure and least adsorbed. The pressure is then reduced for regeneration of adsorbent and purging other gases. The rejected gas, a low-Btu mixture containing methane, CO and CO_2, is burned to provide some pre-heat duty.

Figure 14.2 shows a hydrogen production unit with a PSA. The feed gas, such as methane, and steam pass through a purifier, such as a scrubber, to remove impurities before entering a reformer to form hydrogen. Additional hydrogen is produced in a shift converter through water shift reaction. The gases then enter the PSA to remove all other gases from hydrogen. The PSA unit replaces both the CO_2 removal unit and methanator used in the old hydrogen plants, operating at much higher efficiency and with substantially lower operating costs.

14.2 Sulfur Removal and Recovery [3]

Natural gas can contain significant amounts of hydrogen sulfide. Some gas processing plants in Alberta, Canada, recover more than 2000 tons of sulfur per day using the Claus process. Sulfur can also be recovered by processes such as LOCAT and Selectox. A small Selectox unit converts H_2S to sulfur on Platform Irene off the coast of California [2].

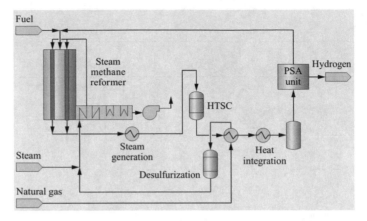

Fig. 14.2 Hydrogen purification by pressure-swing adsorption (PSA) [1] HTSC: high-temperature shift converter

In petroleum refineries, conversion processes, hydrotreaters, and sweetening units remove chemically bound sulfur from organic molecules. The sulfur can end up as SO_x, H_2S, NH_4SH, or $NaSH$.

Fuel-oil fired heaters and the regenerators of FCC units are major sources of refinery SO_x and NO_x emissions. The most obvious way to reduce SO_x emissions from a heater is to use low-sulfur fuels. Unfortunately, although that solution requires no investment, it is probably most expensive due to the relatively high cost of buying low-sulfur fuel oil and/or hydrotreating high-sulfur fuel oil.

A large fraction of the sulfur in the feed to an FCC unit ends up in coke on the catalyst. SO_x are formed in the regenerator when the coke is burned away. Therefore, removing sulfur from the feed, usually by hydrotreating, decreases SO_x emissions. FCC feed pretreating has other substantial benefits. Removing basic nitrogen decreases deactivation of acid sites on the FCC catalyst, which allows the FCC to reach a given conversion at lower temperatures. The saturation of aromatics in the feed pretreater provides the biggest benefit, because it converts hard-to-crack aromatics into easier-to-crack naphthenes. This alone can justify the installation of an FCC feed pretreater.

SO_x transfer additives, first developed by Davison Chemical, react with SO_x in the FCC regenerator to form sulfates. When the sulfated additive circulates to the riser/reactor section, the sulfate is reduced to H_2S, which is recovered by amine absorption and sent to the sulfur plant. In some units, these additives reduce FCC SO_x emissions by more than 70%. Consequently, if a pre-treater or post-treater still must be installed, its size can be reduced. The chemistry of SO_x transfer is summarized below:

FCC Regenerator (Oxidizing Environment)

Coke on catalyst (solid) $+ O_2 + H_2O \rightarrow CO_2 + SO_2, SO_3, HySO_x$ (gases)

$$SO_2, SO_3, HySO_x \text{ (gas)} + MO \text{ (solid)} + O_2 \rightarrow MSO_4 \text{ (solid)}$$

FCC Riser-Reactor (Reducing Environment)

$$MSO_4 \text{ (solid)} + 4H_2 \rightarrow MO \text{ (solid)} + H_2S \text{ (gas)} + 3H_2O \text{ (gas)}$$

Flue-gas scrubbing is a refiner's last chance to keep NO_x and SO_x out of the air. In wet flue-gas desulfurization, gas streams containing SO_x react with an aqueous slurry containing calcium hydroxide $Ca(OH)_2$ and calcium carbonate $CaCO_3$. Reaction products include calcium sulfite ($CaSO_3$) and calcium sulfate ($CaSO_4$), which precipitate from the solution. NO_x removal is more difficult. Wet flue-gas scrubbing removes about 20% of the NO_x from a typical FCC flue gas. To remove the rest, chemical reducing agents are used. In the Selective Catalytic Reduction (SCR) process, anhydrous ammonia is injected into the flue gas as it passes through a bed of catalyst at 500–950 °F (260–510 °C). The chemical reaction between NO_x and ammonia produces N_2 and H_2O.

When sulfur-containing feeds pass through hydrotreaters or conversion units, most of the sulfur is converted into H_2S, which eventually ends up in off-gas streams. Amine absorbers remove the H_2S, leaving only 10–20 wppm in the treated gas streams. H_2S is steam-stripped from the amines, which are returned to the absorbers. The H_2S goes to the refinery sulfur plant.

14.2.1 Amine Treating [3]

Amine gas treating, also known as amine scrubbing, gas sweetening and acid gas removal, refers to a group of processes that use aqueous solutions of various alkanolamines such as monoethanolamine (MEA), diethanolamine (DEA), triethanolamine (TEA), methyldiethanolamine (MDEA), dipropanolamine (DIPA) and diglycolamine (DGA) to remove hydrogen sulfide and carbon dioxide from gases and hydrocarbon streams. Amine scrubbing is common in refineries, and is also used in petrochemical plants, natural gas processing plants and other industries.

The treating chemistry at low partial pressure of the acid gas is:

$$2RNH_2 + H_2S \rightarrow (RNH_3)_2S$$
$$2RNH_2 + CO_2 + H_2O \rightarrow (RNH_3)_2CO_3$$

At high acid gas partial pressure the reactions are:

$$(RNH_3)_2S + H_2S \rightarrow 2RNH_3HS$$
$$(RNH_3)_2CO_3 + H_2O \rightarrow 2RNH_3HCO_3$$

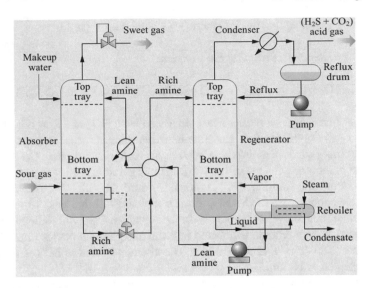

Fig. 14.3 Amine treating unit [3]

As shown in Fig. 14.3, sour gas containing carbon dioxide and/or hydrogen sulfide is charged to a gas absorption tower or liquid contactor where the acid contaminants are absorbed by counter flowing amine solutions (i.e. MEA, DEA, MDEA). The stripped sweet gas is removed overhead, and the amine is sent to a regenerator. In the regenerator, the acidic components are stripped by heat and reboiling action and routed to recovery units. The amine is recycled.

14.2.2   Sulfur Recovery by the Claus Process

Claus sulfur-recovery units, shown in Fig. 14.4, burn hydrogen sulfide in enough air to form a mixture of H_2S and SO_2 in a 2:1 molar ratio. In downstream beds of alumina catalyst, H_2S reacts with SO_2 to form elemental sulfur and water.

$$3H_2S + O_2 + S \rightarrow 2H_2S + SO_2 + H_2O$$
$$2H_2S + SO_2 \rightarrow 3S + 2H_2O$$

In a Claus unit with three catalyst beds, as shown in the figure, overall H_2S recovery in less than 98%. Due to the high concentration of H_2S, the tail gas is toxic and must be incinerated to SO_2 or routed to a tail-gas treating unit (TGTU). Incineration occurs in a high stack to disperse the SO_2. After incineration, no H_2S remains.

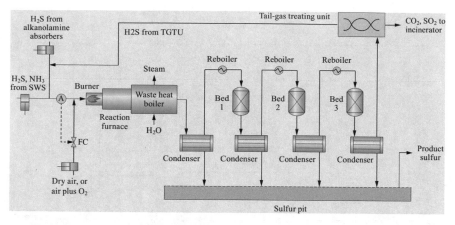

Fig. 14.4 Once-through three-stage Claus sulfur process

In most of the world, air pollution regulations limit SO_x in stack emissions to 10 ppm. In some locales, the limit is as high as 100 ppm. Therefore, tail gas treatment is a necessity for Claus units.

At the sulfur plant, H_2S is combined with sour-water stripper off-gas and sent to a Claus unit. Almost every refinery in the world uses some version of this process to convert H_2S into elemental sulfur. H_2S and a carefully controlled amount of air are mixed and sent to a burner, where about 33% of the H_2S is converted to SO_2 and water. In several units, the combustion air is enriched with oxygen to increase plant capacity.

From the burner, the hot gases go to a reaction chamber, where the reactants and products reach equilibrium. Elemental sulfur is produced by the reversible reaction between SO_2 and H_2S. Ammonia comes in with the sour-water stripper off-gas. In the Claus process, it is thermally decomposed into nitrogen and water.

In the Claus burner, combustion temperatures reach 2200 °F (1200 °C). Much of the heat is recovered in a waste-heat boiler, which generates steam as it drops the temperature to 700 °F (370 °C). Next, the process gas goes to a condenser, where it is cooled to about 450 °F (232 °C). At this temperature, sulfur vapors condense, and the resulting molten sulfur flows through a drain to a heated sulfur-collection pit. At the bottom of the drain, a seal leg maintains system pressure and keeps unconverted gases out of the pit. Uncondensed sulfur and other gases flow to a series of catalyst beds, which recover additional sulfur by promoting the reaction between left-over H_2S and SO_2. With fresh catalyst and a stoichiometric gas composition, the cumulative recovery of sulfur across four condensers is about 50, 80, 95, and 96–98%, respectively.

Claus tail-gas treating (TGT) units bring the total sulfur recovery up to >99.9% in the refineries and over 99.2% in the LNG plants depending feed gas composition [4]. Most tail-gas treating processes send the tail gas to the Beavon Sulfur Removal (BSR) hydrotreater, which converts all sulfur-containing compounds (mercaptans,

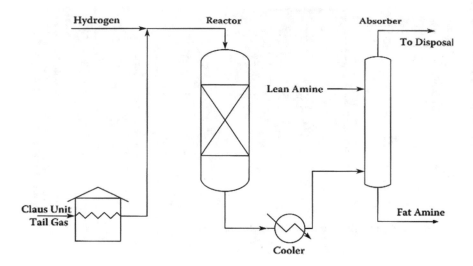

Fig. 14.5   Shell Claus off-gas treatment (SCOT) process [5]

SO_2, SO_3, COS, CS_2 and various vapor forms of S_x) into H_2S. In the Shell Claus Off-gas Treatment (SCOT) process offered by Shell Global Solutions, shown in Fig. 14.5, the H_2S from the BSR section is absorbed by an amine and returned to the front of the Claus furnace [5].

The BSR Stretford process consists of two stages [6]. In the Stretford section, the H_2S is oxidized using an alkaline solution containing vanadium as an oxygen carrier [7]. BSR Stretford removes essentially all of the sulfur compounds from Claus plant tail gases to give elemental sulfur. Chemicals in the Stretford process were toxic, and during process upsets, the sulfur was contaminated by vanadium and could not be sold without purification. Hence, the process is no longer used.

For all sulfur-recovery processes, vapor containing the last traces of unrecovered sulfur go to an incinerator, where they are converted into SO_2 and dispersed into the atmosphere.

14.2.3   Other Sulfur Recovery Processes [8]

At less than about 20% H_2S in acid gas or tail gas mixtures, a stable flame cannot be maintained in a Claus furnace even with a split-flow furnace design. The UOP Selectox process catalytically converts H_2S to elemental sulfur without the need for reaction furnace. It is advantageous for recovering high-purity sulfur from acid gas feed streams containing from less than 5% to about 20% H_2S [9]. As mentioned above, the process has even been used for natural gas processing on an offshore platform. [2] The feed gases are passed over a fixed bed of a proprietary Selectox catalyst at 160–370 °C. It achieves up to 97% sulfur recovery at the start of a catalyst cycle, and is often used for tail gas treatment in conjunction with the BSR process discussed above for overall recovery >99.9%.

In the LO-CAT® process, offered by Merichem, H_2S is air-oxidized to sulfur in an aqueous solution containing a dual chelated iron catalyst under mildly caustic conditions. The system is designed with separate absorber and oxidizer vessels. The absorber removes the H_2S from sour gas, converting it to elemental sulfur. The oxidizer regenerates the catalyst which is pumped between the vessels. It achieves sulfur removal efficiencies up to >99.9% in many different applications and industiies [10].

LO-CAT can be applied to all types of gas streams, including air, natural gas, CO_2, amine acid gas, biogas, landfill gas, refinery fuel gas, etc. The liquid catalyst adapts to the variation in flow and concentration easily. Units require minimal operator attention and are well-suited to removing H_2S from sour gas streams.

14.3 Catalytic NO_x Removal

Selective catalytic reduction (SCR) [11] removes nitrogen oxides (NO_x) by reaction with ammonia to produce nitrogen via the following main reactions.

$$4NO + 4NH_3 + O_2 \rightarrow 4N_2 + 6H_2O$$
$$2NO_2 + 4NH_3 + O_2 \rightarrow 3N_2 + 6H_2O$$
$$NO + NO_2 + 2NH_3 \rightarrow 2N_2 + 3H_2O$$

Secondary reactions involve sulfur oxides and ammonium sulfates.

$$2SO_2 + O_2 \rightarrow 2SO_3$$
$$2NH_3 + SO_3 + H_2O \rightarrow (NH_4)_2SO_4$$
$$NH_3 + SO_3 + H_2O \rightarrow NH_4HSO_4$$

In the process, the gas stream containing NO_x, such as flue gas or exhaust gas, is mixed with a gaseous reductant, typically anhydrous ammonia, aqueous ammonia, or urea, and is adsorbed onto a catalyst. SCR catalysts are made from various ceramic carriers, such as titanium oxide, and active catalytic components of oxides of base metals (such as vanadium, molybdenum, and tungsten), zeolites, or various precious metals. Nitrogen oxides are reduced to nitrogen. If urea is used as the reductant, CO_2 is also produced.

14.4   DEMEX Solvent Extraction

Solvent deasphalting is a process of removing resins and asphaltenes from atmospheric and vacuum residues, which was discussed in Sect. 13.8. Deasphalted oil (DAO) from propane de-asphalting has the highest quality but lowest yield, whereas using pentane may double or triple the yield from a heavy feed, but at the expense of including contaminants (metals and carbon residues) that shorten the life of downstream hydroprocessing catalysts. DEMEX solvent extraction is an extension of propane deasphalting using a less selective solvent to recover not only high quality but also high molecular weight aromatics and other processable constituents from the feedstock. It requires less solvent circulation, thus reducing utility cost and unit size significantly. The demetallized oil (DMO) contains low asphaltene and condensed aromatic contents. The DMO is processed in fixed-bed residue desulfurization units. When metal and carbon residue are sufficient low, the DMO can be processed in FCC units. For hydrocrackers, due to its high endpoint, DMO tends to cause rapid deactivation. The DEMEX vacuum residue can be blended into asphalt or combined with vacuum gas oil to give an acceptable feed for subsequent delayed coking or other residue conversion units.

Figure 14.6 shows a UOP DEMEX solvent extraction unit [12]. The high metal vacuum residuum is mixed with a solvent, which can be a low-carbon alkane or low-boiling naphtha, in an extractor. The overhead is sent to a DMO separator after going through a furnace heater to recover solvent. The bottom is sent to the asphalt stripper to drop out high-metal-content asphalt. The top of the asphalt stripper is sent to a DMO stripper to recover low metal DMO, which is combined with the DMO from the DMO separator as the product.

14.5   Alternatives to Petroleum

Conventional crude oil is getting harder to find, despite the fact that estimates of proven reserves increased substantially with the recent advent of "fracking" technology to produce oil and gas from tight formations and non-conventional reservoirs. Fracking oils, oil sand bitumen, and extraheavy crude oils are called unconventional crude oils.

The primary use of oil and gas is energy production. Declining availability of conventional petroleum can be offset to some extent by non-fossil sources. But unconventional fossil fuels are likely to carry most of the load. Liquid fuels will be produced from natural gas, bitumen, kerogen and coal. Some of these contain daunting amounts of contaminants. Compared to conventional oil, converting these materials will be costly and more difficult. But the technology for doing so already exists.

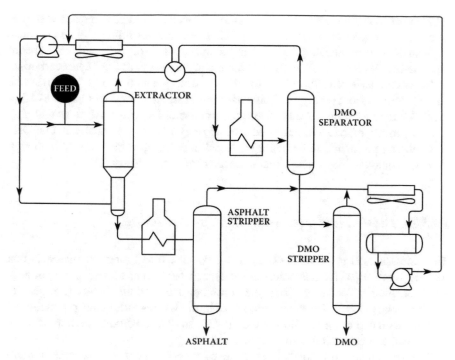

Fig. 14.6 UOP DEMEX solvent extraction unit

14.5.1 Biomass

Biomass is organic matter derived from living or recently living organisms, such as plants, algae, grass, vegetables, animal fats, etc. It can be combusted to produce heat directly or converted into biofuels, most commonly ethanol as a blend component or alternative to gasoline, and as triglycerides as biodiesels. Prior to use in diesel engines, cellulose must be converted into a suitable liquid, and vegetable oil must be stabilized, usually by hydrogenation to remove oxygenates.

In Brazil, ethanol produced from sugar cane replaces a huge percentage of that country's liquid transportation fuel. Growing sugar cane impacts land use, but the lifecycle energy gain is about 8:1.

The United States now requires the blending of ethanol into gasoline. The requirement was established to decrease air pollution and reduce dependence on imported oil. In practice, almost all US ethanol is produced from corn, which is a poor substrate when compared to sugar cane. The estimated energy gain for ethanol from corn ranges is from 1.1 to 1.5. US growers depend heavily on irrigation, which is depleting the Ogallala aquifers and other important sources of ground water. U.S. corn growers consume tremendous amounts of fertilizer, which increases emissions of greenhouse gases, particularly CO_2 and N_2O. Using corn for fuel has increased the price of food nationwide [13–15] Adding 10% ethanol to gasoline decreases the demand for petroleum by less than 10%. However, the gas mileage is less because the heat content of ethanol is lower than the heat content of hydrocarbons. Ethanol increases the RVP of gasoline, thereby increasing emission of light hydrocarbons.

Many countries offer incentives to add oils from plants to diesel fuels. Bio-oils that are derived from pyrolysis of biomass are not suitable for direct use in existing engines due to their high levels of oxygen, immiscibility with fossil fuels and high tendency to polymerize when exposed to air. So they are mixed in small amounts with conventional petroleum to be hydroprocessed in existing refineries. Given enough time, we could develop a new biomass-based chemical industry [16].

14.5.2   Bio-Oil and Petroleum Oil Co-processing

The biomass used to generate biofuels can cover a broad range of materials that include: food crops (1st generation biofuels), non-food crops (energy crops) or non-edible portions of food crops (2nd generation biofuels) and microalgae (3rd generation biofuels) [17, 18]. The 4th generation of biofuels aims to combine production of biofuels with capturing and storing CO_2 (CCS) so that it would be carbon negative rather than carbon neutral with respect to air.

The 1st generation biofuels have achieved a certain degree of success in commercialization, especially under government subsidies or mandates. The bioethanol derived from corn and sugarcane, for example, is unable to replace gasoline completely without major modification to the internal combustion engine, and consequently, it must be blended with gasoline in order to be used within existing engines. Biobutanol, however, can be a good candidate for a gasoline alternative in existing engines.

Many of the energy production and utilization cycles based on lignocellulosic biomass have low or near-zero greenhouse gas (GHG) emissions on a life-cycle basis. A much overlooked concept that should increase the popularity of utilizing these biomass feedstocks is the idea of use within a fully integrated biorefinery, much like what has already been developed within petroleum refining.

One kind of biocrude is fast pyrolysis oil from wood or other biomasses, including municipal wastes that otherwise might go to a sanitary landfill. It has high oxygen content, mostly in the forms of ketones and esters. Intermediates such as alcohols, carboxylic acids or aldehydes may be formed from carbonyl groups upon hydrogenation. Consequently, the carbonyl groups are expected to follow one of the three deoxygenation reactions: hydrodeoxygenation (HDO) of alcohols, decarboxylation of acids and decarbonylaton of aldehydes, to produce pure hydrocarbons. Pyrolysis oil (bio-oil) could serve as a source of certain chemicals.

Biocrude and other biomass oils, such as vegetable oils, can be blended with high acid crude, heavy crudes, or distillation bottoms, to be co-processed. The ideal bio-oil upgrading catalysts need high activity for deoxygenation and must be able to withstand large quantities of coke. The issue of coke can be handled by employing continuous catalyst circulation technology, such as fluid catalytic cracking (FCC) or continuous catalytic reforming (CCR) , but with different catalysts. Furthermore, the applied catalyst must have a high tolerance to water for steady performance during HDO. It is important to note that oxygen removal usually requires hydrogen, which, when produced with steam-methane reforming, generates tremendous amounts of CO_2.

Conventional $NiMo/Al_2O_3$ or $CoMo/Al_2O_3$ hydrotreating catalysts are the most widely used catalyst in the hydroprocessing of biooils. As manufactured, the active metals are oxides—MoO_3 and either NiO or CoO. The oxides are relatively inactive. Prior to use, they are converted to active sulfides—MoS_2 and either NiS or CoS_x,—by reaction with hydrogen and H_2S. With conventional petroleum gas oils, activity retention requires feeds which contain at least 0.05 wt% sulfur. If the feed sulfur is too low, it can be augmented with sulfiding agents such as dimethyldisulfide (DMDS). Too much H_2S inhibits sulfur removal, so in many units excess H_2S is removed with amine treating.

Oxygen-containing feeds, such as those derived from biomass, reduce the performance of conventional hydroprocessing catalysts by slowly but surely converting the active metal sulfides to oxides. Sulfided $NiMo/Al_2O_3$ is more sensitive to this than sulfided $CoMo/Al_2O_3$. Activity maintenance is achieved by blending bio-oils with conventional sulfur-containing petroleum gas oils. Nitrogen has an inhibitive effect on many conventional catalysts, so the low nitrogen content in bio-oil from lignocellulosic biomass is advantageous.

References

1. Crew MA, Shumake BG (2017) Hydrogen production. In: Hsu CS, Robinson PR (eds) Springer handbook of petroleum technology. Springer, New York (2017) (Chapter 24)
2. Bertram RV, Robinson PR (1989) Selectox process for sulfur recovery offshore. In: The 68th annual gas processors association meeting. San Antonio, TX, 14 Mar 1989
3. Robinson PR (2017) Sulfur removal and recovery. In: Hsu CS, Robinson PR (eds) Springer handbook of petroleum technology. Springer, New York (2017) (Chapter 20)
4. Tail Gas Treating: TGT. https://www.chiyoda-corp.com/technology/en/upstream_gasprocessing/tail_gas_treating_tgt.html. Accessed 6 Aug 2014
5. SCOT. http://www.shell.com/business-customers/global-solutions/gas-processing-licensing/gas-processing-technolgies-portfolio/_jcr_content/par/textimage.stream/1444035481012/824ffff53bbcedc4da5daa9094ec8cf1a51451ce5fd2ca8f3021a44b6d2735fb/factsheet-scot-screen.pdf. Accessed 6 Aug 2014
6. Fenton DM, Gowdy HW (1979) The chemistry of the Beavon sulfur removal process. Environ Int 2(3):183–186
7. Stretford process. https://en.wikipedia.org/wiki/Stretford_process
8. LaRue K, Grigson SG, Hudson H (2013) Sulfur plant configuration for weird acid gases. In: Presented at 2013 Laurence Reid gas conditioning conference. The University of Oklahoma, 24–27 Feb 2013. http://www.ortloff.com/files/papers/LRGCC2013.pdf. Accessed 24 Nov 2016
9. http://www.ogj.com/articles/print/volume-99/issue-35/processing/long-term-operating-data-shed-light-on-selectox-process.html Accessed 24 Nov 2016
10. Merichem: LO-CAT® process for cost effective desulfurization of all types of gas streams. http://www.merichem.com/images/casestudies/Desulfurization.pdf. Accessed 6 Aug 2014

11. Wikipedia: selective catalytic reduction. https://en.wikipedia.org/wiki/Selective_catalytic_reduction. Accessed 6 Aug 2014

12. Salazar JR (1986) UOP DEMEX process. In: Handbook of petroleum refining processes. McGraw-Hill, New York, pp 8.61–8.70

13. DeCicco JM, Liu DY, Heo J, Krishnan R, Kurthen A, Wang L (2016) Carbon balance effects of U.S. biofuel production and use, Clim Change 138(3–4):667–680. https://doi.org/10.1007/s10584-016-1764-4. Retrieved 5 Nov 2018

14. Conca J (2014) It's final—corn ethanol is of no use. Forbes, 20 Apr 2014. https://www.forbes.com/sites/jamesconca/2014/04/20/its-final-corn-ethanol-is-of-no-use/#bce8e7767d35. Retrieved 5 Nov 2018

15. Robinson PR (2017) Corn ethanol in gasoline. In: Symposium on biomass to fuels and chemicals: research, innovation, and commercialization. American Chemical Society 254th National Meeting, Washington, DC, 21 Aug 2017

16. Biorefineries and chemical processes. http://www.chemanager-online.com/en/topics/books/biorefineries-and-chemical-processes. Accessed 6 Aug 2014

17. Energy from waste and wood. http://energyfromwasteandwood.weebly.com/generations-of-biofuels.html. Accessed 6 Aug 2014

18. Brodeur G, Ramakrishnan S, Hsu CS (2017) Biomass to liquid (BTL) fuels. In: Hsu CS, Robinson PR (eds) Springer handbook of petroleum technology. Springer, New York (Chapter 38)

Chapter 15
Natural Gas and Petroleum Products

15.1 Natural Gas

Natural gas comes from gas wells, oil wells, and condensate wells. Natural gas at the wellhead contains primarily methane (70–90%) with impurities. Natural gas from oil wells is called associated gas. It can be free gas or dissolved in the oil. Natural gas from a condensate well is produced along with volatile hydrocarbon condensate.

In addition to methane, natural gas mixtures contain ethane, propane, butane and pentanes, water vapor, hydrogen sulfide, mercaptans, carbonyl sulfide, carbon dioxide, ammonia, helium, nitrogen and other components. Natural gas is the major commercial source of helium which is present in natural gas as an inert gas with nitrogen. Helium is thought to form from radioactive decay (α-decay) of uranium and thorium in granitoid rocks of Earth's continental crust. High purity helium is produced by the combination of cryogenic and pressure swing adsorption process. *Natural gas liquids* (NGL) include ethane (35–55%), propane (20–30%), normal butane (10–15%), isobutane (4–8%) and C_5+ natural gasoline (10–15%). They are used in enhanced oil recovery or as raw materials for oil refineries or petrochemical plants, and sources of energy. In the U.S., hydraulic fracturing (fracking) increases the production of natural gas as shale gas.

Natural gas produced at the wellhead, which in most cases contains contaminants and natural gas liquids, must be processed to meet quality specifications before it can be safely delivered to the high-pressure, long-distance pipelines and/or gas tankers that transport the product to the consumers. Natural gas that is not within certain specific gravities, pressures, Btu (heat) content range, dew point, or water content levels will cause operational problems, pipeline deterioration, or even pipeline rupture. Such gas can be especially harmful to equipment in pumping stations. The water content, as defined by water dew point, must be limited to prevent the formation of ice and hydrate in the pipeline. The amounts of entrained hydrocarbons heavier than ethane, as defined by hydrocarbon dew point, must be limited to prevent the accumulation of condensable hydrocarbon liquids to block the pipeline and pumps.

© Springer Nature Switzerland AG 2019   301
C. S. Hsu and P. R. Robinson, *Petroleum Science and Technology*,
https://doi.org/10.1007/978-3-030-16275-7_15

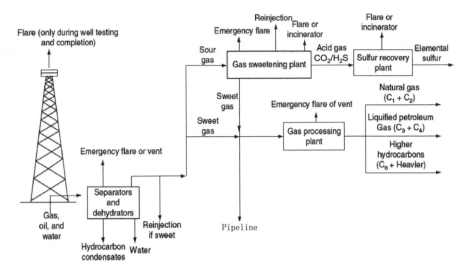

Fig. 15.1 Natural gas from well to the consumer

Figure 15.1 shows the normal procedure to treat natural gas at the well site for subsequent transportation to consumers. The first step is phase separation by gravity to separate water, hydrocarbon condensates, and particulate solids from the gas in a settler. Natural gas pretreatment typically consists of mercury removal, gas sweetening and drying. Mercury is removed by using adsorption processes based on activated carbon or regenerable molecular sieves. Natural gas is dried by absorption in regenerable triethylene glycol (TEG) (glycol dehydration) or molecular sieve adsorbers [1]. It may be necessary to remove H_2S and CO_2 (acid gases) from the natural gas. Sweet gas that contains low concentrations of sulfur compounds can be removed by adsorption along with the removal of water, or directly sent to a gas processing plant. For sour gas that contain high concentrations of sulfur compounds and carbon dioxide, gas sweetening using a solvent, such as amines described in the previous chapter, sulfinol [2] or carbonate washing shown below, is traditionally used. The sweetened gas can then be sent to the gas processing plant or reinjected into the field for enhanced recovery. The sulfur-rich amines solution after sulfur scrubbing is sent to solvent recovering unit to separate the solvent from sulfur, which is sent to a sulfur recovery plant.

$$K_2CO_3 + CO_2 + H_2O \rightarrow KHCO_3$$
$$K_2CO_3 + H_2S \rightarrow KHS + KHCO_3$$

If the natural gas is to be liquefied by cryogenic cooling and stored in the liquid form, carbon dioxide must be removed by amine treatment to prevent the formation of dry ice that would interfere with the refrigeration system and clog the pipeline. Carbon dioxide can be sold as industrial dry ice, and in some locations it is reinjected

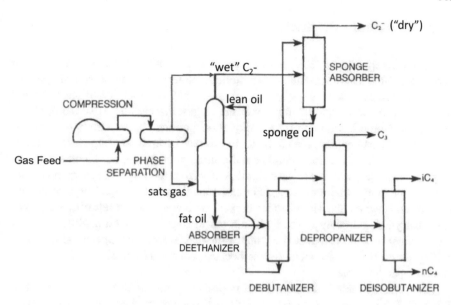

Fig. 15.2 Natural gas processing plant

into the formation. Reinjection brings the added value of reducing releases of CO_2 to the atmosphere; CO_2 reinjection also is known as carbon sequestration.

Natural gas does not contain olefins. Figure 15.2 shows a natural gas processing plant to separate individual light gases [3], which is saturated gas processing plant different than the cracked gas plant for refinery gases. The gas is compressed to approximately 200 psig and fed to an absorber-deethanizer after phase separation, where some ethane and lighter gases (C_2-) evolve. The oil used in the absorber-deethanizer is usually a dehexanized naphtha with an end point of 350–380 °F. The absorber column usually contains 20–24 trays in the top absorption section and 16–20 trays in the stripping section. The lean absorption oil is fed to the top tray to absorb 85–90% of the C_3's and almost all of the C_4's. Significant amounts of C_2- hydrocarbons are vaporized from the lean oil and leave the top of column with the residue gas, from which they are recovered in a sponge absorber. The sponge absorber usually contains 8–12 trays using kerosene or No. 2 fuel oil as the sponge oil, which is derived as a side cut from coker fractionation or catalytic cracking fractionation. The deethanized oil flows to the debutanizer to separate C_3+ from the absorption oil, then to the depropanizer to recover C_3 from C_4's. The C_4 stream is sent to a deisobutanizer to separate isobutane from n-butane. During liquefaction of natural gas, accomplished by reducing its temperature to −260 °F (−160 °C) at atmospheric pressure, ethane can be cryogenically separated from methane or extracted from liquefied natural gas (LNG).

15.2 Petroleum Products

The products produced from various petroleum processes are summarized in Table 15.1. Note that the products from modern refineries come not from single processes but comprise several individual streams, which are blended to meet specific specifications and/or requirements. For example, gasoline can come from straight-run distillation, isomerization, reforming, alkylation, polymerization, catalytic cracking, hydrocracking and coking, blended after hydrotreating to satisfy target octane number. In Table 15.1, the carbon number and boiling point ranges for blend stocks are only for reference, which can vary significantly from refinery to refinery, and from time to time in a given process unit. For example, the yield and quality of reformate from a semi-regen catalytic reformer depend on the age of the reforming catalyst. Also, the yield and quality of straight-run stocks depend on feed quality, which in some refineries can change two or three times per week. The carbon number and boiling point of coke are meaningless because coke is nonvolatile and its carbon number is undefinable.

The consumption of natural gas and petroleum products in the U.S. is listed in Table 15.2. The principal constituents of still gas are methane and ethane, typically used as refinery fuel and petrochemical feedstock. Motor gasoline and diesel fuel account for the majority of consumption. Natural gas liquid, liquefied petroleum gas, ethane/ethylene and propane/propylene are major sources of petrochemical feed-

Table 15.1 Petroleum products

Product	Carbon limit		Lower boiling point		Upper boiling point	
	Lower	Upper	°C	°F	°C	°F
Refinery gas	C_1	C_4	−162	−259	2	35
Liquefied petroleum gas	C_3	C_4	−42	−44	2	35
Naphtha	C_5	C_{17}	40	100	180	350
Gasoline	C_5	C_{12}	40	100	210	400
Kerosene	C_9	C_{16}	150	300	300	550
Jet fuels (icluding avjet)	C_5	C_{18}	40	100	320	600
Diesel fuel	C_{10}	C_{24}	125	250	370	700
Heating oil	C_{14}	$>C_{20}$	250	500	>350	>650
Fuel oil	C_9	$>C_{40}$	150	300	>500	>950
Lubricating oil	C_{14}	$>C_{50}$	250	500	>575	>1050
Wax and Grease	C_{20}	$>C_{50}$	350	575	>575	>1050
Asphalt	$>C_{20}$	$>C_{60}$	>350	>575	>600	>1150
Coke	$>C_{50}$	Unlimited	Nonvolatile	Nonvolatile		

Table 15.2 U.S. consumption of natural gas and petroleum products (in 1000 barrels per day)

	2010	2014
Natural gas liquid	2265	2448
Liquefied petroleum gas	2051	2396
Still gas (principally methane + ethane)	672	693
Ethane/ethylene	880	1048
Propane/propylene	1160	1167
Motor gasoline	8993	8921
Kero-type jet fuel	1462	1470
Distillate fuel (diesel <15 ppm S)	3211	3800
Residual fuel oil	535	257
Lubricants	131	126
Petroleum coke (marketable)	151	112
Petroleum coke (catalyst)	225	233
Asphalt	362	327

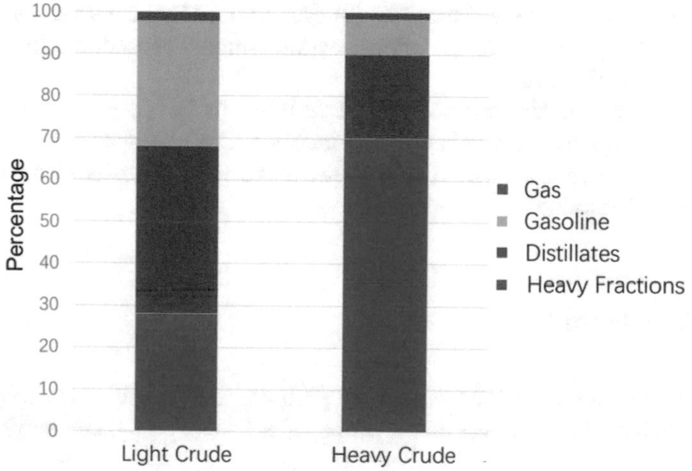

Fig. 15.3 Variation in product distribution

stocks. The consumptions of aviation gasoline, kerosene and wax are less than 20,000 barrels per day, and not listed in Table 15.2.

Figure 15.3 represents the variation of product distribution from a light and a heavy crude. In this figure, gasoline and distillates represent the most valuable products. The heavy oils would need to be converted into these higher-value products, and would increase the costs of refining.

For petroleum products, specifications developed by American Society for Testing Materials (ASTM) International and International Standards Organization (ISO) are

Table 15.3 ASTM specification numbers for hydrocarbon fuels

Product	ASTM specification	Description
Gasoline	D4814	Standard specification for automotive spark-ignition engine fuel
Jet fuel	D1655	Standard specification for aviation turbine fuels
Kerosene	D3699	Standard specification for kerosene
Diesel	D975	Standard specification for diesel fuel oils
Fuel oil	D396	Standard specification for fuel oils

widely used throughout the world. In addition to setting specifications, these institutions develop and publish test methods for analyzing a wide array of materials. They cooperate both with each other and with government regulators. For example, recent low-sulfur gasoline and diesel directives from the U.S. Environmental Protection Agency are incorporated by ASTM into D975 and D4814, respectively. ASTM fuel specifications are listed in Table 15.3.

Other widely used tests and specifications are defined by licensors. For example, UOP's *Laboratory Test Methods* defines several hundred procedures for analyzing catalysts, chemicals and fuels.

Additives are essential components of finished fuels. They increase stability, improve flow properties and enhance performance. Cetane-improvers are routinely added to diesel fuel, and additives that prevent intake-valve deposits are now required in all grades of gasoline in the United States.

15.3   Petroleum Gases

Petroleum gases are gases at ambient temperature and pressure. Gases produced by refineries include methane, ethane, propane, ethylene, propylene, butylenes, and hydrogen.

H_2S, ammonia, and CO_2 also are produced in refineries as by-products. H_2S is collected by adsorption in amine units and converted into elemental sulfur in Claus units equipped with tail-gas recovery. Ammonia, generated from organic nitrogen, ends up in sour water strippers and usually is destroyed in Claus units. In a few refineries, ammonia is collected and sold. CO_2 is a by-product of certain hydrogen generation units, from which it can be collected, purified, and sold as industrial dry ice.

Methane is the main hydrocarbon ingredient of both natural gas and refinery fuel gas. It can be used as a refinery fuel and feedstock for hydrogen production units. The most common of these are steam-methane reformers (SMR units), which make high-purity hydrogen or synthetic gas (syngas), which is a well-defined mixture of hydrogen and CO.

Ethane is used as refinery fuel, as feedstock to SMR units, or as feedstock for production of ethylene, which is an important petrochemical precursor. Propane is used as refinery fuel or as feedstock for propylene manufacturing. Liquefied propane can also be used by other industries and for domestic applications.

Butanes can be blended into gasoline for regulating vapor pressure and promoting better starting during cold weather. In the United States, ethanol must be added to gasoline, and most of the ethanol comes from corn grown in the United States. One of the many environmental problems caused by the ethanol mandate is: ethanol raises the vapor pressure of other gasoline components. To avoid violating vapor pressure regulations, refiners must limit or even eliminate the blending of butane into gasoline. In fact, during the summer, even pentane blending is limited. N-butane can be sold as a component of LPG or serve as a feedstock to produce isobutane in isomerization units. Isobutane is used as a feedstock to alkylation units, where it reacts with C_3–C_5 olefins to produce high-octane isoparaffins (isomerate) for gasoline blending. It can be converted into isobutylene for making polyisobutylenes, methyl tertiary butyl ether (MTBE) and ethyl tertiary butyl ether (ETBE). Before the ethanol mandate, MTBE and ETBE were high-value gasoline blend stocks in the United States. They are still used in other places, especially Europe.

C_2–C_4 olefins are also produced from thermal and catalytic cracking processes. They are feedstocks for poly-olefins, such as polyethylene, polypropylene, poly-isobutylene, etc.

15.3.1 Liquefied Petroleum Gas (LPG)

Liquefied petroleum gas (LPG) is simply propane or a propane-butane mixture obtained from either natural gas liquid or refinery gases. Propylene, butylene and various other hydrocarbons are usually also present in small concentrations if the gas comes from refinery processes, such as coking or catalytic cracking units. LPG is manufactured during the refining of petroleum (crude oil) or extracted from petroleum or natural gas streams as they emerge from the ground. It is used as a fuel in heating appliances, cooking equipment, and vehicles, as well as a feedstock for the petrochemical industry. LPG is stored in salt domes in strategic petroleum reserves.

15.3.2 Liquefied Refinery Gases

Liquefied refinery gases are produced in the refineries from processing crude oils and unfinished oils. They are retained in the liquid state through compression and/or refrigeration. The reported categories are ethane/ethylene, propane/propylene, normal butane/butylene, and isobutane/isobutylene.

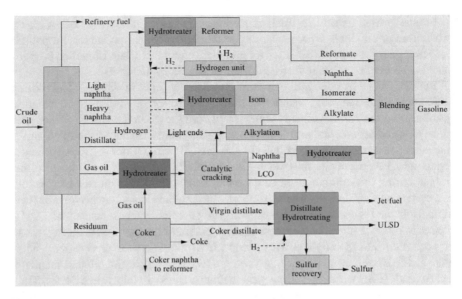

Fig. 15.4 Motor gasoline pool from various refining processes [4]

15.4   Gasoline

Since the birth of modern automobile in 1886, the demand for gasoline to fuel automobiles with internal combustion engines has increased dramatically. The rate of the increase jumped in 1908, when the advent of the Model T made autos affordable by average people. It's estimated the number of automobiles in the world exceeded 1 billion around 2010.

The development of modern refinery processes was driven by rapid growth in demand for transportation fuels. A variety of refining processes have been developed to produce gasoline components, including straight run distillation, isomerization, reforming, alkylation, polymerization, catalytic cracking and hydrocracking, as shown in Fig. 15.4. These components are in a gasoline pool for blending for meeting octane rating requirements and with additives for formulated gasolines. Ninety percent of gasoline produced in the U.S. is used as fuel in automobiles.

15.4.1   Gasoline Engines [5]

The power of a motor vehicle is provided by an internal combustion engine. The essential components of a spark-ignition engine include: gas tank, fuel pump, fuel injection, cylinder, pistons, exhaust valves, spark plugs and catalytic converter. Most common gasoline engine is 4-stroke engine, shown in Fig. 15.5. The 4 strokes are described below:

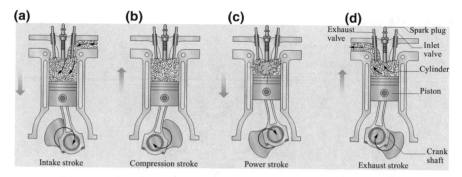

Fig. 15.5 4-Stroke internal combustion engine

1. **Intake** stroke: the piston begins at top dead center (T.D.C.) and ends at bottom dead center (B.D.C.). The intake valve is in the open position while the piston pulls an air-fuel mixture into the cylinder by producing vacuum pressure into the cylinder through its downward motion.
2. **Compression** stroke: the piston begins at B.D.C, or just at the end of the suction stroke, and ends at T.D.C. The piston compresses the air-fuel mixture in preparation for ignition during the power stroke. Both the intake and exhaust valves are closed during this stage.
3. **Power** stroke: The piston is at T.D.C. (the end of the compression stroke) when the compressed air-fuel mixture is ignited by a spark plug, forcefully returning the piston to B.D.C. This stroke produces mechanical work from the engine to turn the crankshaft.
4. **Exhaust** stroke: the piston once again returns to T.D.C. from B.D.C. while the exhaust valve is open. This action expels the spent air-fuel mixture through the exhaust valve.

15.4.1.1 Otto Cycle

The operation of a 4-stroke engine can be expressed by a diagram of pressure versus volume as Otto Cycle, shown in Fig. 15.6. **Intake** (A) stroke is performed by an isobaric (constant pressure) expansion, **Compression** (B) stroke is performed by an adiabatic compression. The combustion upon fuel injection occurs in an isochoric (constant volume) process, which is followed by an adiabatic expansion, characterizing the **Power** (C) stroke. The cycle is complete by an isochoric process and an isobaric compression, characterizing the **Exhaust** (D) stroke.

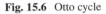

Fig. 15.6 Otto cycle

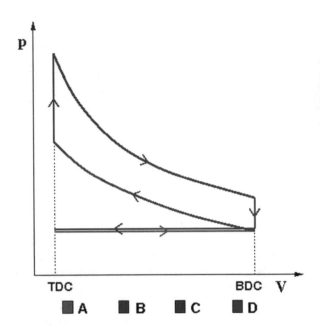

15.4.2   Octane Rating (Octane Number)

In the combustion chamber, vapor heats up during compression. Ideally, the mixture of gasoline vapor and air in a spark-ignition engine is ignited with a spark when the cylinder reaches a predetermined position in the cylinder. However, some compounds can start to ignite before the piston reaches the top end (T.D.C.) near the spark plug. This premature ignition causes engine knock, which pushes piston in wrong direction, reduces the power of the engine, increases engine wear, and can cause serious damage.

Gasoline is a blend of many different components. Different gasoline components have different knock behaviors. The compression ratio (V1/V2), shown in Fig. 15.7, is a factor in knock behavior of a gasoline mixture. Octane rating, or octane number, is an arbitrary unit related to the smallest compression ratio at which an engine starts to knock with a given fuel. It is based on a scale in which the octane number of n-heptane is designated as zero and the octane number of isooctane (2,2,4-trimethylpentane) is 100. When a fuel is tested in a standard single-cylinder engine, mixtures of isooctane and n-heptane of various percentages are used as reference standards for correlating the knocking of the test fuel with the percentage of isooctane in the mixture.

Thus, a gasoline with the same knocking characteristics as a mixture of 94% 2,2,4-trimethylpentane and 6% n-heptane has an octane rating of 94. A rating of 94 does not mean that the gasoline contains just isooctane and n-heptane in these proportions, but that it has the same tendency to knock as this mixture. The blends of known octane ratings can also be used as references as shown in Fig. 15.8. The higher the number, the less likely is a fuel to pre-ignite.

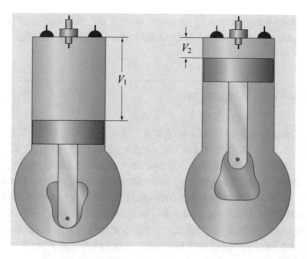

Fig. 15.7 Compression ratio V1/V2

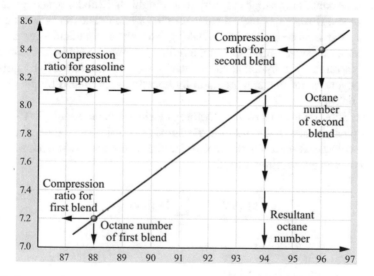

Fig. 15.8 Octane rating (octane number) is determined by comparing compression ratio causing knock using reference blends

ASTM D2699 and ASTM D2700 describe methods for measuring research octane number (RON) at 600 rpm, which is most relevant to low speed city driving with load conditions; and motor octane number (MON) at 900 rpm, which is most relevant to high-speed highway driving conditions.

In North America, the posted octane of gasoline is the arithmetic average of RON and MON: $(R + M)/2$. This is the number displayed on pumps at filling stations. Typical grades are "regular" with a posted octane of 87, "mid-grade or medium" with a posted octane of 89, and "premium" with a posted octane of 91–93. In some locales, customers can dial in any octane they want between 87 and 93.

15.4.3 Gasoline Components

The typical boiling range of gasoline is from 100 °F (~40 °C) to 400 °F (~200 °C). The components have carbon numbers ranging from C_5 to C_{12}. Gasoline is typically blended from several refinery streams including straight-run light naphtha and products from several upgrading processes – isomerization (isomerate), alkylation (alkylate), catalytic reforming (reformate), polymerization (poly gasoline or polymerate), catalytic cracking (cat gasoline), hydrocracking (hydrocrackate) and coking. At some point, every component of gasoline goes through a hydrotreater or hydrocracker, and heavy naphtha must go through a catalytic reformer. Other streams undergo other treatments before they can be blended.

A gasoline containing a high proportion of straight chain alkanes has a greater tendency to knock. However, branched-chain alkanes, cycloalkanes and aromatic hydrocarbons are much more resistant to knocking. Straight-chain alkanes are converted into isoalkanes in several processes in the refinery, as mentioned before. The octane rating of gasolines usually available for cars contain a mixture of straight-chain, branched, cyclic paraffinic and aromatic hydrocarbons, produced by the processes described below.

Another important property of gasoline is Reid vapor pressure (RVP) which is the vapor pressure of a motor gasoline measured at 100 °F in a volume of air 4 times the liquid volume. It indicates the ease of starting and vapor-lock tendency as well as explosion and evaporation hazards.

$$M_t(RVP) = \sum_{i=1}^{n} M_i(RVP)_i$$

where

M_t Total moles of blended product,
$(RVP)_t$ Specification RVP for product, psi
M_i Moles of component i
$(RVP)_i$ RVP of component i, psi or kPa.

In the mid- to late-20th century, making gasoline was a relatively simple task. If a mixture of components met specifications for volatility and octane, it could be shipped to retail outlets and sold as-was. If the octane was low, the problem could be fixed by adding a small amount of tetraethyl lead (TEL). Butanes could be added or left out as needed to adjust volatility, especially during winter time or cold weather. Now, due to environmental regulations on hydrocarbon emissions, refiners must also meet restrictions on RVP, olefins, sulfur content, and oxygen content. The new RVP

Table 15.4 Properties of some gasoline blending components

Component	Density (kg/m³)	RVP (psi)	Boiling range		RON	MON
			°C	°F		
Butanes	0.575	51.6	−12 to −0.5	10–31	93	92
Straight-run gasoline	0.64	11.2	27–80	81–176	78	76
Isomerate	0.65	13.5	34–90	93–194	83	81
Reformate	0.815	3.2	78–197	172–387	100	88
Alkylate	0.705	3.6	39–195	102–383	97	96
Light FCC gasoline	0.66	4.4	25–89	77–192	95	82
Heavy FCC gasoline	0.76	1.4	43–185	109–365	92	77
MTBE	0.746	7.4	48–62	118–144	115	97
Ethanol[a]	0.79	2	78	173	116	112

[a]Data from http://www.txideafarm.com/ethanol_fuel_properties_and_data.pdf

restriction greatly decreases the amount of butanes that can be used when ethanol is present. In some locales during some seasons, not even pentanes can be blended into gasoline without excessive RVP. As a result of these restrictions, especially the limit on RVP, the air is much more breathable in large American and European cities.

As mentioned above, several refinery streams have the right vapor pressure, boiling range, sulfur content and octane to end up in the gasoline pool. Table 15.4 shows properties for blend stocks from which gasoline might be made. Ethanol has very low RVP, compared to gasoline components. However, by blending 10% ethanol with gasoline, it raises the RVP of the hydrocarbon component due to the phobic nature of gasoline to ethanol [6]. RVP restrictions range from 7.8 to 9.0 psi. Depending on location, there can be a 1 psi allowance when the ethanol content is 9–10 vol% [7].

Gasoline used for piston-engine powered aircraft is called aviation gasoline, or avgas, which has high octane number and low flash point to improve ignition characteristics on propeller aircrafts. As mentioned in the jet-fuel section, avgas can mean either Jet B or JP-4.

15.4.4 Gasoline Additives

A formulated (or finished) gasoline contains base fuel and small amounts of additives. Typical additives include metal deactivators, corrosion inhibitors, antioxidants, detergents, demulsifiers, anti-icing additives and oxygenates. Table 15.5 lists the additives used to prepare finished gasoline. Additive packages vary from season-to-season, region-to-region, and retailer-to-retailer. After-market additives contain

Table 15.5 Additives used in gasoline

Additive type	Function
Oxygenates	Decrease emissions
Anti-oxidants	Minimize oxidation and gum formation during storage
Detergents	Clean fuel injector and minimize carbon deposit
Metal passivators	Deactivate trace metals that can accelerate oxidation
Corrosion inhibitors	Minimize rust and corrosion throughout the gasoline supply chain
Demulsifiers	Prevent the formation of stable emulsion
Anti-icing additives	Minimize ice in carburetors during cold weather
IVD control (detergent)	Control deposition of carbon on intake valves
CCD control	Control deposition of carbon in combustion chambers

similar types of ingredients, and usually are more concentrated. They are packaged to be added by consumers to their own vehicles.

Metal deactivators are compounds that sequester (deactivate) metal salts that otherwise accelerate the formation of gummy residues during storage. The formation of these gums is accelerated by metal salts, such as copper salt. These metal impurities might arise from the engine itself or as contaminants in the fuel. Antioxidants, such as phenylenediamines and other amines in 5–100 ppm, are used to prevent the degradation of gasoline. Detergents are used to reduce internal engine carbon buildups, improve combustion, and to allow easier starting in cold climates. Typical detergents include alkylamines and alkyl phosphates at the level of 50–100 ppm.

15.4.5 Reformulated Gasoline (RFG) in the United States

Starting in the 1920s, tetraethyl lead (TEL) was added to gasoline in order to raise engine compression as an octane booster, which improved the fuel economy. However, TEL had damaging effects on catalytic converters and was the main cause of spark plug fouling. There was also a concern about the lead in air which could cause brain damage especially for the younger children to touch surfaces contaminated by lead. Therefore, TEL needed to be phased out, so refiners had to find other ways to provide octane for gasolines. Gasoline blending became more complex in the 1970s.

In the 1990, the Clean Air Act was amended by Congress, requiring the phase-out of TEL. It empowered the Environmental Protection Agency (EPA) to impose emissions limits on automobiles. The reformulated gasoline (RFG) program was mandated. RFG is gasoline blended to burn more cleanly than conventional gasoline and to reduce smog-forming and toxic pollutants in the air, with sufficient octane ratings. Smog or ground level ozone threatens the health and is particularly dangerous to children and individuals with respiratory problems.

RFG was implemented in several phases. The Phase I program started in 1995 and mandated RFG for 10 large metropolitan areas. Several other cities and four entire States joined the program voluntarily. Initially, the oxygen for RFG could be supplied as ethanol or C_5–C_7 ethers. The ethers have excellent blending octanes and low vapor pressures. But due to leaks from filling station storage tanks, methyl-*t*-butyl ether (MTBE) was detected in ground water samples in New York City, Lake Tahoe, and Santa Monica, California. In 1999, the Governor of California issued an executive order requiring the phase-out of MTBE as a gasoline component. That same year, the California Air Resources Board (CARB) adopted California Phase 3 RFG standards, which took effect in stages starting in 2002. The standards include a ban on MTBE and a tighter cap on sulfur content—15 wppm maximum.

Now, ethanol is the only oxygenate allowed to be blended into RFG. Legally, ethanol must be denatured to avoid being classed (and taxed) as a liquor. The most common recognized denaturant is gasoline, at 2–5% by volume. At this point, Almost all of the gasoline in the US is reformulated. However, the blending ethanol into gasoline reduces the total heat content of the RFG respective to an equal volume of conventional gasoline, thus, reduces the fuel efficiency. Butanols, particularly isobutanol and biobutanol (developed by BP and DuPont), have higher heating value and octane ratings. But, only ethanol is used in the U.S. by mandate.

Tier 1 reformulated gasoline regulations required a minimum amount (2%) of chemically bound oxygen, imposed upper limits on benzene and Reid Vapor Pressure (RVP), and ordered a 15% reduction in volatile organic compounds (VOC) and air toxics. VOC react with atmospheric NOx to produce ground-level ozone. Air toxics include 1,3-butadiene, acetaldehyde, benzene, and formaldehyde.

The regulations for Tier 2, which took force in January 2000, were based on the EPA Complex Model, which estimates exhaust emissions for a region based on geography, time of year, mix of vehicle types, and—most important to refiners—fuel properties. As of 2006, the limit on sulfur in the gasoline produced by most refineries in the U.S. was 30 wppm, a 90% reduction from 300 wppm.

In the United States, Tier 3 vehicle emission and fuel standards lowered the allowed sulfur content of gasoline from 30 wppm to 10 wppm, beginning in 2017. The present limit on gasoline sulfur is 10 wppm.

The RFG program, combined with other industrial and transportation controls aimed at smog reduction, is contributing to the long-term downward trend in U.S. smog levels. About 75 million people breathe cleaner air because of RFG [8].

15.4.6 Gasoline (Petrol) in the European Union

Transportation fuel specifications in the European Union are described by EN 228, "Specification for unleaded petrol (gasoline) for motor vehicles." The present version is sometimes called Euro 4 gasoline. Key specifications include the following:

Octane   Minimum RON and MON are 95 and 85, respectively.

Volatility	Minimum percent vaporized at 100 and 150 °C are 46 and 75, respectively.
Hydrocarbon types	The maxima for olefins/aromatics/benzene are 18/35/1 vol%, respectively.
Oxygen	The maximum oxygen content is 3.7% m/m. Oxygenates can include different amounts of methanol, ethanol (with stabilizing agents), isopropyl alcohol, butyl alcohols, ethers such at methyl-t-butylether (MTBE), and other mono alcohols.
Sulfur	10 ppmw.

15.5   Naphtha

Naphtha boils in the same boiling range as gasoline, between 30 and 200 °C, having a similar carbon number range between 5 and 12. Light naphtha is the fraction boiling between 30 and 90 °C and consists of molecules with 5–6 carbon atoms. Some light naphthas, such as those from hydrocracking units, can be used as-is for gasoline blending. Others are sent to isomerization units. Heavy naphthas boil between 80 and 205 °C and consist of molecules with 6–12 carbons. Most heavy naphthas are feeds for catalytic reforming.

Naphtha boiling range products, including reformates, isomerates, and alkylates, are used in high octane gasoline. Straight-run naphtha is used as a diluent in the bitumen mining industry, as feedstock for producing olefins via steam cracking, and as solvents for paints (as diluents), dry-cleaning, cutback asphalt, industrial extraction processes, and in the rubber industry.

15.6   Kerosene and Jet Fuel (Turbine Fuel)

Kerosene, jet fuel, and turbine fuel have similar boiling ranges. The key product properties are flash point, freezing point, sulfur content, and smoke point. The flash point is the lowest temperature at which a liquid gives off enough vapor to ignite when an ignition source is present. The freezing point is especially important for jet aircraft, which fly at high altitudes where the outside temperature is very low. Sulfur content is a measure of corrosiveness. The measurement of smoke point goes back to the days when the primary use for kerosene was to fuel lamps. To get more light from a kerosene lamp, you could turn a little knob to adjust the height of the wick. But if the flame got too high, it gave off smoke. Even today, per ASTM D1322, smoke point is the maximum height of smokeless flame that can be achieved with calibrated wick-fed lamp, using a wick "of woven solid circular cotton of ordinary quality." The smoke point of a test fuel is compared to reference blends. A standard 40%/60% (volume/volume) mixture of toluene with 2,2,4-trimethylpentane has a

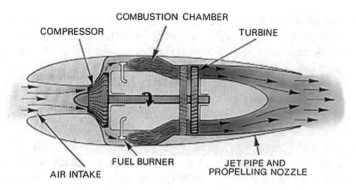

Fig. 15.9 Jet engine [9]

smoke point of 14.7, while pure 2,2,4-trimethylpentane has a smoke point of 42.8. Clearly, isoparaffins have better smoke points than aromatics.

Kerosene boils between 150 and 275 °C (or 350 and 550 °F), with a typical carbon number range between 6 and 16. Prior to the invention of automobiles, kerosene for lighting was the most marketable product of petroleum. Today, the main uses of kerosene include burning in lamps and domestic heaters or furnaces, as a fuel or fuel component for jet engines or rockets, and as a solvent for greases and insecticides.

A jet engine, shown in Fig. 15.9, sucks air in at the front with a fan. The pressure of the air is raised by a compressor, which is powered by a turbine connected to the shaft, to reach a high temperature for fuel to burn. The air/fuel mixture is introduced into fuel burner to combust, and the exhaust from the combustion chamber pushes through the fans of the turbine and exits through the nozzle at the back of the engine. As the jets of gas shoot backward, the engine and the aircraft are thrust forward (per Newton's third law of motion). Turbine engines can operate with a wide range of fuels. Those with higher flash points are less flammable and safer to transport and handle.

The primary sources of jet fuels are hydrotreated straight-run kerosene from an atmospheric distillation unit and hydrocracker products with the right volatility. Jet fuels are used in military and civil jets, with two groups of grading. Jet fuels have two types: naphtha and kerosene jet fuels. Naphtha-type jet fuels are mainly for military usage. The military grades are JP followed by a number, where JP stands for jet propellant. Civil grades start with "Jet" followed by a letter.

In military jet grades [10], JP-1 was specified in 1944. It was pure kerosene, having a freezing point of −60 °C. JP-2 and JP-3 were developed during World War II; they are obsolete today. JP-2 had higher freezing point and JP-3 had higher volatility than JP-1. JP-4 is a wide-cut fuel with a carbon number range of C_4 to C_{16}, for broader availability. JP-5 is specially blended kerosene. JP-6 is a higher cut than JP-4 with fewer impurities. JP-7 is used in supersonic jets requiring higher flash point. JP-8 is kerosene modeled on the Jet A-1 fuels used in civilian aircraft.

For civilian jet fuels [11], Jet A and Jet A-1 are kerosene-type jet fuels. The primary physical difference between the two is freeze point. Jet A specification fuel has been used in the U.S. since the 1950s. Today, the Jet A and Jet A-1 specifications are the same as those published by the International Air Transport Association (IATA). The freezing point of Jet A is <-40 °C versus Jet A-1 at <-47 °C. Both Jet A and Jet A-1 have a flash points higher than 38 °C (100 °F) with autoignition temperatures of 210 °C (410 °F). Kerosene-type jet fuel has a carbon number distribution between 8 and 16, similar to JP-8.

Jet B, also called aviation gasoline, is a lower-boiling fuel used for its enhanced cold-weather performance. Naphtha-type jet fuel, sometimes referred to as "wide-cut" jet fuel, has a carbon number distribution between 5 and 15, similar to JP-4.

The boiling range specifications for commonly used jet fuels are shown in Table 15.6, along with other specifications.

Jet fuel additives include antioxidants, which are usually based on alkylated phenols to prevent gumming; antistatic agents such as dinonylnaphthylsulfonic acid to dissipate static electricity and prevent sparking; corrosion inhibitors; fuel system de-icing agents; biocides to remediate microbial growth in the fuel system; and metal deactivators, such as N,N'-disalicylidine-1,2-propane diamine (MDA), to remediate deleterious effects of trace metals, such as copper even at part-per-billion can catalyze fuel oxidation, on the thermal stability of the fuel.

Table 15.6 Jet fuel Specifications [12]

Property	Jet A	Jet A-1	Jet B	JP-4	JP-5	JP-8
Distillation, D-86, °F, max.						
IBP	342	351	140	140	338	351
10% recovered	400	400	–	–	400	400
20% recovered	–	–	290	290	–	–
50% recovered	450	450	370	370	450	450
90% recovered	–	–	470	470	–	–
FBP	572	572	–	–	572	572
Aromatics, vol%, typical	17.5	18.5	13.2	13.2	19.3	18.5
Olefins, vol%, typical	1.2	0.8	1.5	1.0	0.8	0.8
Naphthalenes, vol%, typical	1.99	1.13	1.20	0.90	1.60	1.13
Sulfur, wt%, maximum	0.3	0.3	0.3	0.3	0.4	0.3
Smoke point, typical, mm	22.5	24.5	25.7	25.7	20.9	24.5
Smoke point. minimum, mm	18	18	18	20	19	19
Flash point, °F, typical	102–148	128–146	Subzero	Subzero	140–158	128–146
Freezing point, °F, typical	−51	−59	<−76	<−80	−56	−59

15.7 Diesel

Diesel fuels are used in diesel or compression ignition engines. Adiabatic compression of a fuel/air mixture provides enough heat for ignition; no spark is needed. Compared to spark ignition engines, a diesel engine is more cost effective because of its greater energy efficiency, higher power output, and better fuel economy under all loads. With previous fuels, diesel engines were noisier and emissions of particulates and nitrogen oxides (NO_x) were considerably higher.

Before 1993, when the United States Environmental Protection Agency (EPA) began regulating diesel fuel, on-road diesel contained as much as 5000 parts per million (ppm) of sulfur. Long-haul truck companies disposed of used crank-case oil, which contained sludge, phosphorous-based additives, and other bad actors, by blending it into diesel at their private terminals. The first step toward cleaner diesel set a cap of 500 ppm sulfur for on-road diesel. Starting in 2006, EPA began to phase-in regulations to lower the allowed sulfur to 15 ppmw. This fuel became known as ultra-low-sulfur diesel (ULSD). At about the same time, similar regulations were imposed in the UK, Europe, Japan, and other developed countries. Now in Europe and the United States, the limit on sulfur content is 10 ppmw. The much stricter diesel fuel standards had a huge impact on cleaning up diesel exhaust. One reason is: sulfates formed by the combustion of organic sulfur compounds comprised a significant proportion of particulates. Another reason: lower sulfur emissions allowed the practical use of particulate traps and catalytic converters in diesel vehicles.

There are three categories of diesel fuels or oils in use: (1) land transportation diesel fuels (officially called 'onroad diesel') used in trucks, buses, trains or other land transportation vehicles that require high variation of speed and load; (2) marine diesel fuels oils, used in ships that have variable speed but relatively high and uniform load; and (3) plant or industrial diesel fuels, used in electric power generation plants that have low or medium speed with heavy load [13]. Onroad diesel fuels used to be known as No.1 diesel (super diesel) and No. 2 diesel (containing cracked oils with a wider boiling range).

Modern automotive diesel specifications resemble those for what used to be called No. 1 diesel. Now, the European Committee for Standardization defines Euro V diesel in European Standard DIN EN 590. The 2017 specifications for Euro V diesel are presented below [14, 15]. Note that the sulfur content must be <10 wppm, the density must be <845 kg/m^3, and the cetane number must be >51. In China, diesel quality is governed by China's State Council. China V, which is similar to Euro V, is mandated for onroad use in major urban areas, must contain <10 wppm sulfur. China IV, which is allowed in other areas, must contain <50 wppm sulfur [16].

Euro V diesel—EN590

Property	Units	low limit	Upper limit	Test-method
Cetane index		46	–	ISO 4264
Cetane number		51	–	ISO 5165
Density at 15 °C	kg/m^3	820	845	ISO 3675, 12185
Polycyclic aromatic hydrocarbons	% (m/m)	–	11	ISO 12916
Sulphur content	mg/kg	–	10	ISO 20846, 20847, 20884
Flash point	°C	55	–	ISO 20846, 20884
Carbon residue (on 90–100% fraction)	% m/m	–	0.3	ISO 2719
Ash content	% (m/m)	–	0.01	ISO 10370
Water content	mg/kg	–	200	ISO 6245
Total contamination	mg/kg	–	24	ISO 12937
Copper strip corrosion (3 h at 50 °C)	Rating	Class 1	Class 1	ISO 12662
Oxidation stability	g/m^3	–	25	ISO 2160
Lubricity. Wear scar diameter at 60 °C	μm	–	460	ISO 12205
Viscosity at 40 °C	mm^2/s	2	4.5	ISO 12156-1
Distillation recovered at 250 °C. 350 °C	% V/V	85	<65	ISO 3104
95% (V/V) recovered at	°C	–	360	ISO 3405
Fatty acid methyl ester content	% (V/V)	–	7	ISO 14078

Truckers use higher density diesel, which used to be called 'No. 2 diesel,' to carry heavy loads for long distances at sustained speeds because it's less volatile than automotive diesel and provides greater fuel economy.

Full-range gas oils boil between 150 and 400 °C (or 335–750 °F), and include molecules with a carbon number range typically between 8 and 21. The boiling range of automotive diesel is narrower. The initial boiling point is set by the flash point, which must be greater than 55 °C. The flash point corresponds to a TBP initial boiling point of about 180 °C (356 °F). The upper distillation limit is determined by the ASTM D86 95% boiling, for which the maximum is 360 °C (680 °F). This corresponds to a TBP final boiling point of 370–380 °C, depending on the nature of the blend components of the fuel.

The most common type of diesel fuel comes from petroleum middle distillate, but alternatives that are not derived from petroleum, such as biodiesel, biomass-to-liquid (BTL) or gas-to-liquid (GTL) diesel, are increasingly being developed and becoming more popular in North America.

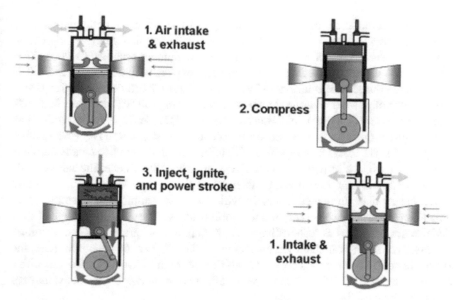

Fig. 15.10 2-Stokes diesel engine [17]

15.7.1 Diesel Engines

Diesel engines use compression heat to ignite a fuel/air mixture in the combustion chamber. A typical 2-stroke diesel engine is shown in Fig. 15.10. At bottom dead center (B.D.C.) the air enters the combustion center above the piston while exhaust moves out of the chamber through open valves. With the exhaust valves closed, the air is compressed to the lowest volume and the highest temperature at top dead center (T.D.C.). This is the moment the fuel is injected and ignited to produce power. The cycle is repeated at the B.D.C.

15.7.2 Cetane Number and Cetane Index

Just as gasoline is rated by its octane number, the ignition quality (knocking tendency) of diesel fuels is measured on a single-cylinder rating engine by matching fuel performance with standard blends. In the historical standard (ASTM D613-10), blends of n-hexadecane with α-methylnaphthalene, where the cetane number is defined as 100 for n-hexadecane (cetane) and 0 for α-methylnaphthalene, that provides the specified standard of 13° (crankshaft angle) ignition delay at the identical compression ratio to that of the fuel sample. For example, a diesel fuel with cetane number of 55 matches the performance of a blend of 55% of n-hexadecane and 45% of α-methylnaphthalene in the cetane engine.

Recently, a modified method has been approved, in which α-methylnaphthalene is replaced by isocetane (2,2,4,4,6,8,8-heptamethylnonane). As before, the cetane number for cetane is defined as 100, but the cetane number assigned to isocetane is 15. Isocetane is preferred because it is stable, less expensive and safer to handle. The cetane is measured in special ASTM variable compression ratio test engine that is closely controlled with regard to temperatures (coolant 100 °C, intake air 65.6 °C), injection pressure (1500psi), injection timing 13° BTDC, and speed (900 rpm) [18]. The compression ratio is adjusted until combustion occurs at TDC (the ignition delay is 13°). The test is then repeated with reference fuels with five cetane numbers difference, until two of them have compression ratios that bracket the sample. The cetane number is then determined by interpolation.

Cetane number correlates inversely with the octane number of gasoline as an indication of ease of self-ignition and ignition delay. The octane number of normal alkanes decreases as carbon chain length increases, whereas the cetane number increases as the carbon chain length increases. The higher the cetane number, the shorter the delay between injection and ignition. Most diesel vehicles use fuel with a cetane rating of 40–55. One of the obvious effects of running on low cetane number fuel is the increase in engine noise.

The molecules with the highest cetane number are the straight chain normal paraffins. The molecules with lowest cetane numbers are those having fewer methylene groups. The presence of double bonds in molecules lowers the cetane number. Hence, the general trend of cetane number among the molecular types is n-paraffins > isoparaffins > cycloparaffins > aromatics.

Typical cetane number ranges of refinery process streams are: virgin distillate streams—light (jet fuel), 35–46, mid (diesel), 35–60, heavy (light virgin gas oil), 46–56; light catalytic cycle oil (LCCO), 14; heavy coker naphtha, 37; light coker gas oil, 40–45; and heavy hydrocracker bottoms, 48–60. Hence, these streams can be blended into diesel fuel pools.

Another measure is Cetane Index (CI), which is a number calculated from the average boiling point and gravity of diesel fuels as $CI = -120.34 + 0.016 \ G^2 + 0.192 \ G \log M + 65.01 \ (\log M)^2 - 0.0001809 \ M^2$, where G = API gravity at 60°F and M = D86 temperature at 50% volume, in °F. Cetane Index calculations and Cetane Number test results are similar, but Cetane Index does not require an expensive engine tests using a large volume of fuel sample over a rather long period of time. American Society for Testing Materials (ASTM) has developed two cetane indices: ASTM D 976 (IP 264) calculated from API gravity and ASTM D 86 mid boiling point (T50), and ASTM D 4737 (IP 380) using density, 10% (T10), 50% (T50) and 90% (T90) recovery temperatures of the fuel as variables.

For Euro V diesel, the minimum Cetane Index is 46 and the minimum Cetane Number is 51. The cetane number can be as high as 60 in premium diesel fuel. Due to its higher heat content, the fuel economy of diesel is better than gasoline.

Railroad and marine diesel fuels can have higher boiling ranges up to 750 °F and lower cetane number in the 30s, as heating oil.

15.7.3 Other Diesel Properties

Other important diesel-fuel properties include flash point, cloud point, pour point, kinematic viscosity, lubricity—and of course sulfur content. Cloud point and pour point indicate the temperatures at which the fuel tends to thicken and then gel in cold weather. In a diesel engine, viscosity not only measures the tendency of a fluid to flow but also indicates how well a fuel atomizes in spray injectors. Lubricity measures the fuel's quality as a lubricant for the fuel system, where it reduces friction between solid surfaces (piston and cylinder) in relative motion. It indicates how the engine will perform when loaded. Table 15.7 lists cetane numbers for selected pure compounds. As with octane, blended cetane numbers can differ significantly from those for pure compounds.

The U.S. EPA now requires that all nonroad, locomotive, and marine (NRLM) diesel fuel must be ULSD, and that all nonroad, locomotive, and marine (NRLM) engines and equipment must use ULSD (with some exceptions for older locomotive and marine engines) [19].

Table 15.7 Cetane numbers for selected pure compounds

Compound	Type	Carbons	Formula	Cetane no.
n-Decane	Paraffin	10	$C_{10}H_{22}$	76
Decalin	Naphthene	10	$C_{10}H_{18}$	48
α-Methylnaphthalene	Aromatic	11	$C_{11}H_{10}$	0[a]
n-Pentylbenzene	Aromatic	11	$C_{11}H_{16}$	8
3-Ethyldecane	Paraffin (iso)	12	$C_{12}H_{26}$	48
4,5-Diethyloctane	Paraffin (iso)	12	$C_{12}H_{26}$	20
3-Cyclohexylhexane	Naphthene	12	$C_{12}H_{24}$	36
Biphenyl	Aromatic	12	$C_{12}H_{10}$	21
α-Butylnaphthalene	Aromatic	14	$C_{14}H_{16}$	6
n-Pentadecane	Paraffin	15	$C_{15}H_{32}$	95
n-Nonylbenzene	Aromatic	15	$C_{15}H_{24}$	50
n-Hexadecane (cetane)	Paraffin	16	$C_{16}H_{34}$	100[a]
2-Methyl-3-cyclohexylnonane	Naphthene	16	$C_{16}H_{34}$	70
Heptamethylnonane	Paraffin (iso)	16	$C_{16}H_{34}$	15[a]
8-Propylpentadecane	Paraffin (iso)	18	$C_{18}H_{38}$	48
7,8-Diethyltetradecane	Paraffin (iso)	18	$C_{18}H_{38}$	67
2-Octylnaphthalene	Aromatic	18	$C_{18}H_{24}$	18
n-Eicosane	Paraffin	20	$C_{20}H_{42}$	110
9,10-Dimethyloctane	Paraffin (iso)	20	$C_{20}H_{42}$	59
2-Cyclohexyltetradecane	Naphthene	20	$C_{20}H_{40}$	57

[a]Used as standards for ASTM D976

Before 2020, marine diesel oil could contain some heavy fuel oil and even waste products such as used motor oil. But as of 2020, MARPOL regulations will require sea-going marine diesel to contain less than 0.5 wt% sulfur [20].

15.7.4   Diesel Additives

Most engines show an increase in ignition delay when the cetane number is decreased from around 50–40. Adding 0.5 vol% of cetane improvers, such as alkyl nitrates, primary amyl nitrates, nitrites, or peroxides, will increase the cetane number by 10 units. The viscosity of the fuel is important because many injection systems rely on the lubricity of the fuel for lubrication. However, fuel lubricity will become poor after usage, so polymeric lubricity agent is needed to maintain the viscosity. Cold flow improvers, or flow-enhancing additives, provide important cold weather properties of the desirable diesel fraction alkanes for cetane have melting points above 0 °C.

Diesel additives may contain cetane number improver, a lubricity agent, detergents, dispersants, metal deactivators, and more. Table 15.8 lists the common additives used in diesel fuel and the reasons they are used.

Table 15.8 Diesel fuel additives

Additive type	Function
Anti-oxidants	Minimize oxidation and gum formation during storage
Cetane improver	Increase cetane number
Dispersants	Improve behavior in fuel injectors for cluster/aggregate formation
Anti-icing additives	Minimize ice formation during cold weather
Lubricity agent	Compensate for poor lubricity of severely hydrogenated diesel fuels
Detergents	Control deposition of carbon in the engine and clean fuel injector
Injector cleanliness agents	Clean deposits from fuel and lubricant
Metal passivators	Deactivate trace metals (e.g., Cu, Fe) that can accelerate oxidation
Smoke suppressants	Reduce black smoke from incomplete combustion
Corrosion inhibitors	Minimize rust throughout the diesel fuel supply chain
Cold-flow improver	Improve flow characteristics in cold weather
Demulsifiers	Allow fuel and water to separate
Stabilizers	React with weakly acidic components
Biocides	Kill bacteria in fuel system that cause clogging

15.8 Heating Oils, Fuel Oils and Marine Diesel Oils

Both heating oil and fuel oil are liquid petroleum products used as fuels for furnaces or boilers to heat buildings, or used in generators to produce power. Larger normal alkanes in heavier distillates burn with a less smoky flame and have higher flash points than gasoline and kerosene, making them desirable for home heating fuels.

No. 1 fuel oil is a volatile distillate similar to kerosene, but with higher pour points and end points. It is intended for ease of vaporization in pot-type burners. It has a carbon number range of 9–16. No. 2 fuel oil is similar to No. 1 but contains cracked stock, having a carbon number range of 10–20. No. 3 fuel oil is for burners requiring low viscosity heating oil, which is described with No. 2 in specifications. No. 4 fuel oil is usually a light residual oil used in a furnace that can atomize the oil and is not equipped with a preheater. It has a carbon number range of 12–70. No. 5 fuel oil has higher viscosities than No. 4, requires preheating to 170–220 °F for atomizing and handling, and has a carbon number range of 12–70. It is also known as Bunker B oil. No. 6 fuel oil is a high viscosity residual oil that requires preheating to 220–260 °F for storage, handling, and atomizing, having a carbon number range of 20–70. It is specified by the U.S. Navy as Bunker C oil for ships [21].

The fuel oils used in marine diesel engines have different classifications [21]. Marine gas oil (MGO) is made from distillate only. Marine diesel oil (MDO) is a blend of heavy gasoil that may contain very small amounts of refinery residue feed stocks, but it needs not be heated for use in internal combustion engines. Intermediate fuel oil (IFO) is a blend of gasoil and heavy fuel oil, with less gasoil than marine diesel oil. Marine fuel oil (HFO) is pure or nearly pure residual oil, roughly equivalent to No. 6 fuel oil (Bunker oil).

Major seaports, including those in the U.S. and the European Union, are imposing tighter emissions limits on sea-going ships and the fuels they burn. In 2012, a global treaty (MARPOL Annex VI) capped fuel sulfur content at 1% in coastal waters and extended the coastal zone to 200 miles offshore around Canada and the United States. The sulfur limit was decreased to 0.1% on January 1, 2015, requiring even tighter limits on the sulfur content of marine fuels. The treaty also imposes limits on SO_x and NO_x emissions, which may require the use of stack scrubbers. Meanwhile, changes in vessels and the use of larger, more fuel-efficient vessels have decreased marine fuel consumption. As mentioned, as of 2020, all marine fuel oil used for international shipping must contain <0.5 wt% sulfur.

15.9 Blending of Fuels

Figure 15.11 gives an overview of product blending of fuels, gasoline, jet fuel (kerosene), diesel and fuel oil, from various refining processes. Lubricant oils are not obtained from distillation followed by various thermal and catalytic processes,

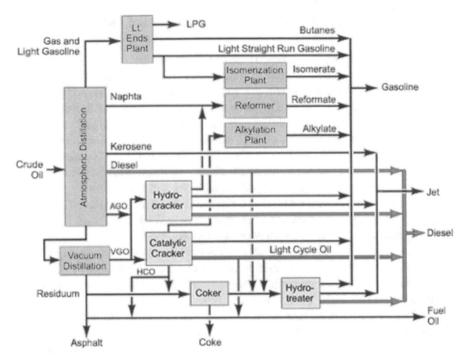

Fig. 15.11 Blending of fuels from various refining processes

as fuels. They are manufactured through various extraction processes; hence, their products are discussed separately.

- Naphtha comes from the following units: distillation, isomerization, alkylation, polymerization (not shown), catalytic reforming, hydrocracking, catalytic cracking and coking. Naphthas from different units have similar boiling ranges but may have contain different components. Some naphtha molecules might end up in gasoline, some might be sold as solvent, and others could go to thermal cracking in an olefins plant.
- Kerosene and jet fuel come from the hydrocracker unit (HCU) and the kerosene hydrotreater (KHT).
- Gas-oil-boiling-range distillates come from the FCC, the HCU, the gas oil hydrotreater (GOHT), visbreaker and coker. FCC gas oil sometimes is used as fuel oil cutter stock, but otherwise it does not meet specifications for diesel. It must undergo further refining before it can be sold. Almost every unit makes off-gas. In the past, offgas streams were collected into a common system, desulfurized and used for fuel gas.

Finished products are comprised of intermediate streams from different units, along with stabilizers, antioxidants, corrosion inhibitors and other additives. To be products, the blends must meet specifications developed and published by international organizations such American Society for Testing and Materials (ASTM), the International Organization for Standardization (ISO), the China State Council, or similar agencies in other countries. Specifications incorporate information from suppliers, users, government agencies and equipment manufacturers.

A material in the kerosene boiling range is a jet fuel if it meets ASTM D1655—15 specifications no matter how it is made. One might assume that a material is gasoline if it meets the specifications of ASTM D4814—14b, no matter how it is made. But in the United States, the assumption is wrong. By law, U.S. gasoline must contain ethanol.

15.10   Lubricant Base Oils, Wax, Grease and Specialty Products

15.10.1   Lube Base Oils

Lubricants were introduced in Chap. 4. Refiners prepare lube base stocks from different oils by removing asphaltenes, aromatics, and waxes. Lube base stocks are hydrofinished, blended with other distillate streams for viscosity adjustment, and compounded with additives to produce finished lubricants. In the past, solvent-based technology was used to prepare lube base stocks. Propane deasphalting was used to remove asphaltenes. Furfural and related substances were used to extract aromatics, and MEK or MIBK were used to remove wax. With the advent of catalytic dewaxing (CDW), some or all of these solvent-based methods can be replaced with hydroprocessing. CDW was developed by Mobil (now part of ExxonMobil) in the 1980s. The Mobil process employs ZSM-5, which selectively converts waxy n-paraffins into lighter hydrocarbons. The Isodewaxing Process, commercialized in 1993 by Chevron, reduces wax catalytically by isomerising n-paraffins into isoparaffins. Isodewaxing also removes sulfur and nitrogen and saturates aromatics. The products have a high viscosity index (VI), low pour point, and excellent response to additives.

Mineral oil-based lubricant base oils can be categorized into 3 groups [22, 23]. Group I is manufactured by solvent extraction, solvent or catalytic dewaxing, and hydro-finishing processes. It contains <90% saturates and >0.03% sulfur with viscosity index (VI) of 80–120. Common Group I base oils are 150 N (solvent neutral), 500 N, and 150BS (bright stock). Group II is manufactured by hydrocracking or catalytic dewaxing processes. It contains over 90% saturates and under 0.03% sulfur with VI of 80–130. Group II base oil has superior anti-oxidation properties since virtually all hydrocarbon molecules are saturated. It has water-white color. Group III is manufactured by special processes such as hydroisomerization and from base oil or slack wax through dewaxing processes. It also contains >90% saturates and <0.03% sulfur, but with VI >120.

Synthetic lubricant base oils are also derived from petroleum with VI >120. In North America, Groups III, IV and V are now described as synthetic lubricants, with group III frequently described as synthesized hydrocarbons, or SHCs. In Europe, only Groups IV and V may be classed as synthetics. Group IV is specific to the base oils derived from poly alpha olefins (PAOs). Group V includes other synthetic oxygen-containing base stock, such as dibasic esters, aromatic esters, polyol esters, etc.

The lubricant industry commonly extends this group terminology to include: Group I+ with a VI of 103–108, Group II+ with a VI of 111–119, Group III+ with a VI of at least 130, and Group IV+ with VI 5–15 higher than conventional PAOs made strictly from 1-decene [24].

15.10.2   Finished (Formulated) Lubricant Oils and Other Products

Finished lubricants are formulated with base oil and special additives. Detergents and dispersants are the dominant performance additives in passenger car motor engine oils. The additives are typically 5–20% of formulated engine oil or 55–70% of performance-package oil. The remaining additives include antiwear agents, ash inhibitors, friction modifiers, etc. [25].

Finished lubricant oils with additives can be used as engine oils, transmission fluids, gear oils, turbine oils, hydraulic oils, metal cutting oils, and paper machine oils. They can be mixed with soaps to make greases.

15.10.3   Typical Additives in Finished Engine Oil

Lubricating oil additives are used to enhance the performance of lubricants and functional fluids. Each additive is selected for its ability to perform one or more specific functions in combination with other additives. Selected additives are formulated into packages for use with a specific lubricant base stock and for a specified end-use application. The largest end use is in automotive engine crankcase lubricants. Other automotive applications include hydraulic fluids and gear oils. In addition, many industrial lubricants and metalworking oils also contain lubricating oil additives.

Additives are organic or inorganic compounds dissolved or suspended as solids in oil. They typically range between 0.1 to 30% of the oil volume, depending on the machine. The formulated lubricant oils in the automotive engine crankcase comprise of up to 5–10% of additives. Oil additives are vital for the proper lubrication and prolonged use of motor oil in modern combustion engines. Some of the most important additives include those used for viscosity and lubricity, for the control of chemical breakdown, and for seal conditioning. Some additives permit lubricants to perform better under severe conditions, such as extreme pressures and temperatures and high levels of contamination.

The major functional additive types in the additive package are dispersants (40–50%), detergents (15–20%), oxidation inhibitors (antioxidants), antiwear agents, ashless inhibitors, friction modifiers, corrosion inhibitors, antifoam agents, demulsifying agents, pour point depressants, metal deactivators, extreme-pressure additives, tackiness agents, and viscosity index improvers. The dispersant and detergent make up about 50–70% of the package. Hence, the chemistry of the total package and the finished formulated oils is greatly influenced by these components.

It should be noted that it is not always better to use more additives. As more additive is blended into the oil, sometimes there isn't any more benefit gained, and at times the performance of the blend actually deteriorates. In other cases, the performance of the additive doesn't improve, but the duration of service does improve.

In addition, increasing the percentage of a certain additive may improve one property of an oil while at the same time degrade another. When the specified concentrations of additives become unbalanced, overall oil quality can also be affected. Some additives compete with each other for the same space on a metal surface. If a high concentration of an anti-wear agent is added to the oil, the corrosion inhibitor may become less effective. The result may be an increase in corrosion-related problems.

Some of the additives are discussed below.

15.10.3.1 Detergents

Corrosion inhibitors protect the metal surface against corrosion in a variety of lubricant applications. Some inhibitors neutralize acids; others form protective films. Detergents are excellent corrosion inhibitors because they protect in both ways. Some detergents can also act as oxidation inhibitors.

The functions of detergents are to solubilize polar components, inhibit corrosion, and prevent high temperature deposits, in part through neutralization of acids. They are composed of two components: surfactants (organic), generally sulfonates, phenates or salicylates; and a colloidal metal phase (inorganic), generally Ca, Mg or Na from the over-basing, where the level of the basic phase is high relative to the amount of surfactant.

The combination of a surfactant molecule with a colloidal inorganic core results in a micellular-type structure shown in Fig. 15.12, with the polar head of the molecule represented by a circle. The long nonpolar alkyl tail (C_{12}–C_{32}) provides good solubility of nonpolar base stock molecules. The amorphous colloidal phase is represented by $CaCO_3$ in the figure. This basic colloidal carbonate neutralizes the acids formed during combustion. The inorganic acids are nitric acid, sulfuric acid and hydrochloric acid, which lead to metal corrosion and wear. Organic acids lead to polymerization, viscosity increase and resin formation. The detergent adheres to dirt and oil insoluble products formed as oxidation by-products during equipment operation. They keep these in suspension, preventing them from depositing onto critical engine surfaces.

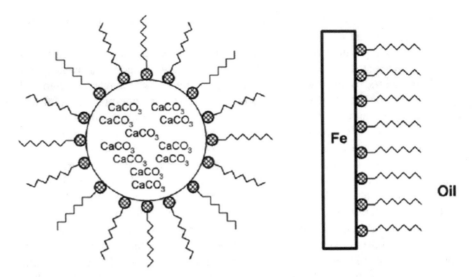

Fig. 15.12 Detergent [25]

Together, the surfactant and the basic components form a protective layer that inhibits corrosion, inhibits oil degradation, reduces high temperature deposits, and solubilizes polar contaminants.

Over-based detergents neutralize acidic combustion products, helping to control corrosion and resinous build-up in the engine. However, the total base number will continue decreasing during the use of oils, leading to an increase in total acid number in the oil. The equivalence point of total base number and total acid number occurs at around 3000–6000 miles of driving. At this point, the acids build up reaches unacceptable levels. It is therefore desirable to change the oil before the total base number and total acid number cross.

15.10.3.2   Dispersants

Dispersants function to suspend soot, thereby mitigating the deleterious effects of large particle agglomerates inside the crankcase. They prevent sludge, varnish and other deposits from forming on critical surfaces. They are primarily used in gasoline engine oils and heavy-duty diesel engine oils, which account for 75–80% of their total use. A dispersant molecule, shown in Fig. 15.13, consists of a polymer backbone component, predominantly polyisobutylene (PIB, MW 500–2000), and a polar group, normally an amine group. Two major classes of dispersants are: succinimide dispersants, which use maleic anhydride, and Mannich dispersants, which use formaldehyde. Both classes use PIB and polar amine groups. Dispersants are designed to have a longer tail than detergents to provide greater steric stabilization

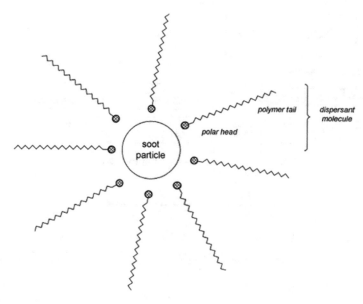

Fig. 15.13 Dispersant [25]

to the dispersed carbon particle or other contaminants in the micellular structure. A polyamine head group is tailored for strong adsorption of soot particles.

Both dispersants and surfactants have similar basic molecular structures. They have a non-polar hydrocarbon backbone for solubility in oils, a polar end group for dissolving or attaching polar contaminants formed during the degradation of the oil, and a functional group as a "hook" to connect the non-polar and polar groups at both ends as shown in Fig. 15.14. The alkyl groups on the detergents can be linear or branched. Their molecular size is in the range of C_{12}–C_{32} and is much longer for dispersants in the range of about C_{35}–C_{50}. The end groups of surfactants are metal salts of organic acids, such as sulfonates and salicylates, which can also function as a hook with alkyl substituents. The end groups of dispersants are polyamines, such as triethylene tetraamine (TETA) and tetraethylene pentaamine (TEPA). A typical hook for a dispersant is maleic anhydride. A representative molecule of succinic anhydride as the "hook" with polyisobutylene (PIB) as oil-soluble backbone and TETA polar end (a PIB succinimide) is also presented.

15.10.3.3   Friction Modifiers, Antiwear Agents and Extreme-Pressure Additives

Extreme pressure (EP) additives are sulfur-phosphorus based chemicals that form organo-metallic salts on the loaded surfaces during operation at high pressures and temperatures. They create a protective layer that reduces wear between two mating

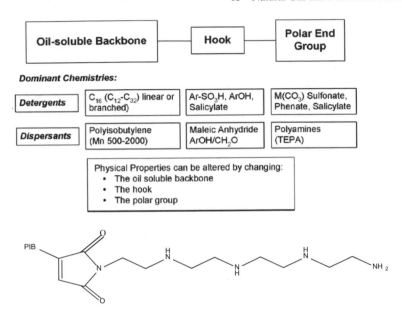

A dispersant molecule: polyisobutylene (PIB) maleic amide with tetraethylene pentaamine (TEPA)

Fig. 15.14 Functional molecules in detergents and dispersants [25]

metal surfaces. Anti-wear additives perform in a similar manner, but tend to operate under lower pressures and temperatures. EP additives are usually supplemented with anti-wear additives to make the oils, especially gear oils and grease. They are effective across a wide range of pressure and temperature conditions. Friction modifiers are mild anti-wear agents that minimize light surface contacts (sliding and rolling). They prevent scoring and reduce wear and noise by forming protective surface layers. They can be esters, natural and synthetic fatty acids as well as some solid materials such as graphite and molybdenum disulfide. The organic molecules have a polar end (head) and an oil-soluble end (tail). Once placed into service, the polar end of the molecule finds a metal surface and attaches itself. These are also called boundary lubrication additives. The inorganic molecules (graphite and MoS_2) are highly stable and are comprised of flat layers that slip easily over each other.

The most important advance in antiwear chemistry was made during the 1930s and 1940s with the discovery of zinc dialkyldithiophosphates (ZDDP). These compounds were found to have exceptional antioxidant and antiwear properties. The antioxidant mechanism of the ZDDP was the key to its ability to reduce bearing corrosion as a corrosion inhibitor. Since the ZDDP suppresses the formation of peroxides, it prevents the corrosion of Cu/Pb bearings by organic acids. Antiwear and extreme-pressure additives function by thermally decomposing to yield compounds that react with the metal surface. These surface-active compounds form a thin layer that preferentially shears under boundary lubrication conditions.

(R = alkyl)

zinc dialkyldithiophosphates (ZDDP)

15.10.3.4   Oxidation Inhibitors/Antioxidants

The rate of oxidation doubles with every 10 °C increase in temperature. In the case of increased temperature, in the presence of oxygen and mechanical load, lubricants age very readily. If not controlled, the radical-induced oxidations generate acids, which lead to the molecular degradation and finally to the failure of the lubricant. The lubricant decomposition will lead to oil thickening and the formation of sludge, varnish, resin and corrosive acids. This leads to an increase in corrosion. Water and polar impurities increase the speed of attack, and internal combustion engines contain plenty of these contaminants. Incorporating an oxidation inhibitor will interrupt and terminate the free radical process of oxidation and thus slow down the ageing of lubricants. Phenolic materials are quite good for this purpose and the two most commonly used inhibitors are 2,6-ditertiary-butylphenol (DBP) and 2,6-di-tertiary-butyl-4-methylphenol or 2,6-di-tertiary-butyl-paracresol (DBPC) which is also known as butylated hydroxytoluene (BHT). The typical recommended value of DBPC and DBP in fresh oil is approximately 0.3% by weight.

15.10.3.5   Viscosity Index Improvers

Mineral oil lubricants become less effective at high temperatures because heat reduces their viscosity and film-forming ability. Viscosity index improvers and thickeners are polymeric molecules, such as olefin copolymers (OCP), and are added to reduce lubricant viscosity changes at high and low temperatures. OCP are copolymers of ethylene and propylene. When polymer coils interact with oil and each other they become increasingly resistant to flow; hence, they can be added to oils to increase their viscosity (thickening). A balance between the thickening efficiency and shear stability of the polymer is important. Higher molecular weight polymers make better thickeners, but tend to have less resistance to mechanical shear. Lower molecular weight polymers are more shear resistant, but do not improve viscosity as effectively at higher temperatures and must be used in larger quantities. Mineral oil lubricants become less effective at high temperatures because heat reduces their viscosity and film-forming ability.

When viscosity index improvers are added to low-viscosity oils, they effectively thicken the oil as temperature increases. This means the lubricating effect of mineral oils can be extended across a wider temperature range. At low temperatures, the molecule chain contracts and does not impact the fluid viscosity. At high temperatures, the chain relaxes and an increase in viscosity occurs. Viscosity improvers are primarily used in multigrade engine oils, gear oils, automatic transmission fluids, power steering fluids, greases and various hydraulic fluids. Most of these uses involve automobiles in which they are subjected to tremendous temperature swings over short periods.

Polymer additives can undergo thermal and oxidative degradation, unzipping back to smaller monomers, which reduces their effect. Hence, the highest possible degree of thermal and oxidative stability is desirable.

15.10.3.6 Metal Deactivators

Metal deactivators or metal passivators, such as benzotrizaole derivatives, create a passivating protective film which prevents the catalytic effects of metals, such as copper and iron, and the transfer of metal ions. These metals are present in fuel-distribution equipment, including pumps, pipeline valves and storage equipment. Dissolved metal in a fuel can accelerate rates of oxidation, resulting in gum and sediment formation and color degradation. Even part-per-billion levels of copper can catalyze fuel oxidation and cause potential stability risks. The use of metal deactivators can prevent the oxidative attack and thus the corrosion of the metal surface and the ageing of the lubricant.

15.10.3.7 Tackiness Agents

Tackiness additives (tackifiers) are a combination of a range of high molecular weight polymers (polyisobutylenes). In selected mineral oils, these additives increase a lubricant's adhesive properties, preventing dripping and splashing during operation over a wide range of conditions. This reduces operational cost and the risk of environmental contamination. They also add an element of corrosion protection.

15.10.3.8 Antifoaming Agents

Most lubricant applications involve agitation, which traps air in the lubricant and encourages the formation of foam. Surface-active materials, such as dispersants and detergents, further increase foaming tendency. Excessive foaming may result in increased lubricant oxidation and decreased operational efficiency.

Antifoaming agents, such as poly dimethyl siloxane and copolymers of polyethylene glycol and polypropylene glycol, alter the surface tension of the oil and help to weaken the structure of air bubbles. The result is better lubricating qualities and reduced maintenance.

15.10.3.9   Pour Point Depressants

Although most of the wax is removed during base oil dewaxing, high-molecular weight species are necessary to achieve the desired target viscosity. A range of pour point depressants based on polymethacrylate (PMA) and styrene ester in lubricant formulation prevent wax fractions in the base oil from forming large crystal networks which inhibit lubricant flow at cold temperatures, while keeping the viscosity benefits at higher temperatures.

15.10.3.10   Demulsifiers

Demulsifiers, or dehydration chemicals, separate water droplets from oil by breaking the emulsion and removing interfacial film. For example, an amine derivative based additive forms high surface tension in the formulated lube oil so as to repel water quickly due to strong cohesive inner forces. A few typical demulsifiers are shown below [26].

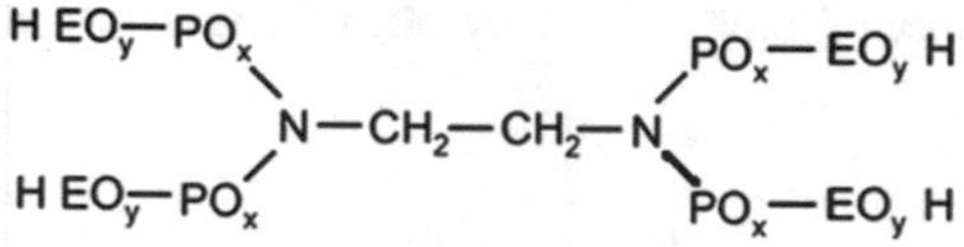

Regular EO+PO complex with polyfunctional amine

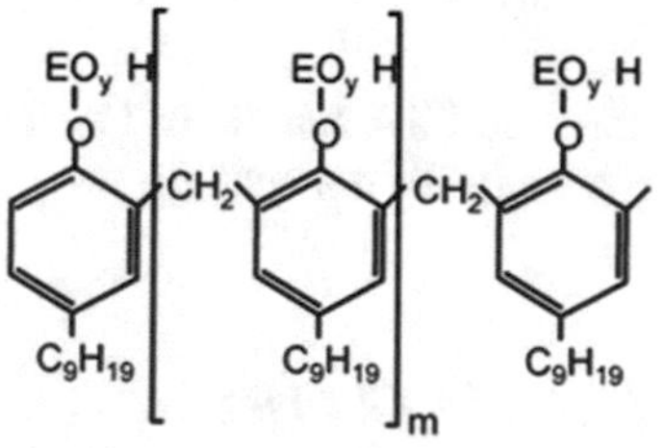

Nonylphenol with EO+PO complex

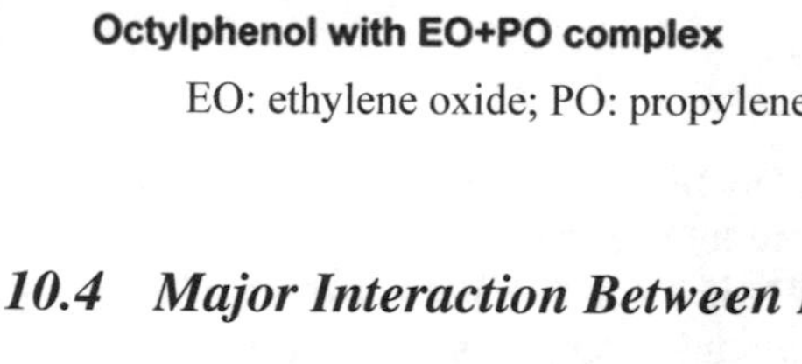

Octylphenol with EO+PO complex

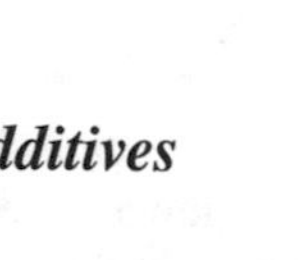

Dodecylbenzene sulphonic acid

EO: ethylene oxide; PO: propylene oxide

15.10.4   Major Interaction Between Base Oil and Additives

There are great improvements in lubricant base oils. However, a good additive package is very important for high-quality performance. To design a lubricant additive package, we need to consider the major interaction between base oil and additives, which is shown in Fig. 15.15. It's not a complete picture, but at least it shows the

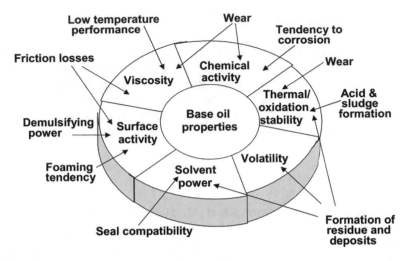

Fig. 15.15 Major interactions of base oil with additives [25]

major additive functions. Compatibility of additives to base oil and with other additives must be considered.

15.10.5 Greases

Greases are made by blending detergents (salts of long-chained fatty acids) into lubricating oils at 400–600 °F (204–315 °C). Antioxidants are added to provide stability. Some greases are batch-produced, while others are made continuously. The characteristics of a grease depend to a great extent on the counter-ion (calcium, sodium, aluminium, lithium, etc.) in the fatty-acid salt.

15.10.6 Specialty Oils

Other specialty products include white oil, insulating (electrical) oils and insecticide. Naphthenic crude oils give white oil products of high specific gravity and viscosity, suitable for pharmaceutical uses. Paraffinic crude oils produce oils with lighter gravity and lower viscosity suitable for lubrication purposes. Technical grade white oils are employed for cosmetics, textile lubrication, insecticides, vehicles, paper impregnation, etc. Pharmaceutical (medical) grade white oils may be employed as laxatives or for lubrication of food-handling machinery. Insulating oils are highly refined fractions of low viscosity and are stable at high temperature. They have excellent electrical insulating properties, so they can be used in transformers, circuit breakers, and oil-filled cables, and for impregnating the paper covering of wrapped cables. Insecticides are petroleum oils in water-emulsion form to be sprayed onto fruit trees and swamp water to killing certain species of insects.

15.11 Petroleum Wax

Wax recovery processes were introduced in Chap. 13. Gas oils or vacuum distillates are the source of paraffin waxes, which forms large crystals or plates with molecular weight between 300 and 450 (C_{17}–C_{35}). Such wax contains mostly normal paraffins and small amounts of isoparaffins, cycloparaffins and minute amounts of aromatics. Vacuum residue is the source of microcrystalline wax with molecular weights between 450 and 800 (C_{35}–C_{60}). Microcrystalline wax contains mostly cycloparaffins with n-alkyl and isoalkyl side chains. Paraffin wax is used for candles, cosmetics, and other purposes. Microcrystalline waxes are used in the tire and rubber industries, where they are sometimes blended with paraffin wax. Food-grade wax is a highly refined product, which is employed for water-proofing beverage containers, for orthodontics and pharmaceuticals, and as food additive.

15.12 Cokes

Green coke is a raw carbonaceous solid of petroleum coke (petcoke) derived from delayed coking. Green coke is calcined to remove volatile hydrocarbons and sulfur compounds. With sufficiently low metal content it can be calcined for use as anode materials (anode grade coke). Green coke with a too-high metal content is used as fuel (fuel grade coke).

Undesirable coke is deposited on catalysts used in certain oil refining processes, such catalytic reforming, hydrotreating and hydrocracking. In the FCC process, coke is not undesirable; it is essential, because burning the coke provides heat to run the unit.

The most common grades of finished coke are fuel grade coke, calcined petroleum coke, and needle coke. Fuel coke comes from delayed coking units in the form of sponge coke, which is named for its sponge-like appearance. It is produced from coker feeds with low-to-moderate asphaltene concentrations. A good fuel coke with high heat value and low ash content is used for power generation. Low ash content is important, because ash can foul boilers. (The low ash content of fuel coke makes it superior to coal for power generation.). When burning fuel coke, some form of sulfur capture is required to meet current North American emission standards. Fuel grade coke can be a feedstock for gasification, during which it is converted into synthesis gas—a mixture of CO and H_2.

Sponge coke is the most common petroleum coke. In its uncrushed state, sponge coke closely resembles coal. If sponge coke meets certain specifications, such as low metals content, it may be calcined and formed into carbon anodes for the aluminum industry. Otherwise, it serves as a fuel. Sulfur recovery is required during calcination, too.

Needle coke, named for its needle-like crystalline structure, is a high-value product made from feeds that contain nil asphaltenes, such as hydrotreated FCC decant oils. Needle coke has a low thermal expansion coefficient, making it suitable for conversion into graphite electrodes for the electric-arc furnaces used in the steel industry. It is less typically to break than the other types of petroleum cokes.

Honeycomb coke is an intermediate between sponge coke and needle coke. It is characterized by ellipsoidal pores uniformly distributed throughout its shape. It has lower electrical conductivity and a lower thermal coefficient compared to needle coke.

Shot coke is undesirable because it tends to be unstable. It forms when the concentration of feedstock asphaltenes and/or coke-drum temperatures are too high. A block of shot coke is a cluster of discrete mini-balls 0.1–0.2 in (2–5 mm) in diameter. The clusters can be as large as 10 in (25 cm) across. If a cluster breaks apart when the coke drum is opened, it can spray a volley of hot mini-balls in every direction. The name "shot" derives from the fact that it can resemble shotgun ammunition. Adding aromatic feeds, such as FCC decant oil, can eliminate shot coke formation. Other methods of eliminating shot coke—decreasing temperature, increasing drum pressure, increasing the amount of product recycle—decrease liquid yields, which is not desired.

Specialty carbon products made from petroleum include recarburizer coke, which is used to make special steels, and titanium dioxide coke, which is used as a reducing agent in the titanium dioxide pigment industry.

Because of the similar coal-like properties, the demand of petroleum coke is rising in China, Mexico, Canada and Turkey. The U.S. is becoming a large net exporter of petroleum coke.

15.13   Asphalt

Asphalt is a sticky, black and highly viscous liquid or semi-solid form of petroleum. It may be found in natural deposits or may be a refined product. Humans used asphalt to make spears and coat baskets as far back as 70,000 years ago near Umm el Tlel, in present-day Syria. As early as 3000–4000 BC, Egyptians used pitch to grease chariot wheels and asphalt to embalm mummies. Mesopotamians used bitumen to line water canals, seal boats, and build roads. Today, the primary use of asphalt/bitumen is in road construction (paving), where it serves as the glue or binder that is mixed with aggregate particles (stone, sand, polymer, etc.) to create asphalt concrete. Other main uses are for waterproof roofing products.

Asphalt base stocks can be produced directly from vacuum residue or by solvent deasphalting. Vacuum residue is used to make road-tar asphalt. To drive off light ends, it is heated to about 750 °F (400 °C) and charged to a column where a vacuum is applied to prevent thermal cracking.

In road-paving, the petroleum residue serves as a binder for aggregate, which can include stone, sand, or gravel. The aggregate comprises about 95% of the final mixture. Polymers are added to the binder to improve strength and durability. The recommended material for paving highways in the United States is Superpave 32 hot-mix asphalt. Superpave was developed in 1987–93 during a US$50 million research project sponsored by the Federal Highway Administration.

Roofing asphalt is produced by bubbling air through liquid asphalt at 260 °C (500 °F) for 1–10 h. During this "blowing" process, organic sulfur is converted to SO_2. Catalytic salts such as ferric chloride ($FeCl_3$) may be added to adjust product properties and increase the rates of the blowing reactions, which are exothermic. To provide cooling, water is sprayed into the top of the blowing vessel, creating a blanket of steam that captures sulfur-containing gases, light hydrocarbons, and other gaseous contaminants. These are recovered downstream. Cooling water may also be sprayed on the outside of the vessel. The length of the blow depends on desired product properties, such as softening temperature and penetration rate. A typical plant blows four to six batches of asphalt per 24-h day. There are two primary substrates for roofing asphalt—organic (paper felt) and fiberglass. The production of felt-based roofing shingles consists of:

- Saturating the paper felt with asphalt
- Coating the saturated felt with filled asphalt
- Pressing granules of sand, talc or mica into the coating
- Cooling with water, drying, cutting and trimming, and packaging.

If fiberglass is used as the base instead of paper felt, the saturation step is eliminated.

15.14 Petrochemicals

Refineries produce a very large quantity of commodity petrochemicals which are often used in subsequent chemical and pharmaceutical production.

Primary petrochemicals can be divided into three groups as olefins, aromatics and synthetic gas, which are discussed in the following chapter.

- Olefins are major source of polyolefins and their derivatives to polyesters.
- Aromatics are mainly benzene, toluene and xylenes. They are solvents and feedstocks for a variety of chemicals.
- Synthesis gas is used for making methanol and ammonia as well as synthetic fuels.

15.15   Overview of Refinery Processes and Products

An overview of refinery processes for making products, including gases, fuels, lubricants, grease, wax, coke and asphalt is shown in Fig. 15.16.

Petroleum products by estimated carbon number ranges are shown in Fig. 15.17. Note that the upper carbon number range is not certain due to the limits of current analytical instruments. The use of high temperature gas chromatography (HT-GC) and field ionization/field desorption mass spectrometry (FI/FD MS), for example, can determine the upper carbon numbers for lubricant oils and asphalts beyond the figure shown. In addition, the ranges shown are only used as reference, which can vary at different refineries and countries to meet products specifications or regulations.

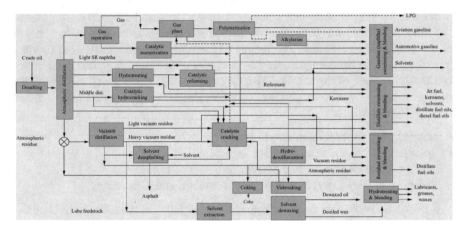

Fig. 15.16 Overall refinery processes and products [27]

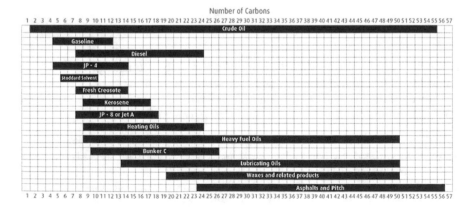

Fig. 15.17 Petroleum products by estimated carbon number with uncertainty in heavy products [28]

References

1. Linde engineering: natural gas processing plants. http://www.linde-engineering.com/internet. global.lindeengineering.global/en/images/HE_1_1_e_12_150dpi19_4271.pdf. Accessed 31 Aug 2014
2. Shell global solution: sulfinol-X. http://www.shell.com/business-customers/global-solutions/gas-processing-licensing/gas-processing-technolgies-portfolio/_jcr_content/par/textimage.stream/1469805944324/ b7412e8c1d28341a55dc73852445a5e7ff45ab36d75b75ce3896c464e29d6048/factsheet-sulfinol-x.PDF. Accessed 31 Aug 2014
3. Processing natural gas. http://naturalgas.org/naturalgas/processing-ng/. Accessed 24 Nov 2016
4. Gonzalez MR (2017) Transitioning refineries from sweet to extra heavy oil. In: Hsu CS, Robinson PR (eds) Springer handbook of petroleum technology. Springer, New York (Chapter 31)
5. Wikipedia: four-stroke engine. https://en.wikipedia.org/wiki/Four-stroke_engine. Access 31 Aug 2014
6. http://www.biofuelsdigest.com/bdigest/2015/06/11/ethanol-and-butanol-symbiotic-partners-for-a-modern-fuel/
7. Gasoline Reid Vapor Pressure (2016) U.S. environmental protection agency: gasoline standards. https://www.epa.gov/gasoline-standards/gasoline-reid-vapor-pressure. Accessed 18 Aug 2016
8. https://www.epa.gov/gasoline-standards/reformulated-gasoline
9. Emergency power: cetane number in diesel fuel. http://www.emergencypower.com/support/diesel-fuel/168-cetane-number-in-diesel-fuel. Accessed 25 Nov 2016
10. How a jet engine works. http://www.youtube.com/watch?v=p1TqwAKwMuM. Accessed 31 Aug 2015
11. Wikipedia: jet fuel. https://en.wikipedia.org/wiki/Jet_fuel. Accessed 6 Aug 15
12. U.S. Air Force, UASF RPT AFAPL-TR-74-71 (1975) Assessment of JP-8 as a replacement fuel for air force standard jet fuel JP-4, June 1975
13. Hsu CS (2000) Diesel fuel analysis. In: Encyclopedia of analytical chemistry. Wiley, pp 6613–6622
14. https://www.dieselnet.com/standards/eu/fuel_automotive.php. Access 30 Nov 2018
15. European Standards, DIN EN 590, Automotive fuels, Diesel—requirements and test methods (includes Amendment: 2017). https://www.en-standard.eu/din-en-590-automotive-fuels-diesel-requirements-and-test-methods-includes-amendment-2017/?gclid= EAIaIQobChMI2oC-ltXj3gIVAqZpCh2FFgjCEAAYASAAEgKBzvD_BwE. Access 30 Nov 2018
16. https://www.theicct.org/sites/default/files/publications/China_6_fuel_Policy_Update_20180410.pdf. Access 30 Nov 2018
17. Explainthatstuff!: diesel engine. http://www.explainthatstuff.com/diesel-engines.html. Accessed 25 Nov 2016
18. ASTM. https://www.astm.org/Standards/D8183.htm. Accessed 15 Oct 2018. Wikipedia: fuel oil. https://en.wikipedia.org/wiki/Fuel_oil. Accessed 6 Aug 15
19. https://www.epa.gov/diesel-fuel-standards/diesel-fuel-standards-and-rulemakings#onroad-diesel. Access 11 Nov 2018
20. DieselNet (2018) IMO marine engine regulations. https://www.dieselnet.com/standards/inter/imo.php. Access 22 Nov 2018
21. Wikipedia: lubricant. https://en.wikipedia.org/wiki/Lubricant. Accessed 6 Aug 15
22. (a) Beasley BE (2006) Conventional lube Basestock manufacturing. In: Hsu CS, Robinson PR (eds) Practical advances in petroleum processing. Springer, New York (Chapter 15). (b) Beasley BE (2017) Conventional lube base stock. In: Hsu CS, Robinson PR (eds) Springer handbook of petroleum technology. Springer, New York (Chapter 33)
23. Lee SK, Rosenbaum JM, Hao Y, Lei GD (2017) Premium lubricant base stocks by hydroprocessing. In: Hsu CS, Robinson PR (eds) Springer handbook of petroleum technology. Springer, New York (Chapter 34)

24. Wu MM, Ho SC, Luo S (2017) Synthetic lubricant base stock. In: Hsu CS, Robinson PR (eds) Springer handbook of petroleum technology. Springer, New York (Chapter 35)
25. Burrington JD, Pudelski JK, Roski JP (2006) Challenges in detergents and dispersants for engine oils. In: Hsu CS, Robinson PR (eds) Practical advances in petroleum processing. Springer, New York (Chapter 18)
26. Petrowiki: oil demulsification. http://petrowiki.org/Oil_demulsification. Accessed 28 Nov 2016
27. Al-Salem SM, Ma X, Al-Mujaibel MM (2017) Carbon dioxide mitigation. In: Hsu CS, Robinson PR (eds) Springer handbook of petroleum technology. Springer, New York (Chapter 32)
28. ALS environmental: petroleum hydrocarbon ranges. http://www.caslab.com/Forms-Downloads/Flyers/PETROLEUM_HYDROCARBON_RANGES_FLYER.pdf. Accessed 23 Nov 2016

Part IV
Petrochemicals, Midstream, Safety and Environment

Chapter 16
Petrochemicals

Petrochemicals are used in subsequent chemical and pharmaceutical production. The raw materials come mainly from two sources: natural gas, which provides methane, ethane, propane, butane, and C_5+ condensate; and crude oil, which yields methane, saturated C_2-C_4 hydrocarbons, C_2-C_4 olefins, and benzenes. Figure 16.1 shows routes for making primary petrochemicals from natural gas and crude oil.

Chemical compounds can be derived from petroleum directly or indirectly through processes, especially cracking. In the discussions of petrochemicals, fuels (gasoline, kerosene/jet fuels, diesel and fuel oil), lubricant oil, wax, asphalt, coke and the like are excluded. General types of petrochemicals are aliphatic compounds (paraffins and isoparaffins), olefinic compounds, aromatic compounds, inorganic compounds (CO_2, elemental sulfur, ammonia), and synthesis gas (CO and H_2) derived from petroleum natural gas and crude oil.

16.1 Chemicals from Methane

The chemicals that can be produced from methane are summarized in Fig. 16.2.

16.1.1 Hydrogen from Methane

Catalytic reforming provides a major source of hydrogen in refinery operations. However, additional hydrogen is required when a refinery has numerous hydrocracking and hydrotreating processes. Additional hydrogen can be obtained from steam methane reforming (SMR) [1]. The two main reactions, as shown below, are reversible.

At 900–1000 °C (1650–1830 °F) over a metal catalyst, such as nickel:

© Springer Nature Switzerland AG 2019
C. S. Hsu and P. R. Robinson, *Petroleum Science and Technology*,
https://doi.org/10.1007/978-3-030-16275-7_16

Feed	Refinery Products	Petrochemicals	Chemicals

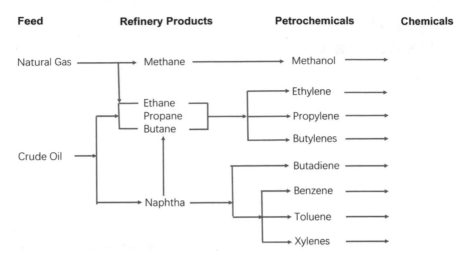

Fig. 16.1 Raw materials and primary petrochemicals

$$CH_4 + H_2O \leftrightarrow CO + 3H_2$$

At 500 °C over a low-temperature shift (LTS) catalyst containing copper or iron:

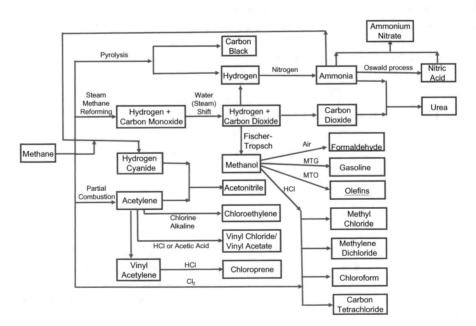

Fig. 16.2 Chemicals from methane

$$CO + H_2O \leftrightarrow H_2 + CO_2$$

The first reaction is highly endothermic, due to the high thermochemical stability of methane and water. Under the conditions shown, the LTS reaction is slightly exothermic.

The carbon dioxide is then removed either by absorption in aqueous ethanolamine solutions, by adsorption into molten potassium carbonate (Benfield process [2]), or by a pressure swing adsorption (PSA) unit. PSA hydrogen usually is 99.99% pure.

For many catalytic processes, CO is a catalyst poison and must be removed. If required, catalytic methanation removes any small residual amounts of carbon monoxide or carbon dioxide from the hydrogen:

$$CO + 3H_2 \rightarrow CH_4 + H_2O \quad \Delta H_{298} = -206 \text{ kJ/mol}$$
$$CO_2 + 4H_2 \rightarrow CH_4 + 2H_2O \quad \Delta H_{298} = -165 \text{ kJ/mol}$$

The methanation of CO_2 over nickel catalysts at elevated temperature (optimally 300–400 °C) was discovered by Paul Sabatier and J. B. Sendersens in 1897 as the "Sabater" reaction. Ruthenium on alumina makes more efficient catalysts. However, nickel is widely used as methanation catalyst due to its high selectivity and low cost.

16.1.2   Methane to Methanol, to Gasoline (by MTG), to Olefins (by MTO) and to Chlorinated Products

Methanol is made from methane (natural gas) in a series of three reactions:

Steam reforming: $CH_4 + H_2O \rightarrow CO + 3H_2 \quad \Delta H = +206 \text{ kJ mol}^{-1}$

Water–gas shift reaction: $CO + H_2O \rightarrow CO_2 + H_2 \quad \Delta H = -41 \text{ kJ mol}^{-1}$

Synthesis: $2H_2 + CO \rightarrow CH_3OH \quad \Delta H = -92 \text{ kJ mol}^{-1}$

Methanol also is made from synthesis gas (syngas), a mixture of CO, H_2 and very often some CO_2, that can be produced by partial oxidation of methane, coal, vacuum residue, and biomass.

16.1.2.1   Methanol to Gasoline (MTG) and Methanol to Olefins (MTO)

The methanol thus formed may be converted to gasoline by the Mobil (ExxonMobil) process developed in the early 1970s [3]. First methanol is dehydrated to give dimethyl ether (DME):

$$2 \ CH_3OH \ \underset{+H_2O}{\overset{-H_2O}{\rightleftharpoons}} \ CH_3\text{-}O\text{-}CH_3$$

$$\downarrow {\scriptstyle -H_2O}$$

Isoparaffins
Aromatics $\longleftarrow$ $C_2 - C_5$ olefins
C_6^+ olefins

The ether is then dehydrated along with methanol over a zeolite catalyst, ZSM-5, to yield an organic product containing up to 80 wt% C_5+ hydrocarbons (paraffins, naphthenes and aromatics). Reactions include polymerization and hydrogenation of the ethylene intermediate. Stabilized gasoline, which produced by removing flue gas and LPG, is split into light and heavy fractions. The heavy gasoline is hydrotreated to reduce durene (1,2,4,5-tetramethylbenzene) and then blended back with the light gasoline. ZSM-5 is deactivated by a carbon build-up ("coking") over time. The catalyst can be re-activated by burning off the coke in a stream of hot air at 500 °C; however, the number of re-activation cycles is limited.

The processes of making gasoline from methane or syngas through methanol by the MTG process and making diesel and jet fuels from syngas by the Fischer-Tropsch process are also known as gas-to-liquids (GTL) processes [4].

At lower temperatures, methanol reacts to form DME. At higher temperatures, olefins are produced and the selectivity for DME decreases. The zeolite catalyst for methanol-to-olefins (MTO) is typically H-SAPO-34. In SAPO-34, some silicon atoms occupying sites in ZSM-5 are replaced by phosphorus atoms so that aluminum and phosphorus occupy the sites alternatively. The pore size is about 3.8 Å, compared to 5.5 Å for ZSM-5. SAPO-34 also has weaker acidity than ZSM-5. Hence, ZSM-5 is used for MTG and SAPO-34 is used for MTO to yield polymer grade ethylene and propylene for the production of polyolefins.

16.1.2.2 Methanol to Formaldehyde

Formaldehyde can be produced by non-catalytic oxidation of methanol, or by dehydrogenation of methanol by catalytic oxidation using an iron–molybdenum oxide catalyst with 99% conversion. The reactions in the shell-and-tube reactor are completely exothermic. The temperature at 250–345 °C is maintained in the reactor by removing the excess heat from reactor tube using Dowtherm™ oil, a heat transfer fluid comprised typically of a eutectic mixture of biphenyl and diphenyl oxide.

Formaldehyde is a building block in the synthesis of many other compounds of specialized and industrial significance: ethylene glycol, acrylates, pentaerythritol, hexamethylenetetramine, ethylene diamine, tetra-acetic acids, methacrylates, etc.

Formaldehyde undergoes a Cannizzaro reaction in the presence of a basic catalyst to produce formic acid and methanol. It can be readily oxidized by oxygen in air into formic acid.

Urea-formaldehyde resin is made by heating formaldehyde and urea in the presence of a base. The resin is a thermosetting resin or plastic.

Formaldehyde polymerizes with melamine to produce a hard, thermosetting plastic material for construction material and kitchen utensils.

Formaldehyde, phenol and melamine reins are thermosetting adhesives having good resistance to high temperatures. Thermoset materials require heat and pressure to form a secure bond.

16.1.2.3 Methanol to Chlorinated Methanes

Various chlorinated methanes—methyl chloride, methylene dichloride, carbon tetrachloride, and chloroform—can be produced by reactions of methanol with hydrogen chloride or concentrated hydrochloric acid catalyzed by zinc chloride by heat. The first three compounds are solvents, as used in degreasing and dry cleaning; the latter is used mainly as a preservative.

16.1.3 Ammonia Production from Natural Gas (Methane)

The first step in ammonia synthesis is to produce hydrogen, usually from natural gas (methane) in SMR as described above. If present, H_2S is first removed from natural gas by amine adsorption. Then, organic sulfur compounds undergo catalytic hydrogenation, which converts them to gaseous hydrogen sulfide:

$$H_2 + RSH \rightarrow RH + H_2S \text{ (gas)}$$

The gaseous hydrogen sulfide is then absorbed and removed by passing it through beds of zinc oxide, where it is converted to solid zinc sulfide:

$$H_2S + ZnO \rightarrow ZnS + H_2O$$

Ammonia is produced from catalytic reactions of hydrogen with nitrogen via the Haber-Bosch process at 400–650 °C and 200–400 bar [5], which is an artificial nitrogen fixation process [6]. The most widely used catalysts are comprised of finely divided iron promoted with K_2O, CaO, SiO_2, and Al_2O_3.

$$3H_2 + N_2 \rightarrow 2NH_3$$

Ammonia can be transported as anhydrous liquid ammonia, which boils at -33 °C. Hence, it must be kept cool and under modest pressure. Ammonia is used primarily for fertilizer. Anhydrous ammonia can be applied directly to soil. Alternatively, ammonia can be converted to ammonium nitrate (NH_4NO_3) or urea (NH_2–O–NH_2) and blended with other compounds to make solid or aqueous fertilizers with defined ratios of nitrogen, potassium, and phosphorous. Ammonia is not only one of the main components of fertilizers, but it also is widely used in other sectors, such as the pharmaceutical, household cleaning, and fermentation industries.

16.1.3.1 Ammonia to Nitric Acid by Oswald Process [7–9]

Anhydrous ammonia is oxidized to nitric oxide, in the presence of Pt or Rh catalysts at a high temperature of about 230 °C and a pressure of 9 bar.

$$4NH_3 \text{ (g)} + 5O_2 \text{ (g)} \rightarrow 4NO \text{ (g)} + 6H_2O \text{ (g)} \quad (\Delta H = -905.2 \text{ kJ})$$

Nitric oxide is then reacted with oxygen in air to form nitrogen dioxide.

$$2NO \text{ (g)} + O_2 \text{ (g)} \rightarrow 2NO_2 \text{ (g)} \quad (\Delta H = -114 \text{ kJ/mol})$$

This is subsequently absorbed in water to form nitric acid and nitric oxide.

$$3NO_2 \text{ (g)} + H_2O \text{ (l)} \rightarrow 2HNO_3 \text{ (aq)} + NO \text{ (g)} \quad (\Delta H = -117 \text{ kJ/mol})$$

The nitric oxide is recycled.
    Alternatively, if the last step is carried out in air:

$$4NO_2 \text{ (g)} + O_2 \text{ (g)} + 2H_2O \text{ (l)} \rightarrow 4HNO_3 \text{ (aq)}$$

The main industrial use of nitric acid is for the production of fertilizers. The other main applications are for the production of explosives, nylon precursors, and specialty organic compounds. Nitric acid is neutralized with ammonia to give ammonium nitrate, a fertilizer. Ammonium nitrate is explosive. It has been responsible for horrendous industrial accidents. When mixed with fuel oil, it becomes ANFO (ammonium-nitrate-fuel-oil), the material used to attack the Murrah Federal Building in Oklahoma City on April 19, 1995 [10].

16.1.3.2 Production of Urea from Ammonia

The production of urea, also known as carbamide, from ammonia is carried out by two steps. In the first step, liquid ammonia reacts with dry ice to form ammonium carbamate ($H_2N–COONH_4$) in an exothermic reaction:

$$2NH_3 + CO_2 \leftrightarrow H_2N-COONH_4$$

In the second step, ammonium carbamate is decomposed into urea and water, an endothermic reaction:

$$H_2N-COONH_4 \leftrightarrow (NH_2)_2CO + H_2O$$

Both reactions combined are exothermic.

More than 90% of world industrial production of urea is destined for use as a time-release nitrogen fertilizer. Urea has the highest nitrogen content of all solid nitrogenous fertilizers in common use [11].

16.1.4 Methane to Acetylene

Acetylene can be produced by the hydrolysis of calcium carbide, a reaction discovered by Friedrich Wöhler in 1862. Since the 1950s, acetylene has mainly been manufactured in the industry by the partial combustion of methane [12]. The heat generated in the initial combustion is used to break the carbon–hydrogen bonds of methane for the formation of acetylene.

Acetylene is highly explosive: it must be stored and handled with great care. When absorbed in diatomaceous earth or dissolved in acetone, it loses its explosive capability, making it safe to transport.

The most common use of acetylene is to produce various organic chemicals including 1,4-butandiol, which is widely used in the preparation of polyurethane and polyester plastics. The second most common use is as the fuel component in oxy-acetylene welding and metal cutting. In the presence of catalysts, acetylene can cyclize to form benzene and cyclooctatetraene or hydroquinone, then reacted with carbon monoxide to yield acrylic acid that further reacted alcohols to make acrylic esters, which can be used to produce acrylic glass.

16.1.4.1 Vinyl Chloride from Acetylene and Polyvinyl Chlorides (PVC)

Vinyl chloride can be produced from acetylene and hydrogen chloride using mercuric chloride as a catalyst. This method has been superseded by the more economical processes based on ethylene in the United States and Europe. It remains the main production method in China. Vinyl chloride is a toxic and carcinogenic gas that has to be handled with special protective procedures.

Polyvinyl chlorides (PVC's) are the third-most widely produced plastics, after polyethylenes (PE) and polypropylenes (PP). PVC is made from vinyl chloride (chloroethane), a product of the reaction between ethylene with oxygen and hydrogen chloride over a copper catalyst.

PVC is used as a corrosion-resistant material for pipes in modern plumbing, siding for houses, decking, and many other applications in construction material supply chains. Adding plasticizers, mostly phthalates, make them flexible for use in clothing, upholstery, electric cable insulation, inflatable products, and rubber replacement. Flexible PVC is also used in medical tubing to replace glass and rubber. PVC is better because it is more resistant to other chemicals than rubber and, unlike glass, it does not break.

The precursor of polyvinyl acetate, vinyl acetate, can also be produced from acetylene via the gas-phase addition of acetic acid in the presence of metal catalysts, such as mercury(II) catalysts. Like vinyl chloride, the industrial route has been superseded by the ethylene-based process which will be discussed later.

16.1.4.2 Chloroprene from Acetylene

In the production of chloroprene, methane is partially oxidized to acetylene. Acetylene is dimerized to give vinyl acetylene, which is then combined with hydrogen chloride to afford 4-chloro-1,2-butadiene (an allene derivative), which in the presence of cuprous chloride, rearranges to the targeted 2-chlorobuta-1,3-diene (chloroprene):

$$HC\equiv C-CH=CH_2 + HCl \rightarrow H_2C=C=CH-CH_2Cl$$
$$H_2C=C=CH-CH_2Cl \rightarrow H_2C=CCl-CH=CH_2$$

Chloroprene is the monomer source of polychloroprene, known as neoprene. Neoprene rubber resists degradation from sun, ozone, and weather and performs well in contact with oils and chemicals. Hence, neoprene is used in a wide variety of applications: gaskets, hoses, corrosion-resistant coatings, adhesives, noise isolation in power transformers, fire-resistant weather stripping, etc.

16.1.5 Methane to Formaldehyde

Formaldehyde is produced industrially by the catalytic oxidation of methanol. In the commonly used Formox process [13], formaldehyde is produced by reacting methanol with oxygen at ca. 250–400°C in presence of iron oxide in combination with molybdenum and/or vanadium:

$$2CH_3OH + O_2 \rightarrow 2CH_2O + 2H_2O$$

Another catalyst is silver-based which usually operates at a higher temperature, about 650 °C. It also undergoes dehydrogenation to produces hydrogen:

$$CH_3OH \rightarrow CH_2O + H_2$$

Formaldehyde is an important precursor to many other materials and chemical compounds, including urea formaldehyde resin, melamine resin, phenol formaldehyde resin, polyoxymethylene plastics, 1,4-butanediol, and methylene diphenyl diisocyanate. The textile industry uses formaldehyde-based resins as finishers to make fabrics crease-resistant. Formaldehyde-based materials are keys to the manufacture of automobiles, and used to make components for the transmission, electrical system, engine block, door panels, axles and brake shoes.

When treated with phenol, urea, or melamine, formaldehyde produces hard thermoset phenol formaldehyde resin, urea formaldehyde resin, and melamine resin, respectively. These polymers are common permanent adhesives used in plywood and carpeting. Production of formaldehyde resins accounts for more than half of formaldehyde consumption.

16.1.6 Production of Hydrogen Cyanide (HCN)

The most important process for producing hydrogen cyanide is the reaction of methane and ammonia in the presence of oxygen at about 1200 °C over a Pt catalyst (Andrussow oxidation [14]):

$$CH_4 + 2NH_3 + 3O_2 \rightarrow 2HCN + 6H_2O$$

In another process of lesser importance, no oxygen is added:

$$CH_4 + NH_3 \rightarrow HCN + 3H_2$$

Hydrogen cyanide (HCN) gas or vapor is highly toxic and fatal in sufficient concentrations. It was used in chemical warfare and is now prohibited for that purpose by international laws. It is a highly valuable precursor to many chemical compounds ranging from polymers to pharmaceuticals. Aqueous sodium cyanide, in concentrations ranging from 100 to 500 parts per million, is used in gold and silver mining. Cyanide salts also are used in electroplating these metals. HCN reacts with aldehydes and ketones to form cyanohydrins ($R_2C(OH)CN$), which are intermediates in many organic syntheses. It reacts with ethylene oxide to form an intermediate which eventually is converted to acrylonitrile.

16.1.7 Chlorine Solvents, Vinyl Chloride and Vinyl Acetate

Methyl chloride, methylene dichloride, chloroform and carbon tetrachloride can be produced from the reactions of methane with chlorine. They can also be produced from methanol, as mentioned above. The reaction of acetylene with chlorine produces chloroacetylenes. Acetylene can also react with hydrogen chloride or acetic acid to form vinyl chloride and vinyl acetate.

16.2   Ethane and Ethylene

The major component of natural gas is methane, accompanied by natural gas liquids (NGL), i.e., ethane, propane, butanes and natural gasoline (C_5+). The components of NGL are separated in a gas processing plant through absorption, condensation, fractionation, or other methods. The separation of ethane, propane, and butanes is accomplished in a series of tall fractionation columns, similar to those shown in Fig. 11.4.

Once the ethane is separated, it is shipped by pipeline to a cracker facility, which is a very sophisticated series of processes that convert the ethane to ethylene. The first step entails using steam to transport ethane or a mixture of ethane with a small amount of propane to a series of cracking furnaces, where it is heated to approximately 1500 °F. This requires a lot of energy. At that temperature, dissociation of C–H or C–C bonds generates free radicals, which through a chain of reactions are converted to ethylene, hydrogen, and heavier products (Fig. 16.3). In commercial units, process conditions are selected to minimize follow-on reactions, such as radical addition to generate C_3+ compounds and cyclization to generate aromatics, polyaromatics, and coke. Typically, ethylene comprises about 80% of the total product. In the same fashion, propane is converted to propylene and other products.

Alternative feeds to cracker furnaces include sulfur-free naphtha and highly upgraded heavy oils, such as hydrocracker unconverted oil. Ethylene and other olefins also come from refinery cracking units. Especially important are propylene and butylenes from FCC units [15].

Due to its highly reactive double bond, ethylene is particularly well-suited for many different chemical reactions. It is one of the most important chemicals in all chemistry. The chemicals derived from ethylene are summarized in Fig. 16.4.

Ethanol can be produced by hydration of ethylene with high pressure steam at 300 °C (572 °F) in a 1.0:0.6 ratio of ethylene:steam, catalyzed by phosphoric acid adsorbed onto a porous support such as silica gel or diatomaceous earth. Ethanol can also be produced by reacting ethylene with sulfur acid to form ethyl sulfate, which in turn reacts with water to form ethanol. Upon partial oxidation over a silver catalyst at about 500–650 °C, ethanol can be converted to acetaldehyde. However, the main production method is the oxidation of ethylene via the Wacker process which involves oxidation of ethylene over a homogeneous palladium/copper catalyst:

$$2CH_2{=}CH_2 + O_2 \rightarrow 2CH_3CHO$$

Ethylene also reacts with other hydrocarbons in a chemical process called alkylation. Benzene reacts with ethylene over a highly acidic zeolite catalyst to form ethylbenzene. Ethylbenzene is a derivative petrochemical building block found in the manufacturing supply chains for styrofoam and automobile tires.

Radical formation

$CH_3CH_3 \rightarrow 2\ CH_3\bullet$

Hydrogen abstraction

$CH_3\bullet\ +\ CH_3CH_3 \rightarrow CH_4\ +\ CH_3CH_2\bullet$

Radical decomposition

$CH_3CH_2\bullet \rightarrow CH_2{=}CH_2 + H\bullet$

$CH_3CH_2CH_2CH_2\bullet\ \rightarrow\ CH_3CH_2CH{=}CH_2$ -or- $CH_3CH{=}CHCH_3 + H\bullet$

Radical addition

$CH_3CH_2\bullet\ +\ CH_2{=}CH_2\ \rightarrow\ CH_3CH_2CH_2CH_2\bullet$

$CH_3CH_2\bullet\ +\ CH_3CH_2CH{=}CH_2\ \rightarrow\ CH_3CH_2CH_2CH_2CH{=}CH_2$ -or- $CH_3CH_2CH_2CH{=}CHCH_3$

Cyclization

$CH_3CH{=}CHCH_2CH_2CH_2\bullet\ \rightarrow\ C_6H_6\,(benzene)\ +\ 2\ H_2\ +\ H\bullet$

$R\bullet\ +\ benzene\ \rightarrow alkylbenzenes, polyring compounds \rightarrow coke$

Termination

$CH_3\bullet\ +\ CH_3CH_2\bullet\ \rightarrow\ CH_3CH_2CH_3$

$CH_3CH_2\bullet\ +\ CH_3CH_2\bullet\ \rightarrow\ CH_2{=}CH_2 + CH_3CH_3$

$H\bullet\ +\ H\bullet\ \rightarrow\ H_2$

Fig. 16.3 Thermal cracking of ethane: mechanism

16.2.1 Polyethylenes

After purification, ethylene goes from the cracker facility to a nearby complex in which it undergoes polymerization to polyethylene.

The polymerization units are custom-designed and operated to control the specific physical properties of the products, which typically are formed into small plastic pellets called polyethylene resins. The different resins can be transformed into diverse products with many shapes, sizes, and colors. Products range from grocery bags, plastic bottles, and plastic toys to sophisticated, high tech military helmets.

In addition to conventional plastics, polymers can be thermoplastics, thermosets, elastomers, rubbers, and fibers. *Thermoplastics* can be shaped and molded easily when heated, and they melt when they are hot enough. Thermoplastics possess a glass transition temperature, Tg. They are soft and pliable above the Tg, and hard and brittle below. The Tg is different for each type of polyethylene. Depending on the crystallinity and molecular weight of a plastic, a melting point and Tg may or may not be observable. Sometimes, additives called plasticizers are introduced to make a plastic softer and more pliable.

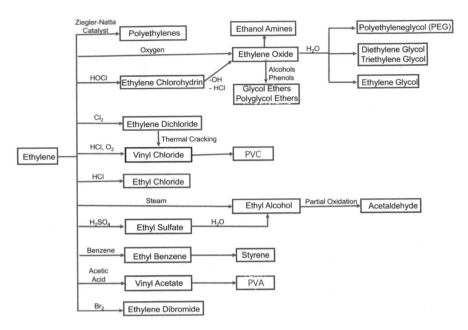

Fig. 16.4 Chemicals from ethylene

Unlike plastics or thermoplastics, crosslinked materials, such as rubbers, don't melt when they are hot. These are called ***thermosets***. Under enough mechanical stress, plastics and thermosets tend to deform permanently under stress or break when they are stretched too hard. ***Elastomers***, on the other hand, can be deformed easily, and they bounce back when the stretching force is released. They are more pliable than *fibers*, which stretch very little when pulled.

Polyethylenes can have the very simple linear structure shown below. Highly linear products are called high density polyethylene (HDPE).

$$CH_2{=}CH_2 \longrightarrow -[CH_2\text{-}CH_2]_n$$

$$\text{www}-\overset{\displaystyle H}{\underset{\displaystyle H}{C}}-\overset{\displaystyle H}{\underset{\displaystyle H}{C}}-\overset{\displaystyle H}{\underset{\displaystyle H}{C}}-\overset{\displaystyle H}{\underset{\displaystyle H}{C}}-\overset{\displaystyle H}{\underset{\displaystyle H}{C}}-\overset{\displaystyle H}{\underset{\displaystyle H}{C}}-\overset{\displaystyle H}{\underset{\displaystyle H}{C}}-\overset{\displaystyle H}{\underset{\displaystyle H}{C}}-\overset{\displaystyle H}{\underset{\displaystyle H}{C}}-\overset{\displaystyle H}{\underset{\displaystyle H}{C}}-\overset{\displaystyle H}{\underset{\displaystyle H}{C}}-\text{www}$$

Other polyethylene chains can have carbon-carbon links to neighboring polyethylene chains. These branched materials are called low density polyethylene (LDPE). Representative structures of HDPE and LDPE are shown below:

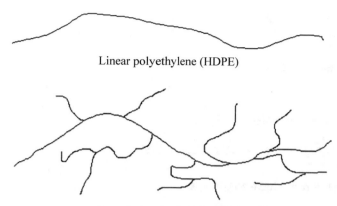

Linear polyethylene (HDPE)

Branched polyethylene (LDPE)

LPDE are made by free radical vinyl polymerization, while HDPE are made catalytically over alkyl aluminum and titanium (III) chloride in the Ziegler-Natta process [16], or silica-supported chromium (VI) oxide in the Phillips process [17]. Both processes are highly exothermic. A simplified free-radical mechanism is illustrated by Fig. 16.5. Reaction conditions are set to maximize radical addition (Reactions 8 and 9) and minimize cyclization (Reactions 10 and 11).

HDPE's are stronger than LDPE's, but LDPE's are cheaper and easier to make. HDPE's have a low degree of branching and thus stronger intermolecular forces and tensile strength. HDPE's can be found in many plastic containers and packaging, such as milk jugs, detergent bottles, margarine tubs, garbage containers and water pipes. One third of all toys are manufactured from HDPE. LDPE's are used for both rigid containers and plastic film applications, such as plastic bags and film wrap.

Ziegler-Natta polymerization can also produce LDPE via incorporation of an alkyl-branched monomer, which reacts with ethylene to form a co-polymer with short hydrocarbon branches.

The density of HDPE is greater than or equal to 0.941 g/cm^3, and the density of LDPE falls in the range of 0.910–0.940 g/cm^3. The molecular weight of linear polyethylene typically ranges from 200,000 to 500,000, but it can be made higher.

Using metallocenes as polymerization catalysts, manufacturers can produce ultra-high molecular weight polyethylenes (UHMWPE) with molecular weights in the millions, usually between 3.1 and 5.67 million [18, 19]. The densities are less than for HDPE (for example, 0.930–0.935 g/cm^3). They are used in a diverse range of applications. These include can and bottle handling machine parts, moving parts on weaving machines, bearings, gears, artificial joints, edge protection on ice rinks, and butchers' chopping boards.

There are many other types of polyethylenes, including linear low-density polyethylenes (LLDPE) and cross-linked polyethylenes (PEX or XLPE). The density range of LLDPE is 0.915–0.925 g/cm^3. LLDPE's are essentially linear polymers with significant numbers of short branches. They commonly are made by copolymerization of ethylene with short-chain alpha olefins (for example, 1-butene, 1-hexene and 1-octene) [20].

Chain reaction initiation

(1) $X_2 \rightarrow 2\,X\bullet$

(2) $R\text{-}X \rightarrow R\bullet + X\bullet$

(3) $X\bullet + H_2 \rightarrow HX + H\bullet$

(4) $X\bullet + CH_2{=}CH_2 \rightarrow X\text{-}\overset{|}{\underset{|}{C}}\text{-}\overset{|}{\underset{|}{C}}\bullet$

(5) $X\text{-}\overset{|}{\underset{|}{C}}\text{-}\overset{|}{\underset{|}{C}}\bullet + H_2 \rightarrow CH_3CH_2\bullet + HX$

Radical decomposition to olefins

(6) $CH_3CH_2\bullet \rightarrow CH_2{=}CH_2 + H\bullet$

(7) $CH_3CH_2CH_2CH_2\bullet \rightarrow CH_3CH_2CH{=}CH_2$ -or- $CH_3CH{=}CHCH_3 + H\bullet$

Chain propagation (radical addition)

(8) $CH_3CH_2\bullet + CH_2{=}CH_2 \rightarrow CH_3CH_2CH_2CH_2\bullet$

(9) $CH_3CH_2CH_2CH_2\bullet + CH_2{=}CH_2 \rightarrow CH_3CH_2CH_2CH_2CH_2CH_2\bullet \rightarrow \rightarrow$

Cyclization

(10) $CH_3CH{=}CHCH_2CH_2CH_2\bullet \rightarrow$ benzene $+ 2\,H_2 + H\bullet$

(11) $R\bullet +$ benzene $\rightarrow$ alkylbenzenes, multiring compounds $\rightarrow$ coke

Chain termination

(12) $CH_3\bullet + CH_3CH_2\bullet \rightarrow CH_3CH_2CH_3$

(13) $CH_3CH_2\bullet + CH_3CH_2\bullet \rightarrow CH_2{=}CH_2 + CH_3CH_3$

(14) $H\bullet + H\bullet \rightarrow H_2$

(15) $X\bullet + H\bullet \rightarrow HX$

Fig. 16.5 Mechanism for free-radical formation of polyethylene

In addition to copolymerization with alpha-olefins, ethylene can also be copolymerized with a wide range of other monomers and ionic compositions to create ionized free radicals. Common examples include a variety of acrylates and vinyl acetate, which yield ethylene-vinyl acetate copolymer, or EVA, widely used in athletic-shoe sole foams. Applications of acrylic copolymer include packaging and sporting goods, and superplasticizers, used for cement production.

Cross-linked bonds introduced into the medium- to high-density ethylene polymer structure change a thermoplastic into an elastomer. They find their way into some potable-water plumbing systems, because tubes made of the material can be expanded to fit over a metal nipple. The tubes then slowly return to their original shape, forming a permanent, water-tight, connection.

Other polymers used as plastics include polypropylenes, polyesters, polystyrenes, polycarbonates, PVC, Nylon and polymethyl(methachrylate). Other polymers used as fibers include polypropylenes, polyesters, Nylon, Kevlar, Nomex, polyacronitriles, cellulose, and polyurethanes.

16.2.2 Ethylene Oxide, Ethylene Glycol and Polyethylene Glycol (Polyethylene Oxide)

If oxygen is added to ethylene, it reacts to form ethylene oxide. Ethylene oxide is used to make surfactants and detergents in the manufacturing supply chain for cleaning products. To take it even further, if water is added to ethylene oxide, it reacts in a process called hydrolysis and forms the product ethylene glycol. Ethylene glycol is used as an ingredient in antifreeze. It is also used to make a fiber called polyethylene terephthalate, otherwise known as polyester, which can be found under the brand name Dacron®. Polyester resin is also used to make beverage bottles and a whole plethora of other useful products.

In earlier times, the production of ethylene oxide involved a chlorohydrin process, which was less efficient than the oxidation process that is used presently. Ethylene oxide is presently industrially produced by direct oxidation of ethylene with pure oxygen in the presence of a silver catalyst.

Ethylene oxide is rather stable in water, but in the presence of a small amount of acid, such as diluted sulfuric acid, it immediately forms ethylene glycol at room temperature. The reaction also occurs in the gas phase, in the presence of a phosphoric acid salt as a catalyst.

$$(CH_2CH_2) + H_2O \rightarrow HO-CH_2CH_2-OH$$

The reaction of ethylene oxide with water does not only end up with ethylene glycol, it can continue to form diethylene glycol (DEG), triethylene glycol (TEG), etc. The formation of these higher glycols is inevitable because ethylene oxide reacts faster with ethylene glycols than with water. The most important variable is the water-to-oxide ration, and in commercial plants the production of DEG and TEG can be reduced by using a large excess of water.

Polyethylene glycol (PEG) is produced by the interaction of ethylene oxide with water, ethylene glycol, or ethylene glycol oligomers, catalyzed by either acidic or basic catalysts. The reactions with ethylene glycol and its oligomers are preferable to produce polymers with a low polydispersity, i.e., narrow molecular weight distribution. Polymer chain length depends on the ratio of reactants.

$$HOCH_2CH_2OH + n(CH_2CH_2O) \rightarrow HO(CH_2CH_2O)_{n+1}H$$

Low molecular weight PEG's are prepared using alkali catalysts, such as sodium hydroxide, potassium hydroxide or sodium carbonate. High molecular weight PEG's, or polyethylene oxides, are synthesized by suspension polymerization to hold the growing polymer chain in solution. The catalysts are magnesium-, aluminum-, or calcium-organoelement compounds.

PEG has been designated as nontoxic and is approved by Federal Drug Administration (FDA) for applications in pharmaceutical formulations, foods and cosmetics. Polyethylene glycol 3350 is in a class of medications called osmotic laxatives to treat occasional constipation. PEG is also used in biomedical research.

16.2.3 Vinyl Acetate and Polyvinyl Acetate (PVA)

The major industrial route of producing vinyl acetate involves the reaction of ethylene and acetic acid with oxygen in the presence of a palladium catalyst.

$$C_2H_4 + CH_3COOH + {}^1/_2O_2 \rightarrow CH_3COOCHCH_2 + H_2O$$

Vinyl acetate can be polymerized to give polyvinyl acetate (PVA). PVA emulsions in water are used as adhesives for porous materials, particularly for wood, paper, and cloth, and as a consolidant for porous building stone, in particular sandstone [21].

With other monomers vinyl acetate can be used to prepare copolymers such as ethylene-vinyl acetate (EVA), vinyl acetate-acrylic acid (VA/AA), polyvinyl chloride acetate (PVCA), etc.

16.2.4 Vinyl Chloride and Polyvinyl Chloride (PVC)

Vinyl chloride is made from ethylene, oxygen and hydrogen chloride (oxychlorination) over a copper(II) chloride catalyst. It can also be produced by ethylene with chlorine over an iron(III) chloride catalyst or by thermal cracking of ethylene dichloride at 500 °C under 15–30 atm (1.5–3 MPa) pressure. Vinyl chloride is a gas with sweet odor. It is highly toxic, flammable and carcinogenic, and therefore must be handled with care.

Upon the action of an initiator, the double bond of vinyl chloride is opened, resulting in thousands of monomer units linked together to form a polymer. Polyvinyl chloride (PVC) is the third-most widely produced plastic, after PE and polypropylene (PP). It's used in construction for pipe and profile applications [22]. By adding plasticizers, mostly phthalates, it becomes flexible for clothing and upholstery, electric cable insulation, inflatable products and rubber replacement.

16.3 Propane, Butane, Propylene and Butylenes

Propane and butanes are by-products of natural gas processing and petroleum refining. The processing of natural gas involves the removal of non-methane natural gas liquids (NGL) including ethane, propane, butanes, condensate, and natural gasoline from the raw natural gas, in order to prevent condensation of these volatiles in natu-

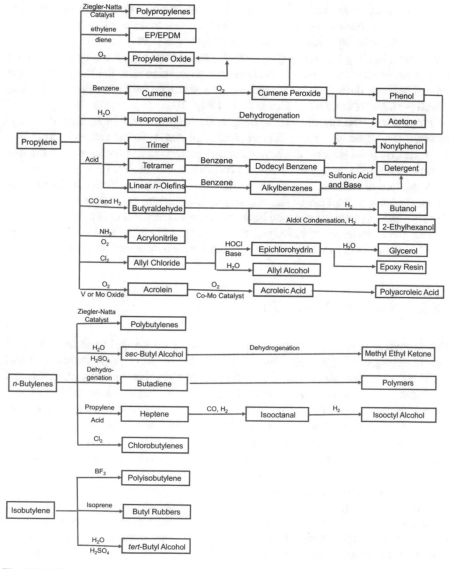

Fig. 16.6 Chemicals from propylene, *n*-butylenes and isobutylene

ral gas pipelines, which would reduce capacity and damage the turbine compressors. Ethane, propane, and butanes (normal butane and isobutane) are separated in a gas processing plant through a fractionation train consisting of three distillation towers in series: a deethanizer, a depropanizer and a debutanizer. The overhead product from the deethanizer is ethane and the bottoms are fed to the depropanizer. The overhead product from the depropanizer is propane and the bottoms are fed to the debutanizer. The overhead product from the debutanizer is a mixture of normal and iso-butane, and the bottoms product is a C_5 + mixture. Isobutane can be further separated from n-butane using a deisobutanizer. Propane and butanes are also by-products of cracking petroleum into gasoline and cycle oils in the FCC and hydrocracking processes.

Propane can be liquefied as liquefied petroleum gas (LPG) and sold in the market for heating, lighting and cooking. It can also be an important feedstock for making propylene. Normal butane can be blended into gasoline to raise the Reid vapor pressure of gasoline during winter in cold weather. It can be isomerized into isobutane, which is an alkylation feed as discussed in Chap. 10.

Propylene can be made from dehydrogenation of propane, with hydrogen as a by-product. Propylene also comes from cracking processes in the refinery. Propylene is the second most important feedstock in the petrochemical industry. It is the raw material for a wide variety of chemical products, shown in Fig. 16.6.

A mixture of butylenes (isobutylene, butene-1 and butene-2) is a by-product of steam cracking of naphtha or gas oils in the manufacture of ethylene. These other light paraffins and olefins can also be minor components of refinery fuel gas.

Propylene, butylene and isobutylene are important chemical feedstocks, with representative products summarized in Fig. 16.6 for propylene and butylenes.

16.3.1   Polypropylenes (PP)

Polypropylene (PP), also known as polypropene, is the second most used thermoplastic polymer, after polyethylene. The structure of polypropylene has a similar backbone as polyethylene with the methyl groups attached to every other carbon. Different polypropylenes have different tacticities. A Ziegler-Natta catalyst is able to restrict the linkage of monomer units to a specific regular orientation, either *isotactic*, when all methyl groups are positioned on the same side with respect to the backbone of the polymer chain, or *syndiotactic*, when the positions of the methyl groups alternate. Isotactic polypropylene is made commercially with two types of Ziegler-Natta catalysts, which will be described in the next section. The first group of catalysts encompasses solid (mostly supported) catalysts or soluble metallocene catalysts containing Ti, Zr, or Hf with an organoaluminum co-catalyst. Isotactic macromolecules coil into a helical shape to line up next to one another to form the crystals. Another type of metallocene catalysts produces syndiotactic polypropylene. These macromolecules also coil into helices (of a different type) and form crystalline materials.

Commercial synthesis of syndiotactic polypropylene is carried out with metallocene catalysts over bridged bis-metallocene complexes of the type bridge-$(Cp_1)(Cp_2)ZrCl_2$, where the first Cp_1 ligand is the cyclopentadienyl group, the second Cp_2 ligand is the fluorenyl group, and the bridge between the two Cp ligands is $-CH_2-CH_2-$, $>SiMe_2$, or $>SiPh_2$ [19, 23]. These complexes are converted to polymerization catalysts by activating them with a special organoaluminum co-catalyst, methylaluminoxane (MAO) [24].

When the methyl groups in a polypropylene chain exhibit no preferred orientation, the polymers are called *atactic*. Atactic polypropylene is an amorphous rubbery material. It can be produced commercially either with a special type of supported Ziegler-Natta catalyst or with some metallocene catalysts.

Most commercial polypropylene is isotatic and has an intermediate level of crystallinity between that of LDPE and HDPE. Polypropylene is normally tough and flexible, especially when copolymerized with ethylene. This allows polypropylene to be used as an engineering plastic, competing with materials such as acrylonitrile butadiene styrene (ABS).

PP is rugged and unusually resistant to many chemical solvents, bases and acids. It's used in a wide variety of applications including packaging, textiles (e.g., ropes, thermal underwear and carpets), plastic parts and reusable containers, laboratory equipment, and automotive components.

16.3.2 Ethylene-Propylene (EP) Copolymers/Ethylene-Propylene-Diene (EPDM) Terpolymer

The polymerization of ethylene and propylene together produces ethylene-propylene (EP or EPM) copolymer or ethylene-propylene rubber. They can also be in combination with a diene (about 5%) – usually ethylidene norbonene or 1,4-hexadiene to make EPDM terpolymer.

Both EP and EPDM copolymers are prepared by dissolving ethylene and propylene in an organic solvent, such as hexane for EP and liquid diene for EPDM, and subjecting them to reaction over Ziegler-Natta catalysts.

In addition to elastic properties, the chemically saturated, stable EP copolymer backbone provides excellent resistance to heat, oxidation, ozone and weathering. The EPDM terpolymer maintains a saturated backbone and places the reactive unsaturation in a side chain available for vulcanization or polymer modification chemistry.

Versatility in polymer design and performance has resulted in broad usage in automotive weather-stripping and seals, glass-run channel, radiator, garden and appliance hose, tubing, belts, electrical insulation, roofing membrane, rubber mechanical goods, plastic impact modification, thermoplastic vulcanizates and motor oil additive applications.

16.3.3 Isopropyl Alcohol (IPA) and Acetone

Isopropyl alcohol (IPA) is easily synthesized from the reaction of propylene with sulfuric acid, followed by hydrolysis of isopropyl sulfate. IPA can also be produced by reacting propylene with water, either in gas phase or liquid phase, at high pressure over a solid or supported catalyst. IPA forms an azeotrope with water. Simple distillation yields 88% IPA and 12% water. Azeotropic distillation with agents such as diisopropyl ether or cyclohexane is required for getting pure IPA.

IPA can be oxidized to acetone using oxidizing agent, such as chromic acid or by dehydrogenation over Raney nickel or a mixture of copper and chromium oxide. Both IPA and acetone are important solvents in the industry. They are important intermediates for making a variety of chemicals. IPA has been used as a mixture with water as rubbing alcohol and other medical sanitary purposes.

16.3.4 Cumene to Acetone and Phenol

Cumene, also known as isopropylbenzene, is produced commercially by Friedel-Crafts alkylation of benzene with propylene using H_3PO_4 as a catalyst. It is a building-block to manufacture other chemicals, such as phenol, acetone, acetophenone, and methyl styrenes. Over 98% of cumene is used to produce phenol and its co-product acetone, which are widely used to make plastics. Cumene hydroperoxide is produced by oxidation of cumene. The oxidation by cumene hydroperoxide affords propylene oxide and the byproduct cumyl alcohol. The largest phenol derivative, biphenol-A (BPA) supplies the growing polycarbonate (PC) sector. PC resins are consumed in automotive applications in place of traditional materials such as glass and metals. Cumene in minor amounts is used as a thinner or solvent in paints, lacquers, and enamels. It is also a component of high-octane motor gasoline.

Phenolic resins are largely used as durable binders and adhesives in wood panels and insulation. The derivative, caprolactam, is mainly the engineering resin sector of nylon market.

Other than as solvent, the acetone derivative methyl methacrylate (MMA) is used to make homopolymers and copolymers in electronic applications, such as TV flat screens.

16.3.5 Acrylic Acid and Acrolein

Acrylic acid is produced by two stages of vapor oxidation of propylene over catalyst. In the first stage, propylene is oxidized to acrolein over a highly active and very selective catalyst (V or Mo oxides). In the second stage, acrolein is further oxidized over a Co-Mo catalyst at 200–300 °C.

Acrylic acid is miscible with water, alcohols, ethers and chloroform. Acrylic acid is a precursor of a wide variety of chemicals in the polymer and textile industries. Examples are manufacture of plastics, latex applications, in floor polish, in polymer solution for coating applications, emulsion polymers, paint formulation, leather finishing and paper coating. Acrylic acid is used to make acrylic esters and resins, chemicals added to protective coatings and adhesives, and the production of polyacrylic acid polymers.

Other than oxidation of propylene, acrolein can also be obtained by thermal decomposition of glycerol at 280 °C. Acrolein is mainly used as a contact herbicide to control submersed and floating weeds, as well as algae, in irrigation canals. In the oil and gas industry, it is used as a biocide in drilling fluids, as well as a scavenger for hydrogen sulfide and mercaptans.

16.3.6 Nonylphenols (Alkylphenyls)

Propylene can be oligomerized to trimers, tetramers and other olefins in the presence of acids. Nonylphenols are produced by acid-catalyzed alkylation of phenol with a mixture of propylene trimers. They are a family of closely related organic compounds called alkylphenols, and used in manufacturing antioxidants, lubricating oil additives, detergents, emulsifiers, cosmetics, and insecticides. These compounds are important non-ionic surfactants alkylphenol ethoxylates and nonylphenol ethoxylates, which are used in detergents, paints, pesticides, personal care products, and plastics.

16.3.7 Dodecylbenzene and Linear Alkylbenzenes

Dodecylbenzene and some related linear alkylbenzenes are produced industrially by alkylation of benzene with dodecane (propylene tetramer) and corresponding linear mono-olefins in the presence of hydrogen fluoride, aluminum chloride or related acid catalysts. The resulting linear dodecylbenzenes are sulfonated to yield corresponding allkylbenzene sulfonic acids. The acids are then neutralized with base to give biodegradable alkylbenzene sulfonates, which are subsequently blended with other components to give detergents and various cleaning products.

16.3.8 Alkyl Chloride, Epichlorohydrin and Glycerol

Allyl chloride is produced by the chlorination of propylene. At lower temperature, the main product is 1,2-dichloropropane, but at high temperature (500 °C), allyl chloride is predominant through a free-radical reaction. The great majority of allyl chloride is converted to epichlorohydrin by treating with base. Other derivatives include allyl alcohol, allylamine, allyl isothiocyanate and allyl silane.

Epichlorohydrin is mainly converted to bisphenol A diglycidyl ether (see below) which is a building block of epoxy resin. The other product is glycerol. This source competes with the rapid increase in production of glycerol from biodiesel, which is creating a glut in the market.

Synthetic glycerol is now used only in sensitive pharmaceutical, technical and personal care applications where quality standards are very high.

$$CH_2CHOCH_2Cl + 2H_2O \rightarrow HOCH_2CH(OH)CH_2(OH) + HCl$$

16.3.9 Acrylonitrile and Acetonitrile

Most industrial acrylonitrile (ACN) is produced by catalytic ammoxidation of propylene, known as the SOHIO process [25], with acetonitrile and hydrogen cyanide as bi-products.

$$2CH_3-CH=CH_2 + 2NH_3 + 3O_2 \rightarrow 2CH_2=CH-C\equiv N + 6H_2O$$

The reactants pass through a fluidized bed reactor containing bismuth phosphomolybdate supported on silica at 410–510 °C and 50–200 kPa. The reaction product is quenched by aqueous sulfuric acid, yielding a mixture containing acrylonitrile, acetonitrile, hydrocyanic acid and ammonium sulfate. After removal of water-soluble components, acrylonitrile and acetonitrile are separated by distillation.

Acylonitrile (ACN) is highly flammable and toxic. It's also suspected to cause lung cancer through exhaust emission and cigarette smoking.

ACN is used mainly as a monomer or comonomer in the production of synthetic fibers, plastics and elastomers. It is the monomer source of polyacrylonitrile (PAN), a homopolymer, and precursor of high quality carbon fiber; styrene-acylonitrile (SAN); acylonitrile butadiene styrene (ABS); acrylonitrile styrene acrylate (ASA) and other rubbers, such as acrylonitrile butadiene rubber (NBR).

Nitrile rubbers are used in the automotive and aeronautical industries to fabricate hoses for fuel and oil handling, and to make seals and grommets. Other uses include gloves, footware, adhesives, sealants, sponges, floor mats, etc.

Acrylonitrile is also a precursor in the industrial manufacture of acrylamide and acrylic acid.

16.3.10 Butaldehyde, Butanol and 2-Ethylhexyl Alcohol

Higher carbon number alcohols can be manufactured from lower carbon number olefins with carbon monoxide and hydrogen through hydroformulation via the oxo process, which is discussed in Sects. 16.4 and 16.5.

16.3.11 Polybutyenes

Isotactic polybutylene-1 (PB-1), as is the case for isotactic polypropylene and other poly(olefin-1), is synthesized commercially using two types of heterogeneous Ziegler-Natta catalysts [26]. The first type of catalyst contains two components, a solid pre-catalyst (the δ-crystalline form of $TiCl_3$), and a solution of an organoaluminum cocatalyst, such as $Al(C_2H_5)_3$. The second type of pre-catalyst is supported. The active ingredient in the catalyst is $TiCl_4$ and the support is microcrystalline $MgCl_2$.

Polybutylenes are used in piping systems, plastic packaging, hot melt adhesives (for its compatibility with tackifier resins), domestic water heaters, wire and cable, shoe soles, and polyolefin modification (for improving thermal bonding, enhancing softness, increasing temperature resistance, etc.).

Polybutylene water pipes are no longer accepted by US and Canada because they leak due to corrosion caused by chlorinated water, but they are still available in Europe and Asia (e.g., UK, South Korea and Spain).

16.3.12   Polyisobutylenes (PIB)

Isobutylene was discovered by Michael Faraday in 1825. Polyisobutylene (PIB) was first developed by BASF in 1931 using a boron trifluoride catalyst at low temperatures. In the industry today, polyisobutylenes are produced by ionic polymerization of the monomer, isobutylene, at temperatures from -80 to $-100\,°C$; they are processed using the ordinary equipment of the rubber industry [27, 28].

Polyisobutylenes are either viscous liquids with molecular weights of 10,000–50,000 or rubbery, amorphous products with molecular weights of 70,000–225,000. Their softening point is 185–200 °C and they do not decompose up to 350 °C, although their mechanical properties deteriorate significantly even at 100 °C; they retain their elasticity down to $-50\,°C$.

The characteristic features of PIB are low gas permeability and high resistance to the action of acids, alkalis, and solutions of salts, as well as high dielectric indexes (loss tangent, 0.0002 at 50 Hz). They degrade gradually under the action of sunlight and ultraviolet rays. The addition of carbon black slows this process.

PIB are soluble in hydrocarbons, chlorinated hydrocarbons, and ether. They combine easily with natural or synthetic rubbers, polyethylene, polyvinyl chloride, and phenol-formaldehyde resins. PIB's are used for electrical insulation and as anticorrosion coatings for chemical apparatus and pipelines, adhesives, waterproof fabrics, and hermetic compounds. Polyisobutylenes with molecular weights of 10,000–20,000 are used as additives and thickeners in lubricants.

16.3.13   Butyl, Chlorobutyl and Bromobutyl Rubbers

Butyl rubber [27, 28] was invented in 1937 by Standard Oil of New Jersey (now ExxonMobil). During World War II, this synthetic rubber enabled Canada and the United States to manufacture tires after supplies of natural rubber from Southeast Asia were cut off. Butyl rubber is a copolymer of isobutylene with isoprene, produced by polymerization of about 98% of isobutylene with about 2% of isoprene for crosslinking by vulcanization, with the following representative structure. It is impermeable to air and used in many applications requiring an airtight rubber.

$$\sim\sim CH_2 - \underset{\underset{CH_3}{|}}{\overset{\overset{CH_3}{|}}{C}} - \left[CH_2 - \underset{\underset{CH_3}{|}}{\overset{\overset{CH_3}{|}}{C}} \right]_n CH_2 - \underset{\overset{CH_3}{|}}{C} = CH - CH_2 - CH_2 - \underset{\underset{CH_3}{|}}{\overset{\overset{CH_3}{|}}{C}} \sim\sim$$

PIB and butyl rubber are also used in adhesives, agricultural chemicals, fiber optic compounds, ball bladders, caulks and sealants, cling film, electrical fluids, lubricants (2-cycle engine oil), paper and pulp, personal care products, pigment concentrates, for rubber and polymer modification, for protecting and sealing certain equipment for use in areas where chemical weapons are present, as a gasoline/diesel fuel additive, and even in chewing gum.

As a fuel additive, PIB has detergent properties. When added to diesel fuel, it resists fouling of fuel injectors, leading to reduced hydrocarbon and particulate emissions. It is blended with other detergents and additives to make a "detergent package" for gasoline and diesel fuel to resist buildup of deposits. Such deposits can catalyze pre-ignition, so preventing their formation reduces engine knock.

Butyl rubber and halogenated rubber are used for the inner liner ("inner tube") that holds air in tires. Bromobutyl has air impermeability and resists UV radiation properties which make it ideal for tubeless tires.

16.3.14   Butadiene

The butadiene isomers of importance for production of synthetic rubbers is 1,3-butadiene while 1,2-butadiene has no industrial significance. 1,3-butadiene is a by-product in steam cracking of liquid hydrocarbon mixtures to produce ethylene and other light olefins. It is isolated from other C_4-olefins by extractive distillation using a polar aprotic solvent such as acetonitrile, N-methylpyrrolidone, furfural, or dimethyl-formamide, from which it is then stripped by distillation. 1,3-Butadiene can also be produced by catalytic oxidative dehydrogenation of normal butenes or normal butane in the Houdry process.

Most butadiene is polymerized to polyisobutylene and with other co-monomers to produce tough and/or elastic synthetic rubber, such as acrylonitrile butadiene styrene (ABS), acrylonitrile butadiene rubber (NBR) and stryrene-butadiene rubber (SBR). SBR is the material most commonly used for the production of automobile tires. ABS is one of the most widely-used thermoplastics for basic everyday items such as computer keyboards, kitchen appliances, toys, the plastic guards on wall sockets, auto parts, etc.

Smaller amounts of butadiene are used to make the nylon intermediate, adiponitrile by hydrocyanation, or adipic acid by carbonylation (see below). Carbonylation of butadiene (reaction with CO and water) yields adipic acid. Other synthetic rubber materials such as chloroprene, and the solvent sulfolane are also manufactured from butadiene.

$$CH_2=CH-CH-CH_2 + 2CO + 2H_2O \rightarrow HOOC(CH_2)_4COOH$$

16.4   Hydroformylation (Oxo Process)

Hydroformylation, also known as oxo synthesis or the oxo process, is an industrial process entailing net addition of the formyl group (CHO) and hydrogen atom to a

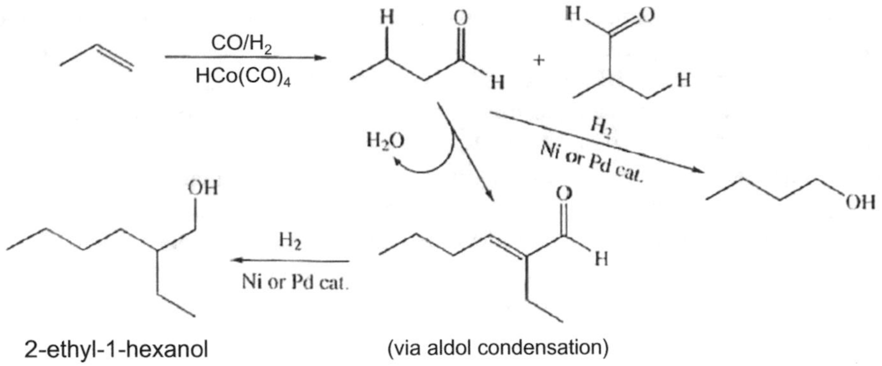

2-ethyl-1-hexanol   (via aldol condensation)

Fig. 16.7 Hydroformylation

carbon-carbon bond. Thus, it converts linear 1-olefins to aldehydes and alcohols, shown in Fig. 16.7. The alcohols produced are called oxo alcohols.

In the process, the olefin is mixed with a 1:1 ratio of CO and H_2 in the presence of a cobalt homogeneous catalyst, $HCo(CO)_4$, at 110–180 °C and 200–300 bar pressure to produce a ratio of normal and branched aldehydes of about 4:1; these can be hydrogenated over a Ni or Pd catalyst to form corresponding alcohols [29]. The addition of an alkyl phosphine ligand, such as tributyl phosphine, requires less CO pressure (50–100 bar) and increases hydrogenation capacity to combine the hydroformylation and hydrogenation stages into one. But the rate is 5–10 times slower. Alternatively, the use of rhodium–phosphine complexes as catalysts reduces the pressure requirements to 50–120 bar with lower temperatures (80–100 °C) because the rhodium complex is more active than the cobalt complex. The linear to branch ratios can be as high as 15:1. The aldehydes can also undergo aldol condensation to form dimer hydroxyl aldehydes, which are hydrogenated to form "dimer" alcohols. To make dimer or higher alcohols through aldol condensation, linear structures are preferred. Hydroformylation is an important process for making plasticizers (diesters of phthalic acid) for plastics and long alkyl alcohols for detergent and surfactants.

16.5 Butanols and Methyl Ethyl Ketone (MEK)

The most common process for making butanol starts with propylene, which undergoes hydroformylation to form normal- and iso-butyraldehydes, which are then reduced with hydrogen to 1-butanol and/or 2-butanol. Secondary and tertiary butanols can be made from butane-1 and isobutylene, respectively, with water in the presence of sulfuric acid. Butanol can also be produced by fermentation of biomass by bacteria (as biobutanol).

Butanol is considered as a potential biofuel (butanol fuel). At 85%, it can be used in cars designed for gasoline without any change to the engine (unlike 85% ethanol), and it contains more energy for a given volume than ethanol and almost as much as gasoline. The consumption of butanol as a fuel in a vehicle would be more comparable to gasoline than ethanol. Butanol can also be added to diesel fuel to reduce soot emissions.

Butanols are used as solvents in a variety of textile processes and many industrial applications. Normal butanol is a solvent for paints, resins and other coatings. A large proportion of n-butanol is converted to esters for various applications, such as butyl acetate as paint solvent and dibutyl phthalate as a plasticizer. Isobutanol is used as a solvent and in plasticizers. s-Butanol is used in solvents and esters to a limited extent. Large amounts are oxidized to methyl ethyl ketone (MEK), an important dewaxing solvent, and in the manufacturing of plastics, fabrics and explosives. t-Butanol is used as a solvent; smaller amounts are used in flavoring and in perfumes.

16.6 BTX Derivatives

In refineries, in addition to hydrogen and other products, catalytic reforming generates benzene, toluene, xylenes (grouped as BTX), and ethylbenzene (EB). EB is normally converted to xylenes as a chemical feedstock. Figure 16.8 shows the pathways by which BTX become petrochemical intermediates.

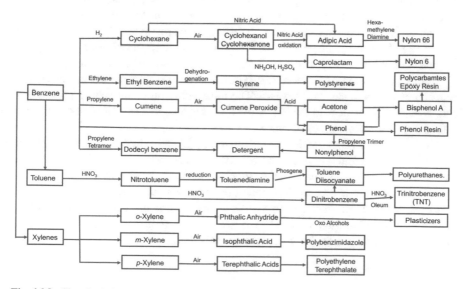

Fig. 16.8 Chemicals from BTX

16.6.1   Benzene Derivatives

Benzene can't be used in gasoline due to health and environmental concerns, but other reformer products—toluene, xylenes, and EB—are high-octane blend stocks. Planners often must decide, based on prices and relative demand, whether to blend these products into gasoline or send them to a chemicals plant.

The reaction of benzene with propylene yields cumene, which produces a mixture of acetone and phenol via cumene peroxide. The condensation of phenol and acetone in a 2:1 molar ratio produces bisphenol A (BPA). BPA is a starting material for the synthesis of plastics, primarily polycarbonates for drinking water bottles and epoxy resins.

Nonylphenol is produced by the acid catalyzed alkylation of phenol with a mixture of nonenes. Nonyl phenol is used in manufacturing antioxidants, lubricating oil additives, laundry and dish detergents, emulsifiers and solubilizers.

Partial hydrogenation of phenol or hydrogenation of benzene gives cyclohexane which yields cyclohexanol and cyclohexanone upon oxidation. The mixture of cyclohexanol and cyclohexanone is oxidized with nitric acid to give adipic acid. The reaction of the acid with hexamethylenediamine produces Nylon 66, a widely used fiber and plastic.

16.6.1.1   Nylons

Nylon is a thermoplastic, silky material and was the first commercially successful synthetic thermoplastic polymer produced by DuPont in the 1930s. Commercially, Nylon 66 is made by reacting adipic acid with hexamethylenediamine, shown on top of Fig. 16.9. Another polymer of Nylon 6 is made from only one monomer of caprolactam through ring opening polymerization, shown in the bottom of Fig. 15.9.

Both adipic acid and caprolactam can be derived from benzene through cyclohexane and cyclohexanone/cyclohexanol. The most common manufacturing process of adipic acid is the nitric acid oxidation of a cyclohexanol and cyclohexanone mixture, called KA (for ketone-alcohol) oil.

Cyclohexane is produced from benzene by liquid phase hydrogentation at a pressure of 1.34 MPa and a temperature of 210 °C using a Raney nickel catalyst. The mixture of cyclohexanol and cyclohexanone is produced by continuous oxidation of cyclohexane by air using a trace of cobalt naphthenate as catalyst.

Caprolactam is produced from cyclohexanone by being converted into oxime with hydroxylamine first, which is then treated with oxime with acid to introduce Beckmann rearrangement for yielding ε-caprolactam.

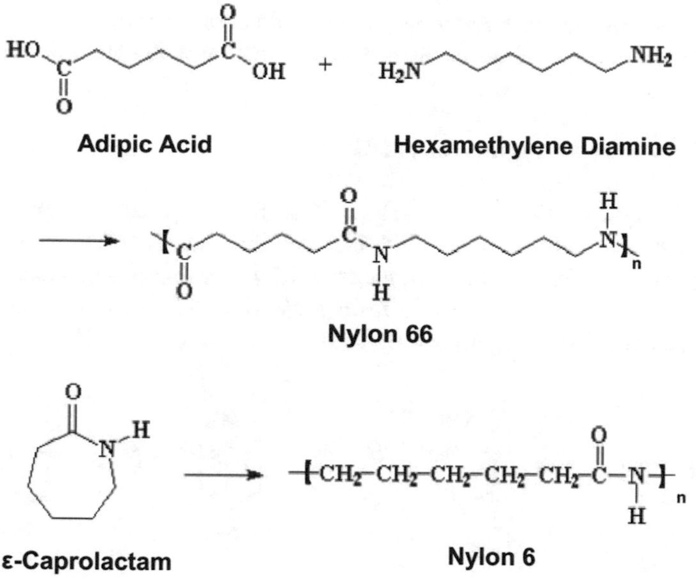

Fig. 16.9 The manufacturing of Nylon 66 and Nylon 6

Nylon polymers have found significant commercial applications in fibers (apparel, flooring (carpet fibers), ropes, conveyor belts, and for rubber reinforcement). It comes in shapes (molded parts for cars, electrical equipment, etc.), and in films (mostly for food packaging) [30].

16.6.2 Toluene Derivatives

16.6.2.1 Trinitrotoluene (TNT)

Trinitrotoluene, specifically 2,4,6-trinitrotoluene, is best known as an explosive material. Its explosive yield is considered to be the standard measure of bombs and other explosives.

In industry, TNT is produced in a three-step process [31]. First, toluene is nitrated with a mixture of sulfuric and nitric acids to produce mononitrotoluene (MNT). The MNT is separated and then renitrated to dinitrotoluene (DNT). In the final step, the DNT is nitrated to trinitrotoluene (TNT) using an anhydrous mixture of nitric acid and oleum. Nitric acid is consumed by the manufacturing process, but the diluted sulfuric acid can be reconcentrated and reused. After nitration, TNT is stabilized by a process called sulfitation, where the crude TNT is treated with aqueous sodium sulfite solution to remove less stable isomers of TNT and other undesired reaction products.

TNT is one of the most commonly used explosives for military, industrial, and mining applications. TNT has been used in conjunction with hydraulic fracturing.

16.6.2.2   Toluene Diisocyanates (TDI)

Toluene diisocyanate (TDI), $CH_3C_6H_3(NCO)_2$, has six possible isomers. Two of them are commercially important: 2,4-TDI and 2,6-TDI. 2,4-TDI are prepared in three steps from toluene via dinitrotoluene and 2,4-diaminotoluene (TDA). Finally, the TDA is treated with phosgene to form TDI, with HCl as a co-product that is a major source of industrial hydrochloric acid.

Distillation of the raw TDI mixture produces an 80:20 mixture of 2,4-TDI and 2,6-TDI, known as TDI (80/20). Differentiation or separation of the TDI (80/20) can be used to produce pure 2,4-TDI and a 65:35 mixture of 2,4-TDI and 2,6-TDI, known as TDI (65/35) [32]. All TDIs are reactive chemicals and potentially hazardous to humans.

TDIs are important building blocks to produce polyurethanes by reacting with hydroxy groups, such as a wide range of polyols, with formation of carbamate links. Polyurethanes are used to produce countless products, particularly in the transportation and construction industries, such as rigid and flexible thermoset foams, coatings, adhesives, sealants, elastomers and lighter automobile parts for energy savings.

16.6.2.3   Toluene Disproportionation

Disproportionation is said to occur when a reactant is transformed into two or more dissimilar products. In the UOP Tatoray process [33], for example, toluene is converted by disproportionation into benzene and xylenes, as shown. The process also achieves transalkylation of C_9 and C_{10} aromatics to form additional benzene and xylenes. Selective toluene disproportionation with the UOP PX Plus™ Process yields mixed xylenes, shown in Fig. 16.10, in which the p-xylene content is nearly 90% [34].

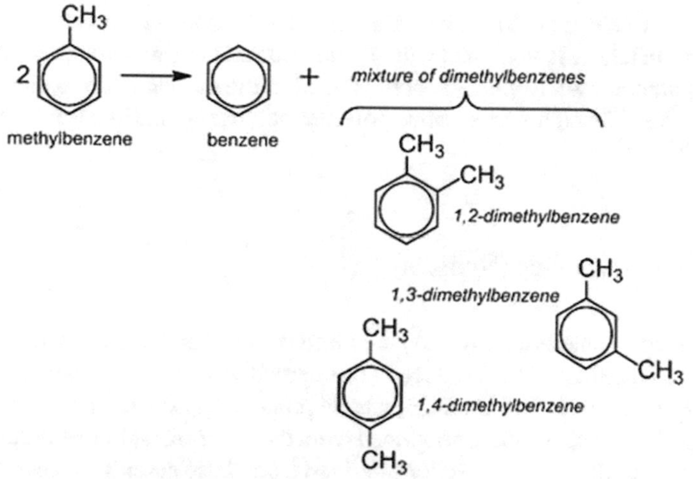

Fig. 16.10 Toluene disproportionation

16.6.3 Xylenes Derivatives

There are three xylene isomers—*ortho*, *meta*, and *para*. The most valuable is purified *p*-xylene, which undergoes oxidation to form terephthalic acid (TPA), shown below. TPA is a fundamental building block for polyethylene terephthalate (PET), commonly known as polyester.

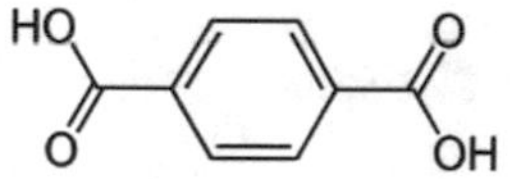

Several refinery processes are devoted to converting other small aromatics into *p*-xylene. The UOP Parex process recovers high-purity *p*-xylene from mixtures of C_8 aromatics [35]. In the UOP Isomar process [36], *p*-xylene-free mixtures of *o*-xylene and *m*-xylene are allowed to equilibrate, forming a new mixture which contains all three isomers along with ethylbenzene. At one set of conditions, an equilibrium mixture contains 60% *m*-xylene, 14% *p*-xylene, 9% *o*-xylene, and 17% EB. Therefore, it is important that the Isomar process also converts ethylbenzene by disproportionation into benzene and xylenes. EB itself can also be converted to styrene (vinyl benzene) by dehydrogenation. Styrene is then polymerized to form polystyrenes, one of the most widely used plastics.

Phthalic anhydride is a principal commercial form of phthalic acid. It is produced from oxidation of *o*-xylene over a vanadium pentoxide (V_2O_5) catalyst between 320–400 °C. Phthalic anhydride is an important ingredient to make plasticizers with oxo alcohols.

Isophthalic acid is produced by oxidizing m-xylene using oxygen over a cobalt-manganese catalyst. It is a precursor of the fire-resistant material Nomex. Mixed with terephthalic acid, isophthalic acid is used in the production of resins for drink plastic bottles. The high-performance polymer polybenzimidazole is produced from isophthalic acid.

16.7   Synthesis Gas (Syngas)

As mentioned, synthesis gas (syngas) is a mixture of CO and hydrogen and very often some carbon dioxide. CO is produced commercially by partial oxidation of natural gas, coal or biomass. Both CO and H_2 can be produced by steam/methane or naphtha reforming. Additional H_2 can be obtained from the water-gas shift reaction, with CO_2 as a byproduct. Many of the reactions derived from these gases have been discussed in Sect. 16.1. The main use of syngas is to produce alkanes, olefins and alcohols through the Fischer-Tropsch process. In South Africa during apartheid, when due to an embargo the country could not import enough petroleum to meet its needs, syngas from coal was the main source of commercial hydrocarbons. A similar situation developed in Germany during World War II. The country supplemented its limited supplied of petroleum by converting coal into syngas, which in turn was converted through Fischer-Tropsch synthesis into fuels and lubricants for war machines (planes, tanks, ships, etc.). Syngas can also be produced from biomass for the production of renewable biofuels and biochemicals.

16.7.1   Fischer-Tropsch Process

Liquid transportation hydrocarbon fuels and various other chemical products can be produced from syngas via Fischer-Tropsch synthesis (FTS), named after the two German inventors, Franz Fischer and Hans Tropsch, in the 1920s. The Fischer–Tropsch (F-T) process involves a series of chemical reactions that produce a variety of hydrocarbons:

$$(2n + 1)H_2 + nCO \rightarrow C_nH_{2n+2} + nH_2O$$

where n is typically 10–20. The alkanes are linear and fall in the diesel range. Higher hydrocarbons are produced at low temperatures (150–300 °C) with cobalt catalysts. In addition to alkane formation, competing reactions give small amounts of alkenes, as well as alcohols and other oxygenated hydrocarbons. The technology is often referred to as a gas-to-liquid (GTL) process for natural gas.

The Fischer–Tropsch process starts with partial oxidation of methane (natural gas) or other hydrocarbons to carbon dioxide, carbon monoxide, hydrogen gas and water. The ratio of carbon monoxide to hydrogen is adjusted using the water gas shift reaction, while the excess carbon dioxide is removed with aqueous solutions of alkanolamines. Removing the water yields syngas, which is chemically reacted over an iron or cobalt catalyst to produce liquid hydrocarbons and other byproducts.

For the F-T Process, the feed must be free of sulfur since sulfur compounds can poison the catalyst. The process is operated in the temperature range of 150–300 °C (302–572 °F). Higher temperatures lead to faster reaction and higher conversion, but also tend to favor methane production. Hence, the temperature is usually maintained at the low and middle part of the range. Typical pressure ranges from one to several tens of atmospheres. Higher pressures favor the production of long-chain paraffins but lead to catalyst deactivation from coke formation. The optimal H_2:CO ratio is around 1.8 to 2.1. Different ratios can lead to different product distributions. Cobalt catalysts are highly active, especially when the feedstock is natural gas. Iron-based catalysts may be more suitable for certain applications, which are preferred for lower quality feedstocks, such as coal and biomass, which tend to have low H_2:CO ratios (<1), due to their intrinsic water shift activity. For iron, promoters (alkali metals) are essential to obtain high basicity and to stabilize high specific metal surface areas. Nickel can also be used, but it tends to favor methane formation. Ruthenium is another commonly used catalyst, with a much higher cost. Its activity at lower temperatures is better than conventional F-T catalysts. At low temperatures and high pressures, large amounts of very high molecular weight waxes are produced.

The F-T reaction is highly exothermic, therefore heat removal is essential in reactor designs. The three types of reactors used commercially include fixed-bed reactors, fluidized bed reactors and slurry bed reactors.

The hydrocarbon product distribution from the F-T process generally followed an Anderson-Schulz-Flory distribution as:

$$W_n/n \rightarrow (1 - \alpha)^2 \alpha^{n-1}$$

where Wn is the weight fraction of hydrocarbons containing n carbon atoms, and α is chain growth probability. The latter is largely determined by the catalyst and process conditions. Higher alkanes can be produced at lower temperatures. Normal alkanes having carbon numbers greater than 20 are waxy and solid at room temperature. They can be hydrocracked into fuel range products, or hydrodewaxed/hydroisomerized into lubricant base oils. At high temperature (nearly 340 °C) over alkalized fused iron catalysts, low molecular weight unsaturated hydrocarbons are produced in a fluidized bed that is adopted from catalytic cracking technology. The olefins can form gasoline from olefins processes, such as oligomerization, isomerization, reforming and hydrogenation to produce gasoline. Figure 16.11 summarizes various fuel and lubricant products through high-temperature and low-temperature Fischer-Tropsch synthesis (FTS) routes.

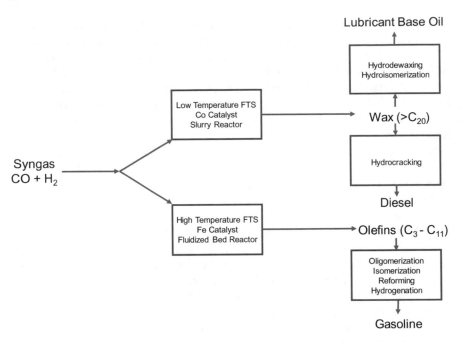

Fig. 16.11 Various fuel and lubricant products from high-temperature and low-temperature Fischer-Tropsch synthesis (FTS) routes

16.7.2 Syngas to Gasoline Plus (STG+) Process

The syngas to gasoline plus (STG+) process consists of four fixed bed reactors in series in which syngas is converted to synthetic fuels. For producing high-octane gasoline, syngas is fed to first reactor to convert most of the syngas (CO and H_2) to methanol by passing through the catalyst bed. The methanol-rich gas enters the second reactor to be converted to dimethyl ether (DME) by catalytic dehydration. The gas from the second reactor is next fed to the third reactor containing a catalyst to convert DME to hydrocarbons, including paraffins, naphthenes, aromatics and small amounts of olefins. Then, the fourth reactor provides transalkylation and hydrogenation to reduce durene/isodurene and trimethylbenzene leading to desirable viscometric properties. Finally, the mixture enters a separator to separate non-condensed gas from liquid product. The gas is recycled back to the first reactor.

16.7.3 Syngas to Methanol

Methanol is produced when the CO and CO_2 produced in steam methane reforming and water shift reactions are reacted with hydrogen in a ratio of $CO_2:CO:H_2 = 5:5:90$ at 50–100 bars and 225–275 °C over $Cu/ZnO/Al_2O_3$ catalysts. The predominant reactions are:

$$CO + 2H_2 \leftrightarrow CH_3OH \quad \Delta H_{298\ K, 5\ MPa} = -90.7 \text{ kJ mol}$$
$$CO_2 + 3H_2 \leftrightarrow CH_3OH + H_2O \quad \Delta H_{298\ K, 5\ MPa} = -40.9 \text{ kJ mol}$$

As these reactions are exothermic and lead to a decrease of the amount of molecules present, the productivity can be increased by increasing pressure and reducing temperature.

Methanol can serve as a fuel that has a high octane rating. It has low volatility. However, it is used on a limited basis to fuel internal combustion engines, in part due to its toxicity. It is primarily used as a feedstock for manufacture of chemicals, such as gasoline, olefins, formaldehyde, chlorinated methanes discussed in Sects. 16.1.2.1, 16.1.2.2 and 16.1.2.3,

About 40% of methanol is converted to formaldehyde, and from there into diverse products: plastics, plywood, paints, explosives and permanent press textiles. A broad range of formaldehyde applications are discussed in Sect. 16.1.2.2.

16.8 Integrated Gasification Combined Cycle (IGCC) [37]

Syngas can also be produced from integrated gasification combined cycle (IGCC) where a high-pressure gasifier with sub-stoichiometric oxygen turns heavy petroleum residue, coke, coal and biomass into pressurized gas. Impurities such as sulfur dioxide, particulates, mercury, and in some cases carbon dioxide are removed prior to power generation. Syngas cleanup includes the removal of bulk particulates by filters, fine particulates by scrubbing and mercury by solid adsorbents. Carbon monoxide emission can be reduced by adding a water-gas shift reactor. Carbon dioxide for the shift reaction can be separated, compressed and stored through sequestration. After the shift reactor, the syngas is fairly pure hydrogen.

The syngas and hydrogen are used for power generation to drive generators rather than making chemicals. The IGCC plant is called integrated because (1) the syngas produced in the gasifier is used as fuel for the gas turbine and (2) the steam generated by syngas cooler in the gasifier is used to drive a steam turbine in the combined cycle for electricity generation. A third function is sometimes added: co-generation of hydrogen for refineries and other industrial facilities. This process could increase values for undesirable cokes that cannot be sold in the market.

16.9 Links Between Refineries and Petrochemical Plants

Refineries supply several feedstocks for petrochemical plants (Table 16.1). Primary petrochemicals can be divided into three groups: olefins, aromatics, and synthesis gas.

As mentioned, olefins are major sources of polyolefins, such as polyethylene, polypropylene, and ethylene/propylene co-polymers. Ethylene and propylene are sources of other polymers, too. Ethylene reacts with benzene to make ethyl benzene, which leads to styrene and polystyrene. Propylene is oxidized to acrylic acid, used to make polyacrylates. Oxidation of propylene in the presence of ammonia produces acrylonitrile, which leads to polyacrylonitrile and several derivatives. Butyl rubbers made from C_4 olefins are basic ingredients for tires. Bromobutyl (and Chlorobutyl) is the key ingredient for tubeless tires due to its excellent air impermeability, UV resistant and antioxidation properties. High polymers can be further processed into plastics and engineering plastics.

Aromatics from catalytic reformers are mainly benzene, toluene and xylenes. They are solvents and feedstocks for a variety of chemicals. p-Xylene is a major feed for polyester production.

Synthesis gas is used for making methanol and ammonia as well as synthetic fuels and lubricants.

Figure 16.12 illustrates that it's a two-way street: petrochemical plants also feed refineries. The olefin plant is essentially a steam cracking plant to produce olefins. Hydrogen is, for the most part, a by-product of olefins plants, but it has high value in refineries. Pyrolysis gasoline (py gas) from steam cracking can be stabilized by hydrotreating and blended into the gasoline pool. Similarly, excess butanes from olefins plants can be converted into high-value gasoline blendstocks or into isobutane as a feedstock for alkylation in refineries.

Feedstocks make up anywhere from 60 to 70% of the cost to manufacture petrochemicals; therefore, the feedstock flexibility of domestic petrochemical processes has enabled US producers to remain competitive. As both oil and gas became more expensive in the US during the early 2000s, however, coupled with increased energy

Table 16.1 Sources of petrochemical feedstocks from refineries

Process unit	Petrochemical or feedstocks
Refinery gas processing	C_2-C_4 paraffins and olefins → olefin plant → polymers
Naphtha hydrotreating	Clean naphtha → olefin plants → polymers
FCC	Propylene and butylene → polymers
Hydrocracking	Unconverted oil → olefin plants → polymers
Catalytic reforming	Aromatics → solvents, polymers

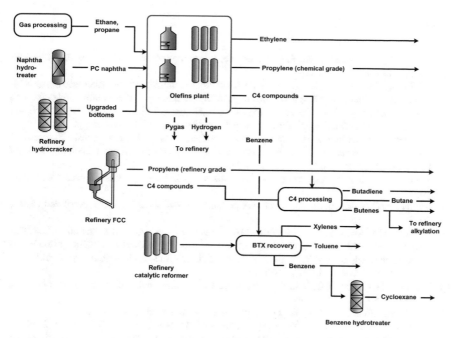

Fig. 16.12 Synergies between a refinery and chemical plants

costs, the competitive advantage that came with feedstock flexibility all but disappeared. Petrochemical companies began to seek joint ventures in other parts of the world where energy and feedstock costs were cheaper. All that changed around 2008, when production of oil and gas from shale "fracking" led to a dramatic decrease in the cost of industrial energy and feedstocks in the US.

Shale development is providing manufacturers with an unprecedented opportunity in low cost energy; however, that is only part of the picture. Along with the natural gas and oil coming from shale plays like the Marcellus, Bakken, Utica and Eagle Ford, these areas contain an abundance of NGLs. Ethane is a major component of NGLs, especially in the Marcellus, Utica and Eagle Ford plays. The economic law of supply and demand states that an increase in supply, other things being equal, results in a decrease in price. This new-found bounty of NGLs has significantly reduced the price of ethane in the US. The low cost of energy and ethane has prompted American petrochemical manufacturers to expand ethane cracking capacity. Over the past three years, petrochemical companies have announced plans to invest over $80 billion in new manufacturing infrastructure.

References

1. Chen I-T (2010) Steam reforming of methane. http://large.stanford.edu/courses/2010/ph240/chen1/. Accessed 6 Aug 2014
2. Oli: Benfield Process. http://wiki.olisystems.com/wiki/Benfield_process. Accessed 6 Aug 2015
3. Helton T, Hindman M (2014) Methanol to gasolsine technology: an alternative for liquid fuel production. Presented at GTL Technology Forum, Houston, TX, 30–31 July 2014. http://cdn.exxonmobil.com/~/media/global/files/catalyst-and-licensing/2014-1551-mtg-gtl.pdf
4. Wikipedia: Gas to Liquids. https://en.wikipedia.org/wiki/Gas_to_liquids. Accessed 6 Aug 2014
5. Encyclopaedia Britannica: Haber-Bosch Process. https://www.britannica.com/technology/Haber-Bosch-process. Accessed 15 Aug 2015
6. Wikipedia: Haber Process. https://en.wikipedia.org/wiki/Haber_process. Accessed 15 Aug 2015
7. Wikipedia: Oswald Process. https://en.wikipedia.org/wiki/Ostwald_process. Accessed 15 Aug 2015
8. Elyna M. The Oswald Process. http://chemistryhomework.weebly.com/. Accessed 15 Aug 2015
9. World of Chemical: Manufacturing of nitric acid by Oswald process. http://www.worldofchemicals.com/449/chemistry-articles/manufacturing-of-nitric-acid-by-ostwald-process.html. Accessed 15 Aug 2015
10. Wikipedia: Oklahoma City bombing. https://en.wikipedia.org/wiki/Oklahoma_City_bombing. Accessed 15 Aug 2015
11. Wikipedia: Urea. https://en.wikipedia.org/wiki/Urea. Accessed 15 Aug 2015
12. Wikipedia: Acetylene. https://en.wikipedia.org/wiki/Acetylene. Accessed August 6, 2014
13. Johnson Matthey Process Technologies: the Formox process. http://www.formox.com/formox-process. Accessed 6 Aug 2014
14. Andrussow L (1935) The catalytic oxidation of ammonia-methane-mixtures to hydrogen cyanide. Angew Chem 48(37):593–595
15. https://www.researchgate.net/publication/264509790_Propylene_Production_Methods_And_FCC_Process_Rules_In_Propylene_Demands. Accessed 15 Aug 2015
16. Ziegler-Natta Vinyl Polymerization. http://pslc.ws/macrog/ziegler.htm. Accessed 6 Aug 2014
17. Phillips Process. http://wwwcourses.sens.buffalo.edu/ce435/Polyethylene/CE435Kevin.htm. Accessed 6 Aug 2014
18. Benedikt GM, Goodall BL (eds) (1998) Metallocene catalyzed polymers. ChemTech, Toronto
19. Metallocene Catalysis Polymerization. http://pslc.ws/macrog/mcene.htm. Accessed 15 Aug 2016
20. Wikipedia: Linear Alpha Olefins. https://en.wikipedia.org/wiki/Linear_alpha_olefin. Accessed 15 Aug 2016
21. Vikipedia: Polyvinyl acetate. https://en.wikipedia.org/wiki/Polyvinyl_acetate. Accessed 15 Aug 2015
22. PVC Profiles: https://www.kellerplastics.com/pvc-profiles/. Accessed 15 Jan 2019
23. Wikipedia: Polypropylene. https://en.wikipedia.org/wiki/Polypropylene. Accessed 6 Aug 2014
24. Kaminsky W (2012) Discovery of methyaluminoxane as cocatalyst for olefin polymerization. Macromolecules 45(8):3289–3297
25. Wikipedia: Acrylonitrile. https://en.wikipedia.org/wiki/Acrylonitrile. Accessed 15 Aug 2015
26. Wikipedia: Polybutylene. https://en.wikipedia.org/wiki/Polybutylene. Accessed 6 Aug 2016
27. Wikipedia: Butyl Rubber. https://en.wikipedia.org/wiki/Butyl_rubber. Accessed 15 Aug 2015
28. Polyisobutylene. http://pslc.ws/macrog/pib.htm. Accessed 6 Aug 2015
29. Sanchez S, Cho J (2015) Hydroformylation. http://www.chem.tamu.edu/rgroup/marcetta/chem462/lectures/Cho-Sanchez-Hydroformylation.pdf. Accessed 6 Aug 2015
30. Wikipedia: Nylon. https://en.wikipedia.org/wiki/Nylon. Accessed 15 Aug 2015

31. Wikipedia: Trinitrotoluene. https://en.wikipedia.org/wiki/Trinitrotoluene. Accessed 15 Aug 2015
32. Wikepedia: Toluene Diisocyanate. https://en.wikipedia.org/wiki/Toluene_diisocyanate. Accessed 9 Oct 2018
33. http://www.treccani.it/export/sites/default/Portale/sito/altre_aree/Tecnologia_e_Scienze_applicate/enciclopedia/inglese/inglese_vol_2/591-614_ING3.pdf. Accessed 15 Aug 2015
34. The Essential Chemical Industry Online: Cracking and related refinery processes. http://www.essentialchemicalindustry.org/processes/cracking-isomerisation-and-reforming.html. Accessed 15 Aug 2015
35. Honeywell UOP: Aromatics. https://www.uop.com/processing-solutions/petrochemicals/benzene-para-xylene-production/ Accessed 24 Nov 2016
36. Choice of the Processes: http://www.sbioinformatics.com/design_thesis/Orthoxylene/orthoxylene_Methods-2520of-2520Production.pdf. Accessed 6 Aug 2015
37. Wikepedia: Integrated Gasification Combined Cycle. https://en.wikipedia.org/wiki/Integrated_gasification_combined_cycle. Accessed 15 Aug 2015

Chapter 17
Midstream Transportation, Storage, and Processing

The petroleum industry is often divided into three business sectors: Upstream, Midstream and Downstream [1, 2]. Upstream and downstream have been well defined and adopted in petroleum companies and communities.

Upstream involves the discovery (exploration) and recovery (production) of petroleum resources, including natural gas, condensate, and crude oil, from underground or underwater fields. The recently developed in situ steam-assisted gravity drainage (SAGD) and hydraulic fracturing (fracking) to recover trapped gas and oils from shales and other tight formations are upstream operations. The open mining of oil sand bitumen deposits belongs to upstream.

On the other hand, downstream operations include the refining of crude oils, production of petrochemicals from petroleum, Fischer-Tropsch production of synthesis gas (syn gas), and distribution of petroleum-based products. Refining produces fuels, lubricant oils, and petrochemicals. Petrochemicals include light olefins and BTX (benzene, toluene and xylenes), steam-cracker feedstocks such and light naphtha and hydrocracker unconverted oil. In turn, petrochemicals are converted into thousands of chemicals, including solvents, polymers, fibers, coatings, plastics, synthetic rubbers, and pharmaceuticals.

Generally speaking, "midstream" connects upstream with downstream. Midstream operations and processes generally include transportation by pipeline, tanker, barge, rail and trucks; storage; and trading of crude oil and refined products [2]. Trading of refined products is often considered to be a downstream business. Trading natural gas, crudes and other refinery feedstocks can be included in upstream.

The midstream sector is huge. Estimates of total pipeline mileage differ widely. According to the United States Department of Transportation, in 2014 the United States had the largest oil and gas pipeline network in the world, accounting for ~40% of the world total [3]. According to Pipeline 101, in 2018 the U.S. has more than 2.4 million miles of gas pipeline pipelines and more than 190,000 miles of liquid petroleum pipelines, of which 72,000 miles are for crude oils [4]. The cost of installing a pipeline ranges from $300,000 to $1.5 million per mile, so the cost of replacing this infrastructure would be enormous. The United States Central Intelli-

© Springer Nature Switzerland AG 2019
C. S. Hsu and P. R. Robinson, *Petroleum Science and Technology*,
https://doi.org/10.1007/978-3-030-16275-7_17

gence Agency reports that there are 4295 ocean-going oil tankers in the world, each with a deadweight tonnage (DWT) greater than 1000 long tonnes [5].

17.1   Relationship with Upstream and Downstream

"Midstream," generally starts after wellhead treatments or at the point where oil, gas, or other produced hydrocarbons enter a transportation system—a pipeline or a vehicle, such as a barge, tanker or rail car.

There are some gray areas. The liquefaction of natural gas to LNG, extraction/dilution of bitumen, and production of synthetic crude oil (syn crude) for transportation seem to fit best into midstream rather than upstream. The processing of natural gas includes recovering sulfur, carbon dioxide, helium, and hydrocarbon liquids before the gas goes to a product pipeline. CO_2 will form dry ice upon refrigeration to clog tubes in pipelines and tankers. Hence, it has to be removed prior to natural gas liquefaction for transportation. The acid gases (CO_2 and H_2S) are removed with alkanol amines, such as diethanol amine (DEA); usually, the H_2S is carried by alkanol amine to a Claus plant, where it is converted into sulfur. The physical separation of gas, oil, water and particulates in a sedimentation vessel at or near well head, however, is an upstream process.

The dilution of bitumen by froth treatment, and the upgrading of the oil sand bitumen into synthetic crude oil via coking prior to pipeline transportation, is called midstream operations by some companies and downstream operations by others. Salt dome storage of natural gas or crude oil, including in strategic reserves, can also be considered as a midstream operation.

Mid-stream ends at the point where a hydrocarbon stream enters refinery storage tanks (i.e., the tank farm). Hence, storage tank management at a transshipment point is part of midstream, but refinery tank farms are a part of downstream operations, including crude oil blending.

17.2   Transportation

Large quantities of crude oils are transported by pipelines, either on or beneath land or underwater, or in tankers. Seagoing tanker transportation has been heavily used where pipeline transportation becomes difficult, uneconomical or impossible, such as for intercontinental transportation. The larger the tanker, the lower is the unit cost of transportation. Viscous oil sand bitumen produced in Canada is diluted by condensates, naphtha or middle distillates, or converted to synthetic crude oil on-site for pipeline transportation.

For transportation of crude oil through pipeline, the excess quantities of water have to be removed to meet specification. The maximum tolerable water content for pipeline transportation is from 0.5 to 2.0%.

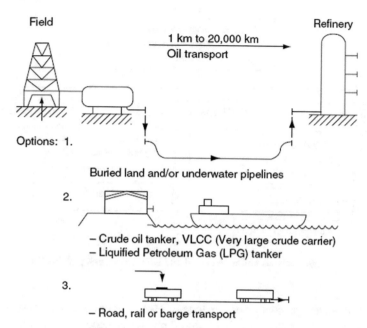

Fig. 17.1 Transportation of natural gas, crude oils and their products [6]

Figure 17.1 summarizes transportation means for crude oil and natural gas as well as their products. Not shown are the storage facilities (tank farms) needed at transportation ports and refining centers for the large quantities of petroleum. Crude oils may also be stored in such geological features as salt domes, used by the U.S. Strategic Petroleum Reserve.

17.3   Transportation of Heavy Oil and Bitumen

Pipelines are the most convenient and economical means of transporting crude oil from the producing field to the refinery. For long distance or over a vast of water body, such as an ocean, where construction of pipelines becomes impractical, tankers are used. However, pipelines are still needed for transporting the oils to marine terminals for loading onto tankers and unloading at destination ports for shipping to the refineries.

In recent years, heavy oil, extraheavy oil and bitumen resources are more than double the conventional oil reserves worldwide. Very large reserves are mainly located in Venezuela and Canada. Because of their extremely low mobility due to high viscosity (>1000 cP), prior reduction in viscosity is necessary for transportation via pipeline. In addition to the flow properties, asphaltene deposition, heavy metals, sulfur and

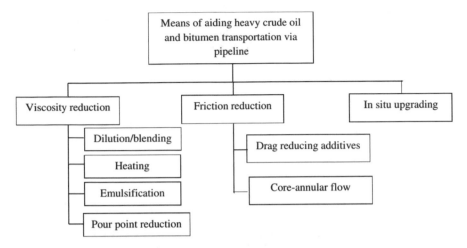

Fig. 17.2 Technologies of transporting heavy oil and bitumen through pipeline [7]

brine or salt content present other challenges. Hence, upgrading heavy crude oil and bitumen to meet conventional light crude oil properties become an option.

Common technologies of transporting heavy oil and bitumen through pipeline include preheating, blending or dilution, emulsification, partial upgrading and core-annular flow, summarized in Fig. 17.2. Each of these techniques is aimed at reducing viscosity as well as energy required for pumping, to enhance flowability of the oil via pipelines.

For viscosity reduction, dilution/blending is a commonly used technique. The most widely used diluents include condensate, naphtha, kerosene and light crude oil. The resulting blend of heavy crude oil and diluents, i.e., diluted bitumen or "dilbit", has lower viscosity and therefore easier to pump. However, instability of the blend due to asphaltene precipitation should be avoided by choosing the proper diluent.

Preheating followed by subsequent heating of the pipeline enhances the flow properties of heavy crude oil and bitumen. But heating pipelines is expensive due to high heat loss, a greater number of pumping/heating stations, expansion of pipelines, and greater corrosion inside the pipe due to the higher temperature. Consequently, heated pipes are used only for short distance transportation within a processing facility.

The emulsion method consists of dispersing heavy crude oil in water in the form of droplets stabilized by surfactants. The surfactants stabilize the emulsion to prevent drop growth and phase separation.

The use of polymers (polyacrylates, polymethacrylates, etc.) as pour point depressants inhibits precipitation of high density and high viscosity asphaltenes.

Drag-reducing additives, such as silicones, polymers, fibers and surfactants suppress turbulent flow and reduce friction near the pipeline wall. In core-annular flow, the heavy crude oil flows through the pipeline with a film layer of water or solvent near the pipe wall, which acts as a lubricant, thus, reducing the pump pressure drop over a long distance as comparable to the one for water alone. Due to their relatively low cost and ease of handling, silicones are the most widely used friction reducers, but they are problematic because they poison refinery catalysts.

In situ upgrading is achievable by thermal recovery of the oil and bitumen, including SAGD, CSS, THAI, etc. These upstream processes involve thermal cracking underground, which yield lighter hydrocarbons with lower viscosity.

17.4   Oil Sand Upgrading and Transportation

Oil sand or tar sand is a mixture of bitumen, sand, water and clay. It does not act like conventional crude oil. It must be mined in an open pit or in situ underground. Then its viscosity must be reduced prior to transportation to refineries for upgrading into finished products. For open mining, froth treatment is needed to separate bitumen from the aqueous and solid contaminants (sand, water and clay) [8]. This extraction process is not required for in situ processing, such as SAGD (Chap. 7, Sect. 7.9.2.3), because the separation happens in the ground.

In surface-mined oil sands operations, bitumen is extracted from oil sand ore using a warm-water extraction process that produces bitumen froth typically containing 60 wt.% bitumen, 30 wt.% water, and 10 wt.% mineral solids. The froth is first diluted with a naphthenic or paraffinic hydrocarbon solvent to reduce the viscosity and density of the oil phase, thereby accelerating the settling of the dispersed phase impurities by gravity or centrifugation. Figure 17.3 shows the froth formation after mixing bitumen with solvent for 2, 4, 8, 12 and 16 min. The mineral and inorganic impurities settled down to the bottom. The diluted product is called "diluted bitumen"

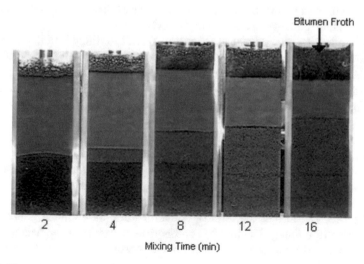

Fig. 17.3   Formation of bitumen froth after mixing [11]

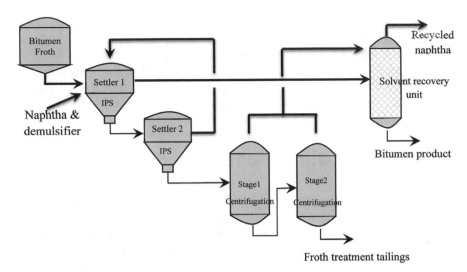

Fig. 17.4 2-Stage naphthenic bitumen treatment process [13]

or "dilbit", a marketable product, which can be shipped through an oil pipeline to an oil refinery or oil terminal. The solvent is recovered from the diluted bitumen to be used again.

The separation of oil from sand is never complete. The leftover sands and clays are discharged into vast tailings ponds. There are more than 176 km² of tailings ponds near Fort McMurray, Alberta [9]. Tailing pond water is contaminated with significant concentrations of polycyclic aromatic hydrocarbons (PAH) and trace elements, including arsenic. Moreover, the hydrocarbons are being oxidized to CO_2, making tailings ponds are a major source of greenhouse gases. Efforts are underway to use hot-house plants to recover a fraction of the CO_2 [10].

At the refineries, the solvent is recovered from the diluted bitumen and sold. When distances are short, i.e., when a refinery is nearby, the solvent is sent back to the production site to be used again. The challenge is to remove emulsified water and colloidal mineral particles. The bitumen can be further upgraded by means of a coker, a hydrotreater, or a residue hydrocracking process such as LC-Fining, H-Oil, or VCC.

There are two major technologies used for froth treatment: naphthenic and paraffinic [12]. In the naphthenic solvent froth treatment, shown in Fig. 17.4 for a 2-stage process, the solvent naphtha is used at relatively low solvent-to-bitumen ratio. The separation is enhanced by addition of chemicals to destabilize the emulsion. Typical diluted bitumen products from this process contain 1–2% of water and about 0.5% mineral solids. Because of the level of contamination, the bitumen is not suitable for pipelining or refining. The bitumen is therefore upgraded using heat to produce synthetic crude oil by means of a coker unit. The water (~80%) and mineral solid

(~15%) fraction obtained after separation in the treatment is commonly referred to as froth treatment tailings.

In paraffinic solvent froth treatment, light alkanes, such as pentane or hexane, are used as solvent, leading to precipitation of some of the asphaltenes which contribute to the high viscosity in the bitumen. The precipitated asphaltenes form agglomerates with the mineral solid particles and emulsified water droplets. These agglomerates quickly settle, producing very clean diluted bitumen. Thus, the bitumen product obtained has lower levels of contaminates, reaching almost 2 orders of magnitude.

17.5 Synthetic Crude Oil

Synthetic crude oil is a blend of naphtha, distillate, and gas oil range materials, with little residuum 1050 °F + (565 °C+) material. It is a high-quality intermediate product from a bitumen/extra heavy oil upgrader facility used in connection with oil sand production. Synthetic crude oil also is produced from oil shale, which comes from kerogen-bearing rock. Oil in shale is recovered by pyrolysis (retorting) [14]. The properties of the synthetic crude depend on the processes used in the upgrading. Typically, it is low in sulfur and has an API gravity of around 30, with properties close to conventional crude oil.

In the syncrude plant, the bitumen is heated and sent to drums where excess carbon, in the form of petroleum coke, is removed. The superheated hydrocarbon vapors from the coke drum are sent to fractionators where vapors condensed into naphtha, kerosene and gas oil. These products are blended into synthetic crude oil which is shopped by pipeline to refineries. Diesel can be sold as a final product. Shale oil products often contains significant amounts of arsenic, which is another severe catalyst poison in refineries [15].

The largest producers of synthetic crude in the world are Syncrude Canada, Suncor Energy Inc., and Canadian Natural Resources Limited, with a cumulative production of approximately 600,000 barrels per day (95,000 m^3/d).

17.6 Natural Gas Liquefaction

Much natural gas comes from production sites far from the market place. Natural gas is mostly methane. After removal of water, H_2S, CO_2, mercury, and higher hydrocarbons (condensate), natural gas is transported by pipeline or by tankers. For tanker transport,, natural gas is converted into liquefied natural gas (LNG), a process called liquefaction.

Figure 17.5 shows the sequence of transportation of LNG. Transportation from the field to the liquefaction facility is accomplished with gas pipelines. Treatment in the facility involves the removal of acid gases, H_2S and CO_2, and mercury. The condensate and water are also removed; condensate is a valuable byproduct. Precooling

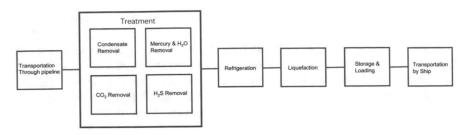

Fig. 17.5   Natural gas liquefaction sequence

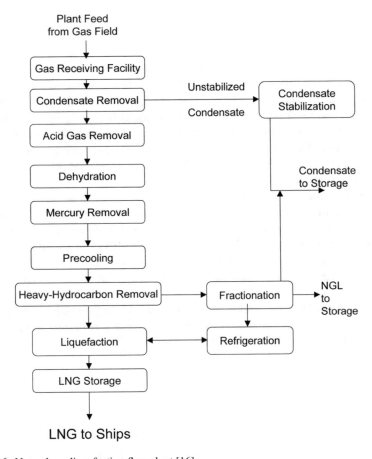

Fig. 17.6   Natural gas liquefaction flow chart [16]

removes heavy hydrocarbons, which can be combined with condensate for storage. The remaining gas is refrigerated. The refrigerant gas is compressed, cooled, condensed and dropped in pressure through a valve to reduce temperature by the Joule-Thomson effect. The temperature of the feed gas is eventually reduced to -161 °C. At this temperature, all hydrocarbon gas including methane liquefies. Constituents of natural gas (propane, ethane and methane) are typically used as refrigerant gases.

Since carbon dioxide normally co-exists with natural gas in gas wells, the removal of carbon dioxide by diethanolamine (DEA) or equivalent is necessary to avoid dry ice formation clogging the pipes, tubing, and other transferring components, such as valves. Several options for amine treating were discussed in Chap. 14, Sect. 14.1.2. In sour gas fields, such as those in Western Canada, both CO_2 and H_2S are removed in the same acid gas scrubbing facility.

Mercury is present in natural gas predominantly as elemental mercury. It is highly corrosive to LNG aluminum alloy cryogenic exchangers, causing equipment failure. Adsorption is a primary method for its removal. Sulfur impregnated carbon captures mercury as mercuric sulfide and fixes the compound due to its high energy adsorption into the pores of the adsorbent. Water can be removed simultaneously. The mercury level of the treated gas can be less than 0.01 µg per normal cubic meter.

The final product is LNG, ready for storage and shipping. A flow chart for the LNG process is shown in Fig. 17.6 [16]. Hydrogen and helium remain as gases to be recovered. As mentioned earlier, natural gas plants are the major commercial sources of helium.

References

1. Global Energy: Challenges and Solutions in an Upstream and Downstream Oil and Gao Operation: http://globalenergy.pr.co/65678-challenges-and-solutions-in-an-upstream-and-downstream-oil-and-gas-operation. Retrieved 6 Aug 2016
2. Wikepedia: Midstream. https://en.wikipedia.org/wiki/Midstream. Retrieved 6 Aug 2016
3. Wikipedia: https://www.bts.gov/content/us-oil-and-gas-pipeline-mileage. Retrieved 8 Oct 2018
4. http://www.pipeline101.org/where-are-pipelines-located. Retrieved 8 Oct 2018
5. Retrieved November 8, 2016
6. United States Central Intelligence Agency: The World Factbook, https://www.cia.gov/library/publications/the-world-factbook/fields/2108.html
7. Speight JG (2007) The chemistry and technology of petroleum, 4th edn. CRC Press, Boca Raton
8. Hart A (2014) A review of technologies for transporting heavy crude oil and bitumen via pipeline. J Petrol Explor Prod Technol 4:327–336
9. https://thenarwhal.ca/environment-canada-study-reveals-oilsands-tailings-ponds-emit-toxins-atmosphere-much-higher-levels-reported. Access 30 Nov 2018
10. https://www.cbc.ca/news/business/fort-mcmurray-greenhouse-to-turn-garbage-into-veggies-1.2984722. Access 30 Nov 2018
11. https://www.oilsandsmagazine.com/technical/mining/froth-treatment. Retrieved 7 Oct 2018
12. Froth Treatment: https://www.nrcan.gc.ca/energy/oil-sands/5873. Natural Resources Canada. Retrieved 15 Sept 2015

13. Rao F, Liu Q (2013) Froth treatment in Athabasca Oil Sands Bitumen Recovery Process: a review. Energy Fuels 27(12):7199–7207
14. Frost CM, Paulson RE, Jensen HB (2018) Production of synthetic crude shale oil produced by in-situ combustion retorting. https://web.anl.gov/PCS/acsfuel/preprint%20archive/Files/19_2_LOS%20ANGELES_04-74__0156.pdf. Retrieved 7 Oct 2018
15. https://web.anl.gov/PCS/acsfuel/preprint%20archive/Files/23_4_MIAMI%20BEACH_09-78_0018.pdf. Access 30 Nov 2018
16. https://petrowiki.org/File:Vol6_Page_369_Image_0001.png. Retrieved 7 Oct 2018

Chapter 18
Safety and Environment

18.1 Introduction

In addition to desired products, refineries and chemical plants also produce significant amounts of toxic, harmful and environmentally damaging organic and inorganic
materials. Hence, environmental considerations are addressed during design, processing, product handling, and disposal. Various treatment, capture and disposal units
are required. Fugitive emissions from evaporation, sedimentation, seeping, drifting,
spreading and natural dispersion are constantly monitored. Air and water qualities
after disposal have to meet specifications and standards. Oil spills into water and
onto soil threaten environment. Inadvertent discharges can cause tremendous loss
in life and environmental damages on a large scale, such as that which happened
after Texas City refinery explosion, the *ExxonValdez* oil spill and the recent deadly
blowout of the *Deepwater Horizon* in the Gulf of Mexico.

Air Quality. In the 1970s and 1980s, environmental laws compelled oil and
gas companies to reduce emissions of SO_x, NO_x, CO_2 and hydrocarbons. In the
atmosphere, SO_x reacts with water vapor to make sulfurous and sulfuric acids and
sulfate particulates, which return to earth as wet or dry acid deposition; wet deposition
is often called acid rain. Volatile hydrocarbons react with NO_x to make ozone. CO_2
is a major "green-house" gas. To reduce these pollutants, the industry tightened its
operations by:

- Reducing fugitive hydrocarbon emissions from valves and fittings
- Removing sulfur from refinery streams and finished products
- Adding tail-gas units to sulfur recovery plants
- Reducing the production of NO_x in fired heaters
- Scrubbing SO_x and NO_x from flue-gases
- Reducing the production of CO_2 by increasing energy efficiency.

In several oil fields, companies keep CO_2 out of the atmosphere by pumping it
into oil wells to maintain reservoir pressure, thereby enhancing production.

© Springer Nature Switzerland AG 2019
C. S. Hsu and P. R. Robinson, *Petroleum Science and Technology*,
https://doi.org/10.1007/978-3-030-16275-7_18

Waste Water Treatment. Waste water treatment is used to purify process water, runoff, and sewage. As much as possible, purified waste-water streams are re-used in the refinery. Wastewater streams may contain suspended solids, dissolved salts, phenols, ammonia, sulfides, and other compounds. The streams come from just about every process unit, especially those that use steam, wash water, condensate, stripping water, caustic, or neutralization acids.

Primary treatment uses a settling pond to allow most hydrocarbons and suspended solids to separate from the wastewater. The solids drift to the bottom of the pond, hydrocarbons are skimmed off the top, and oily sludge is removed. Difficult oil-in-water emulsions are heated to expedite separation. Acidic wastewater is neutralized with ammonia, lime, or sodium carbonate. Alkaline wastewater is treated with sulfuric acid, hydrochloric acid, carbon dioxide-rich flue gas, or sulfur.

Some suspended solids remain in the water after primary treatment. These are removed by filtration, sedimentation or air flotation. Flocculation agents may be added to consolidate the solids, making them easier to remove by sedimentation or filtration. Activated sludge is used to digest water-soluble organic compounds, either in aerated or anaerobic lagoons. Steam-stripping is used to remove sulfides and/or ammonia, and solvent extraction is used to remove phenols.

Tertiary treatment processes remove specific pollutants, including traces of benzene and other partially soluble hydrocarbons. Tertiary water treatment can include ion exchange, chlorination, ozonation, reverse osmosis, or adsorption onto activated carbon. Compressed oxygen may be used to enhance oxidation. Spraying the water into the air or bubbling air through the water removes remaining traces of volatile chemicals such as phenol and ammonia.

Solid Waste Handling. During drilling and production operations, oil-bearing cuttings from wells present environmental challenges. Onshore, the cuttings can be stored onsite in clay-lined pits. Offshore, the cuttings are returned to the sea. Other wastes are transported to waste treatment facilities. In the oil sands fields in Alberta, Canada, bitumen is washed way from the inorganic substrate with hot water. The water goes into huge tailings ponds. The ponds emit considerable amounts of CO_2, and they contain significant amounts of trace contaminants.

Refinery solid wastes may include spent catalysts and catalyst fines, acid sludge from alkylation units, and miscellaneous oil-contaminated solids. All oil-contaminated solids are treated as hazardous and sent to sanitary landfills.

Recently, super-critical extraction with carbon dioxide has been used with great success to remove oil from contaminated dirt, including cuttings from oil-well drilling.

Improving Energy Efficiency and Safety with Automation. The ever-increasing development and use of better instruments and analyzers, engineering models, and real-time online optimization continue to improve the efficiency and safety of petroleum processing plants. Subsequent sections provide examples. Implementing advanced process control increased production by 2.5% at the Chemopetrol ethylene plant in Litvinov, Czech Republic. Other examples are provided on the web sites of companies such as Applied Manufacturing Technologies, Aspen Technology,

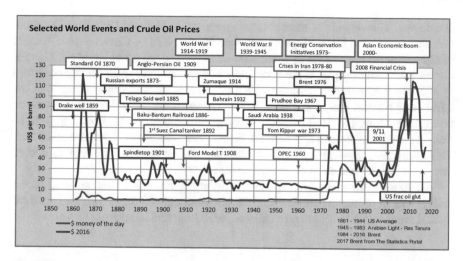

Fig. 18.1 Impact of world events on crude oil prices. After the 2008 financial crisis, prices dropped considerably due to lower consumption. Similarly, in 1973, conservation initiatives stalled the decades long upward trend in consumption

Emerson Process Management, Honeywell Process Solutions, Invensys Operations Management, Yokogawa and others.

Protecting the environment. The push to control anthropogenic CO_2 emissions is gaining momentum. Government agencies around the world continue to tighten regulations on sulfur in fuels, air quality, waste water quality, and the disposal of hazardous solids. Such regulations are now being adopted in less developed countries. They improve the quality of air, water and land, but they also require investment.

Conservation. Conservation also decreases the demand for petroleum and other fossil hydrocarbons while reducing pollution. According to the Rocky Mountain Institute modest investment in proven, widely applied technology, the United States could increase the efficiency of power production by 30%.

Figure 18.1 demonstrates that prices have a tremendous effect on energy consumption. We can save energy not just be traveling less and turning off lights and driving less, but also through innovation. Never underestimate the ingenuity of an energy-hungry human.

In 1984, a distraught oil company CEO was being grilled by reporters about an explosion, which had killed 19 workers and injured at least 23 others. When asked to explain the cause, his purported response was: "What more can I say? It's a dangerous business." His answer outraged many, especially employees who believed that his old-school attitude was part of the problem. (The same CEO was fond of saying in private meetings: "The solution to pollution is dilution.")

Health, safety, reliability, and protecting the environment are inextricably linked. In today's litigious society, they are prerequisites to profit. Modern executives behave accordingly. "It's a dangerous business" has been replaced by Zero-Zero-Zero campaigns aimed at achieving zero injuries, zero loss of containment, and zero unplanned shutdowns. The slogan for ExxonMobil's program is Nobody Gets Hurt [1]. Shell preaches Goal Zero [2]. KBR promotes Zero Harm 24/7 [3]. Chevron's ten Tenants of Operation [4] are prefaced by two statements: "Do it safely or not at all. There is always time to do it right." These particular companies have departments dedicated to HSE (health, safety, environment) led by executives who report directly to the CEO or COO, not just to someone in human resources. The 0-0-0 initiatives are not just for show. They reflect a sincere belief that our business doesn't have to be deadly. HSE rules are strictly enforced. Employees who violate them are dismissed. This relatively recent shift in attitude requires investment, but the return on such investments is enormous in many measureable ways.

Vox populi—the voice of the people—instigated the regulations that play such a critical role in protecting workers and the environment. Worker safety and environmental movements [5, 6] are international, with long and interesting histories.

18.2   History of Labor Legislation and Decrees

Worker-protection rules go back to 15th Century England, when provisions were made to prevent abuses in the "truck system"—the paying of wages in goods instead of money. Similar reforms were implemented in Switzerland in the 17th Century.

Efforts to improve the plight of coal miners go back as far as 1775, when an Act of Parliament partially emancipated colliers, coal-bearers, and salters in Scotland [7]. According to the Act, the workers were in a "state of bondage." They had been since 1606, when an Act of the Scottish Parliament bound them to the mines. In 1799, another Act explained and amended the 1775 Act, abolishing bondage totally.

In 1802 in England, the Health and Morals of Apprentices Act limited working hours for apprentices to twelve per day and abolished night work. It required basic education and the provision of clothing and adequate sleeping accommodations. The Factory Act of 1819 limited working hours to twelve per day and prohibited employment of children under nine years of age.

In 1841 in France, a law prohibited the employment of children under 8 years of age and prohibited night labor for any child under 13. In 1848, the working day was limited to 12 h per day for adults in factories. The law was amended in 1885 to include factories with motor power or furnaces and workshops employing more than 20 people.

The first labor law in the Netherlands, passed in 1874, applied only to children working in factories. In 1889, the law was extended to all industries except agriculture, forestry, and fishing. It prohibited the employment of children under 12. In 1895, a law established standards for space, lighting, ventilation, and sanitation. And in 1897, a decree limited working hours according to age and sex, with special provisions for certain industries and pregnant women.

The American Colonies were largely agrarian. Slaves provided labor in southern plantations. Industry was conducted for the most part by master and journeyman craftsmen, supported by apprentices and indentured servants [8]. Labor disputes included a 1636 fishermen's strike on an island near Maine, a 1677 strike by car-men in New York City, and a 1746 strike by carpenters in Savannah, Georgia.

The United States Congress passed the first federal law governing mine safety in 1891 [9]. The law established ventilation requirements for underground coal mines and prohibited owners from employing children under 12. Owners had resisted acting voluntarily, arguing that doing so would put them at a competitive disadvantage.

In 1910, after a decade in which coal mine fatalities exceeded 2000 annually, Congress established the Bureau of Mines to conduct research and reduce accidents in the coal mining industry. Regulations were implemented, but violations were rampant until 1941, when Congress empowered federal inspectors to enter mines.

Over the years, regulations improved, culminating in the Federal Mine Safety Act of 1977, which was amended by the MINER Act of 2006.

US child labor laws lagged the rest of the world by decades. According to the 1890 U.S. Census, about 1.12 million children under ten years old were engaged in "gainful occupations" out of a total population (including adults) of 63.0 million [10–12]. According to the 1900 US Census, the number had increased to 1.75 million children out of a total population of 76.2 million. About 16.7% of children under ten were working, and 18% of children between 10 and 15 were employed full time. Child labor on family farms was expected, but the increase of 50% in under-10 employment for wages, mostly in factories, alarmed many Americans. Attempts at reform were thwarted by the Supreme Court. In 1912, a law established the United States Children's Bureau in the Department of Commerce and Labor. In 1916, the Keating-Owen Act prohibited interstate shipment of goods manufactured or processed by child labor. The bill passed 337 to 46 in the House and 50 to 12 in the Senate. But in 1918, in a 5-4 decision, the law was ruled unconstitutional by the Supreme Court. In 1938—after years of court decisions overturning attempts at labor reform—and 118 years after the first child labor law in England—Congress passed the Fair Labor Standards Act (FLSA) [13]. The FLSA banned oppressive child labor, set a minimum wage of 25 cents per hour, and limited the work-week to 44 h. The FLSA survived Supreme Court challenges, including the *United States v. Darby Lumber Co.* case in 1941. The FLSA is in force today.

18.3 History of Environmental Laws and Decrees

As humans living on an ever-more-crowded planet, our main reason for abating pollution is self-preservation. While addressing waste in *What Is Life?* Margulis and Sagan [14] remind us that "every species or organism produces wastes that are incompatible with its own existence."

In *The Closing Circle* [15], Barry Commoner generalized this principle with his Four Laws of Ecology. The first two are especially relevant:

1. Everything is connected to everything else. What affects one organism affects all.

2. Everything must go somewhere. There is no "waste" in nature and there is no "away" to which things can be thrown.

Fallout from open-air nuclear testing appeared all across the planet. Radioactive isotopes showed up in food, water, and the bones of animals. For the first time, it became clear that certain events have long-lasting global impact. On January 15, 1958, Nobel Laureate Linus Pauling presented to the United Nations a petition signed by 9235 scientists from many countries protesting further nuclear testing [16]. The nuclear powers, including the Soviet Union, the United States, and the UK, voluntarily stopped testing that same year. The Soviet Union announced the resumption of open-air testing in August 1961, adding urgency to efforts to conclude a test-ban treaty. In July 1963, the Nuclear Test Ban Treaty was signed, outlawing atmospheric testing but allowing underground testing.

In 1962, Rachel Carson published *Silent Spring* [17]. Among other things, she presented evidence linking DDT to declines in the bald eagle population, due to the thinning of egg shells. She used the DDT linkage to tell a broader story: What affects one organism can affect all. According to data recently cited by PAN [18], an anti-pesticide organization, DDE breakdown products are still being found in food supplies and human blood.

In response to the efforts of outspoken organizations and people such as Carson, Commoner, and Pauling, the United States formed the Environmental Protection Agency (US EPA). Signed into law by President Nixon in 1970 [19], EPA was an amalgamation of departments, bureaus, agencies, etc., taken from the Interior Department; the Department of Health, Education, and Welfare; the Agriculture Department; the Atomic Energy Commission; the Federal Radiation Council; and the Council on Environmental Quality. One of the first actions promoted by EPA was the 1972 ban on DDT.

The history of environmental protection goes back (at least) to the 17th Century [20].

After the Great London Fire of 1666, King Charles II proclaimed that all buildings must be constructed from stone, and streets must be widened. The London Cooking Fire Bylaw of 1705 specifically prohibited open fires in the attics of thatched buildings [21].

In 1842, the Madras Board of Revenue started local forest preservation efforts in British India. In 1855, under Governor-General Lord Dalhouise, a large-scale, state-managed forest conservation program was established.

Other early examples of British legislation are the 1863 Alkali Acts [22], implemented to regulate emissions of hydrochloric acid from the Leblanc process, the leading process for making soda ash. Pressure for the Acts came from the urban middle-class.

Parliament passed the Public Health Act of 1875 to combat the spread of diseases such as cholera and typhus, which percolated in the sewage that flowed through public streets [23]. The Act required all new construction to include running water and internal drainage. It also regulated housing construction and required every public health authority to employ a medical officer and an inspector. Significantly, it required all furnaces and fireplaces to consume their own smoke.

Previous attempts at reform were inhibited by the reluctance of factory owners to improve conditions voluntarily. The Act itself included a "grandfather" clause, which exempted existing facilities. Both factors have played roles in safety and environmental programs ever since.

The emergence of great coal-burning factories and the concomitant growth in coal consumption increased air pollution in industrial centers enormously. Smoke, soot, and sulfur oxides led to the Great Smog of 1952, which is described in Sect. 18.3.

Conservation in the United States began with the establishment of national parks [24], such as the Hot Springs Reservation (1832) [25], Yellowstone (1872), Sequoia (1890), Yosemite Valley (1890), Mount Rainier (1899), and Crater Lake (1902).

Major U.S. environment-related legislation includes the following:

- The National Park Service Organic Act (1916)
- Land and Water Conservation Act (1964)
- National Trails System Act (1968)
- National Wild and Scenic Rivers System Act (1968)
- National Environmental Policy Act (1969)
- Clean Air Act (1970).

18.4 Pollution Control and Abatement Technology

This section describes the predominant gaseous, liquid and solid pollutants generated by the coal, petroleum, and petrochemical industries.

In response to environmental regulation, the petroleum industry reduced pollution by the actions listed in Sect. 18.1 under "Air Quality". The technology behind these actions is explained below.

18.4.1 Particulate Matter (PM)

In refineries, coking operations and FCC regenerators are the main sources of PM emissions. Coke-derived PM10 can be reduced by building enclosures around coke-handling equipment—conveyor belts, storage piles, rail cars, barges, and calciners.

For flue gas from FCC regenerators, many licensors offer scrubbing technology. ExxonMobil offers wet-gas scrubbing (WGS), which removes particulates, SO_x and NO_x [26]. UOP and Shell Global Solutions offer third-stage separator (TSS) technology, which removes PM in conjunction with flue-gas power recovery [27].

The main sources of air pollution from petroleum refineries are listed in Table 18.1. Refineries can be significant sources of particulate matter (PM), which can irritate

Table 18.1 Main sources of air pollution in hydrocarbon processing

Source	PM	SO$_2$	CO	VOC	NO$_x$	Hg	VTC
Coal handling	x	x					
Coal combustion		x	x	x	x	x	
Fluid catalytic cracking (FCC) units	x	x	x	x	x		
Coking units	x	x	x	x	x		
Compressor engines		x	x	x	x		
Vapor recovery and flare systems		x	x	x	x		
Vacuum distillation unit and condensers				x			
Sulfur recovery units		x	x		x		
Waste water treatment plants				x			
Boilers and process heaters		x	x		x		
Storage tanks				x			
Petrochemical production			x		x		x

PM = particulate matter
SO$_2$ = sulfur dioxide
CO = carbon monoxide
VOC = volatile organic compounds
NO$_x$ = nitric oxide (NO) and nitrogen dioxide (NO$_2$)
Hg = mercury
VTC = volatile toxic chemicals

the respiratory tract. PM is especially harmful when it is associated with sulfur and nitrogen oxides (SO$_x$ and NO$_x$).

PM from coal is coal dust. PM from refining comes mainly from two sources—delayed coking units and the regenerators of fluid catalytic cracking (FCC) units. FCC regenerators also emit ammonia, which combines with SO$_x$ and NO$_x$ in the air to form ammonium sulfates and nitrates. According to the South Coast Air Quality Management District (AQMD) [28] in Southern California, 1 ton of ammonia can generate 6 tons of PM10—airborne particulates with particle diameters less than 10 microns. PM2.5 stands for airborne particulates with diameters less than 2.5 microns.

18.4.2 Carbon Monoxide

Carbon monoxide (CO) is formed by incomplete combustion in boilers, process heaters, power plants, and FCC regenerators. CO is toxic because it binds strongly to the hemoglobin in blood, displacing oxygen. It is colorless and odorless, so without a special analyzer, it is hard to detect. This adds to its danger. Under particular conditions in refinery hydroprocessing units, the nickel on NiMo catalysts can react with CO to form nickel carbonyl, an exceptionally hazardous gas.

CO from partial combustion in FCC regenerators is converted to CO_2 in CO boilers. Flue gas from other boilers, process heaters, and power plants can also contain some CO, which can be diminished by the installation of high-efficiency burners and/or the implementation of advanced process control.

Fugitive emissions (leaks) from storage tanks, sewers, process units, seals, valves, flanges, and other fittings [29] can contain both CO and volatile organic compounds (VOC). Floating roofs can be added to open tanks, and tanks that already have a roof can be fitted with vapor recovery systems. Open grates above sewers can be replaced with solid covers. Emissions from seals, valves, etc., can be pin-pointed with portable combustible-gas detectors. Repairs can then be made at convenient times, e.g., during a maintenance shutdown.

18.4.3 Hydrogen Sulfide and Sulfur Oxides

Hydrogen sulfide (H_2S) is a pervasive hazard in petroleum production, petroleum refining, and natural gas processing facilities. Sour natural gas can contain several percent H_2S, which is recovered by amine adsorption and converted into elemental sulfur. In developed countries, H_2S recovery must be more than 99%—more than 99.8% in most instances.

Workers are most likely to be exposed due to leaks when taking samples of sulfur-containing gases or liquids. Sample-taking stations should be well-ventilated and equipped with several in-place toxic-gas monitors. Sample-taking procedures include the buddy system: one person takes the sample while another watches. In the past, workers died when rushing forward to help a stricken companion, so the watchers stand at a safe distance with quick access to a radio and/or SCBA. The impact of H_2S depends on its concentration and duration of exposure [30]. At 0.01 to 1.5 ppm, the characteristic rotten-egg odor is detectable. The odor becomes offensive at 3 to 5 ppm, where prolonged exposure causes nausea, tearing, asthma, and headaches. Above 20–30 ppm, the odor is said to be sweet or sickly sweet. Above 100 ppm, olfactory nerves are paralyzed and H_2S can no longer be smelled. Exposure to more than 700–1000 ppm causes immediate collapse and death after one or two breaths.

H_2S is flammable. The explosive range is 4.5–45.5% in air. When sulfur-containing feeds pass through hydrotreaters or conversion units, some or most of the sulfur is converted into H_2S, which eventually ends up in off-gas streams. Amine absorbers remove the H_2S, leaving only 10–20 ppmw in the treated gas streams. H_2S is steam-stripped from the amines, which are returned to the absorbers. The H_2S is conveyed by the amine system to the refinery sulfur plant, where it is converted into sulfur.

Fuel-oil fired heaters and FCC regenerators units are major sources of refinery SO_x and NO_x emissions. The most obvious way to reduce SO_x from a heater is to burn low-sulfur fuels. Switching fuels requires no capital investment, but it is probably the most expensive solution due to the relatively high cost of low-sulfur fuels.

A large fraction of the sulfur in the feed to an FCC unit ends up in coke on the catalyst. SO_x are formed in the regenerator when the coke is burned away. Removing sulfur from FCC feed with a pretreater decreases SO_x emissions.

SO_x irritate the respiratory tracts of people and other animals. When incorporated into particulate matter (PM 2.5), SO_x are especially bad. While gaseous SO_x molecules are trapped by mucous in the upper respiratory tract, inhaled particulates can penetrate deep into lungs.

In the atmosphere, SO_x react with water vapor to make sulfurous and sulfuric acids. The acids return to earth as "acid rain," which poisons trees and contaminates lakes and rivers. Experts estimate that SO_2 can remain in the air for 2–4 days. During that time, it can travel 600 miles (1000 km) before returning to the ground. Consequently, SO_2 emissions have caused a number of international disputes. Before 1990, acid rain from neighboring countries caused the death of about 1/3 of the trees in Germany forests. In the past, the United States and Canada argued bitterly about the cross-border impact of SO_x emissions from U.S. power plants and copper smelters. Fortunately, the passage and enforcement of clean-air legislation has reduced the problem dramatically.

Processes for removing sulfur oxides from stack gases include dry absorption, wet absorption, carbon adsorption, and catalytic oxidation.

Historically, wet flue-gas desulfurization processes used aqueous slurries of lime, dolomite, and/or sodium hydroxide. Sulfur oxides react with lime or limestone ($CaCO_3$) to produce calcium sulfate ($CaSO_4$) and calcium sulfite ($CaSO_3$), which precipitate from the scrubbing solution. The products move to a settling tank, in which the solid calcium salts separate from the solution as the scrubbed gas goes up the stack. After some time, the solids are removed and sent to a sanitary landfill. The solution is recycled, and fresh lime is added as needed.

The "dual alkali" approach starts with solutions or slurries of sodium hydroxide ($NaOH$), sodium carbonate (Na_2CO_3), or sodium bicarbonate ($NaHCO_3$). These compounds react with SO_2 to give sodium sulfite (Na_2SO_3) and sodium bisulfite ($NaHSO_3$), which stay dissolved in the solution. Some of the sodium sulfite reacts with excess oxygen in the flue gas to give sodium sulfate. Sulfate and sulfite are removed by reaction with lime or limestone ($CaCO_3$). The sodium hydroxide solution is recycled. Make-up hydroxide is added as needed to compensate for losses. Selected dual-alkali reactions are shown below:

$$2NaOH + SO_2 \rightarrow Na_2SO_3 + H_2O$$
$$Na_2CO_3 + SO_2 \rightarrow Na_2SO_3 + CO_2$$
$$Na_2SO_3 + {}^1/_2O_2 \rightarrow Na_2SO_4$$
$$Na_2SO_3 + CaCO_3 \rightarrow CaSO_3 + Na_2CO_3$$
$$Na_2SO_4 + Ca(OH)_2 \rightarrow CaSO_4 + 2NaOH$$

In a carbon adsorption process developed by Lurgi, hot flue gas first goes through a cyclone or dust collector for particulate removal. The gas is cooled with water and sent to an adsorption tower packed with activated carbon. The carbon adsorbs SO_x. Water is sprayed into the tower intermittently to remove the adsorbed gas as a weak aqueous acid. The scrubbed gas goes out the stack. The acid goes to the gas cooler, where it picks up additional SO_x by reacting with incoming flue gas. Cooler discharge is sold as dilute sulfuric acid.

In the Reinluft carbon adsorption process, the adsorbent is a slowly moving bed of carbon. The carbon is made from petroleum coke and activated by heating under vacuum at 1100 °F (593 °C).

Flue-gas scrubbing with catalytic oxidation (Cat-Ox) is an adaptation of the contact sulfuric acid process, modified to give high heat recovery and low pressure drop. In the Monsanto process, particulates are removed from hot flue gas with a cyclone separator and an electrostatic precipitator. A fixed-bed converter uses solid vanadium pentoxide (V_2O_5) to catalyze the oxidization of SO_2 to SO_3. Effluent from the converter goes through a series of heat exchangers into a packed-bed adsorption tower, where it contacts recycled sulfuric acid. The tower overhead goes through an electrostatic precipitator, which removes traces of acid mist from the scrubbed gas. Liquid from the tower (sulfuric acid) is cooled and sent to storage. Some of the acid product is recycled to the absorption tower.

In a flue-gas desulfurization process from Mitsubishi Heavy Industries (MHI), manganese dioxide (MnO_2) is the absorption agent. The final product is ammonium sulfate $(NH_4)_2SO_4$, which is an excellent fertilizer. MHI claims better than 90% removal of SO_x with this process.

The Wellman-Lord process uses a solution of potassium sulfite (K_2SO_3) as a scrubbing agent. K_2SO_3 adsorbs SO_2 to give potassium bisulfite ($KHSO_3$). The bisulfite solution is cooled to give potassium pyrosulfite ($K_2S_2O_5$). This can be stripped with steam to release SO_2, which is fed to a sulfuric acid plant.

Arguably, SO_x transfer additives are the most cost-effective way to lower SO_x emissions from a full-combustion FCC unit. These materials, first developed by Davison Chemical, react with SO_x in the FCC regenerator to form sulfates. When the sulfated additive circulates to the riser/reactor section, the sulfate is chemically reduced to H_2S, which is recovered by amine absorption and sent to a sulfur plant. Sulfur that would have gone up the stack as SO_x goes instead to the sulfur plant as H_2S.

FCC Regenerator (Oxidizing Environment)
Coke on catalyst (solid) $+ O_2 + H_2O \rightarrow CO_2 + SO_2, SO_3, H_ySO_x$ (gases)
SO_2, SO_3, H_ySO_x (gas) $+$ M (solid) $+ O_2 \rightarrow$ M·SO_4 (solid)
FCC Riser-Reactor (Reducing Environment)
M·SO_4 (solid) $+ 5 H_2 \rightarrow$ M (solid) $+ H_2S$ (gas) $+ 4 H_2O$ (gas)

In some units, SO_x additives can reduce FCC SO_x emissions by more than 70%. This can have a dramatic effect on the design and/or operation of upstream and downstream equipment—FCC feed pretreaters, FCC gasoline post-treaters, and flue-gas scrubbers for FCC regenerators.

18.4.4   Nitrogen Oxides, VOC, and Ozone

Like CO and SO_x, nitric oxide (NO) and nitrogen dioxide (NO_2) are emitted by fired heaters, power plants, and FCC regenerators. NO_x also damage respiratory tissues and contribute to acid rain.

Ozone, a nasty component of smog, is generated by reactions between oxygen and NO_x. The reactions are initiated by sunlight. In the troposphere, ozone reacts with volatile organic hydrocarbons (VOC) to form aldehydes, peroxyacetyl nitrate (PAN), peroxybenzoyl nitrate (PBN) and a number of other substances. PAN irritates nasal passages, mucous membranes, and lung tissue. Collectively, these compounds are called "photochemical smog." They harm humans, animals, and plants, and they accelerate the degradation of rubber and construction polymers. In some areas, smog looks like a brownish cloud just above the horizon. It makes for spectacular sunsets, but nothing else about it is good.

In refineries and petrochemical plants, NO_x is formed in several ways. In high-temperature fired heaters, NO_x is produced by the reaction of nitrogen with oxygen. In FCC regenerators, NO_x is produced from the nitrogen deposited with coke on spent catalysts. FCC NO_x emissions go up when (a) the catalyst contains more combustion promoter, (b) when oxygen in the flue gas goes up, (c) at higher regenerator temperatures, and (d) at higher feed nitrogen contents. Combustion promoter is a noble-metal material that accelerates the reaction between CO and O_2 to form CO_2. By removing CO, the promoter inhibits the following reactions:

$$2NO + 2CO \rightarrow N_2 + 2CO_2$$
$$2NO_2 + 4CO \rightarrow N_2 + 4CO_2$$

Dual-alkali flue-gas scrubbing only removes about 20% of the NO_x from a typical flue gas. Therefore, instead of simple scrubbing, chemical reducing agents are used. In selective catalytic reduction (SCR) processes, anhydrous ammonia is injected into the flue gas as it passes through a bed of catalyst at 500–950 °F (260–510 °C). The reaction between NO_x and ammonia produces N_2 and H_2O.

The MONO-NO_x process offered by Huntington Environmental Systems employs a non-noble metal catalyst. For SO_x, NO_x and VOC removal, Ducon uses ceramic honeycomb or plate-type catalysts in which titanium dioxide is the ceramic and the active coatings are vanadium pentoxide and tungsten trioxide (WO_3). The working catalyst temperature ranges from 600 to 800 °F (315–427 °C). For NO_x abatement, Ducon provides complete ammonia injection systems with storage tanks, vaporizers and injection grids. Either anhydrous or aqueous ammonia can be used.

NO_x-removal catalysts are offered by Haldor-Topsøe, KTI, and others. The Thermal DeNO$_x$ process offered by ExxonMobil is non-catalytic.

18.4.5   Chemicals that React with Stratospheric Ozone [31]

The previous section mentions the harmful effects of ground-level ozone. The stratosphere, located about 6–30 miles (10–50 km) above the ground, contains a layer of ozone that is beneficial, because it protects organisms from harmful ultraviolet-B (UV-B) solar radiation.

In the 1980s, scientists noticed that the stratospheric ozone layer was getting thinner [32]. Researchers linked several man-made substances, mostly volatile halogenated compounds, to ozone depletion, including carbon tetrachloride (CCl_4), chlorofluorocarbons (CFCs), halons, methyl bromide, and methyl chloroform. These chemicals leak from air conditioners, refrigerators, insulating foam, and some industrial processes. Winds carry them through the lower atmosphere into the stratosphere, where they react with strong solar radiation to release chlorine and bromine atoms. These atoms initiate chain reactions that consume ozone. Scientists estimate that a single chlorine atom can destroy 100,000 ozone molecules.

In 1987, twenty-seven countries signed the Montreal Protocol, which recognized the international consequences of ozone depletion and committed the signers to limit production of ozone-depleting substances. Today, more than 190 nations have signed the Protocol, which now calls for total elimination of certain ozone-depleting chemicals.

The 1998 and 2002 Scientific Assessments of Stratospheric Ozone firmly established the link between decreased ozone and increased UV-B radiation. In humans, UV-B is linked to skin cancer. It also contributes to cataracts and suppression of the immune system. The effects of UV-B on plant and aquatic ecosystems are not well understood. However, the growth of certain plants can be slowed by excessive UV-B. Scientists suggested that marine phytoplankton, which are the foundation of the ocean food chain, were already under stress from UV-B.

In the United States, production of halons ended in January 1994 [33]. In January 1996, production virtually ceased for several other ozone-depleting chemicals, including CFCs, CCl_4, and methyl chloroform. January 1, 2010 saw a ban on production, import, and use of HCFC-22 and HCFC-142b, except for continuing servicing needs of existing equipment. In 2015, the production, import, and use of all HCFCs was banned, except for continuing servicing needs. In 2020, there will be no production or imports of HCFC-22, and HCFC-142b, achieving a total reduction of 99.5%.

New products less damaging to the ozone layer have gained popularity. For example, computer makers now use ozone-safe solvents to clean circuit boards, and automobile manufacturers use HFC-134a, an ozone-safe refrigerant, for air conditioners in new vehicles.

Studies indicate that the Montreal Protocol has been effective. Stratospheric concentrations of methyl chloroform are falling, indicating that emissions have been reduced. Concentrations of other ozone-depleting substances, such as CFCs, are also decreasing. According to the U.S. EPA [34], the ozone layer has not grown thinner since 1998 and appears to be recovering. Antarctic ozone is expected to return to pre-1980 levels sometime between 2060 and 2075.

18.4.6   Greenhouse Gases [35, 36]

In a greenhouse or in an automobile with the windows rolled up, sunlight comes in through the glass and gets absorbed by various objects inside. These objects reradiate the adsorbed energy as infrared heat, which can't go back out through the glass, at least not very quickly. Consequently, on a sunny day the inside of a greenhouse and inside of a sealed car is hotter than the outside air.

In the earth's atmosphere, gases such as CO_2, CH_4, and N_2O are called "greenhouse gases," because they are nearly transparent to visible sunlight, but they absorb some of the re-emitted infrared (IR) radiation. In essence, they warm the atmosphere by slowing the release of heat into space.

On average, every person in the world is responsible for just over a ton of CO_2 emissions each year. Most of this is due to the burning of fossil fuels in power plants and vehicles. According to some estimates, to prevent dangerous climate change while allowing for some increase in population, that number must be reduced to about 0.3 tons per person per year. Per capita CO_2 emissions are orders of magnitude greater in developed countries than they are in poor developing countries. In 2014, Qatar had the highest emission (45.4 metric tons per person) while Somalia and 16 other countries had emissions <0.1 metric tons per person. The values for the United States and China were 16.5 and 7.5, respectively [37].

18.4.6.1   Global CO_2 and Temperature Balance

Life, as we know, requires liquid water and nutrients. Macronutrients include compounds containing carbon, hydrogen, sulfur and nitrogen. Arguably, the most fundamental food-chain macronutrient is carbon dioxide. Water and carbon dioxide play key roles in regulating planetary temperatures.

Water is an effective heat buffer, with both a high heat of fusions (freezing and melting) and a high heat of vaporization (vaporizing and boiling). It takes a lot of heat to vaporize water completely at its boiling point, and it also takes a lot of heat to melt ice at water's freezing point. On earth, water is present in three phases: vapor, liquid, and ice—mostly liquid. In the atmosphere, water vapor condenses and falls to the surface as rain, snow, etc. In cold regions, snow collects into ice sheets, which partially or completely melt during the summer. Rain and melt water collect into rivers, lakes, and oceans, where evaporation returns water vapor to the atmosphere.

Table 18.2 provides estimates of the global carbon distribution [38]. On per-mole basis, sedimentary rock contains more than 99.8% of the carbon in the earth crust. Most of the rest is found in the sea. In oceans and lakes, water serves as a carbon dioxide buffer. On average, the oceans hold 60 times more CO_2 than the atmosphere.

Table 18.2 Global distribution of carbon

Source	Moles of Carbon $\times 10^{18}$	Relative to the atmosphere
Sediments		
Carbonates	1530	28,500
Organic carbon	572	10,600
Land		
Organic carbon	0.065	1.22
Oceans		
CO_2 and H_2CO_3	0.018	0.3
HCO_3^-	2.6	48.7
CO_3^{2-}	0.33	6.0
Dead organic	0.23	4.4
Living organic	0.0007	0.01
Atmosphere		
CO_2	0.0535	1.0

Dissolved in water, CO_2 forms carbonic acid (H_2CO_3), bicarbonate (HCO_3^-) and carbonate (CO_3^{2-}). Sea water is slightly alkaline, with a surface pH of 8.2, so it readily reacts with H_2CO_3. However, rapid exchange with the atmosphere only occurs in the upper wind-mixed layer, which is about 300 feet (100 m) thick. This layer contains roughly one atmosphere equivalent of CO_2. CO_2 in its various forms is removed from the sea by foraminifera, coral reefs, and other marine organisms, which produce solid calcium carbonate, the main component of sea shells.

About 30–50% of the CO_2 released into the air stays there. The rest goes into the hydrosphere and biosphere.

Only 0.0025% of the earth's carbon is present as atmospheric CO_2. But even small changes have a dramatic impact on surface temperature balance. The air retains about 30–50% of the excess CO_2 it receives from combustion of fossil fuels, cement manufacturing, and other human activities. The rest goes into the hydrosphere and biosphere. An alarming trend is ocean acidification. Coral reefs moderate the impact of CO_2 on ocean pH, but as they so, they slowly dissolve [39].

18.4.6.2 Climate Change

Atmospheric CO_2 concentrations are increasing faster than ever, and most of the increase is due to industrial growth. According to the Goddard Institute for Space Studies (GISS), which is administered by the U.S. National Aeronautics and Space Administration (NASA), the average global temperature has increased 1.4 °C between 1850 and 2005 [40]. ModelE2 is a GISS application which indicates that man-made CO_2 is responsible for most of the change.

If warming continues, it may do the following:

- Increase the number and intensity of dangerous heat waves
- Increase severe storm activity
- Damage certain crops
- Raise the average sea level by expanding the oceans (the density of water decreases as it warms) and by melting ice sheets in polar regions and elsewhere.

These in turn will increase weather-related deaths, damage coastal cities and towns, and ruin coastal ecosystems.

Global warming sceptics base their objections on the fact that other phenomena affect temperatures, and that predictions are based on models. Models are complex because they must include a plethora of natural and anthropogenic factors [41]. The most important factors are summarized in Fig. 18.2. Man-made factors include aerosols (which decrease temperature), human-generated greenhouse gases, man-made ozone, and land use (deforestation). Natural factors include periodic changes in the earth's orbit, variations in solar temperature, and volcanic gases (which include CO_2). The 1.4 °C change since 1850 is "only" 0.1 °C per decade, but this is significant, especially when one considers that vast quantities of energy are required to accomplish a relatively minor increase in average ocean temperature. Human-generated greenhouse gases correlate with temperature changes more strongly than all other factors.

18.5 Waste Water from Petroleum Production and Processing

Upstream and midstream operations, particularly SAGD (steam-assisted gravity drainage) and cyclic steam stimulation (CSS) require tremendous amounts of water [42]. SAGD produces waste water contaminated with fines, trace metals, and hydrocarbons. At present, only a fraction of waste water is recovered. Most goes to large tailing ponds, where the oil undergoes oxidation, generating significant amounts of CO_2 [43]. Environment Canada tolerates the tailing ponds.

The recent increase in U.S. oil production is due to hydraulic fracturing (fracking) requires immense amounts of water. Over 100 billion gallons are used for this purpose in the US annually. There are still many unresolved questions and issues about its impact to the environment. Experience shows that fracking itself causes little if any damage to groundwater, but the wastewater from fracking is problematic [44]. A *Texas Tribune* article by Kate Galbraith and Terrence Henry [45] provides a succinct discussion of problems associated with fracking wastewater. In fracking, million gallons of water, combined with sand and chemicals, are pumped into a well at high pressure. The pressure shatters the reservoir rock, and the sand props the cracks open, increasing the flow of oil and gas into the well. Some of the water comes back up, along with additional underground water. Typically, fracking wastewater is a dark, thick liquid that smells of sulfur and contains a cocktail of chemicals and minerals.

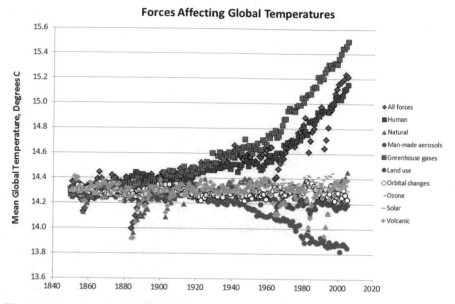

Fig. 18.2 Data from ModelE2, developed by NASA's GISS group. Shown are forces affecting global temperatures, 1850–2005. Man-made forces include aerosols (which decrease temperature), human-generated greenhouse gases, man-made ozone, and land use (deforestation). Natural factors include periodic changes in the earth's orbit, solar temperature variations, and volcanic gases (which include CO_2)

It generally has high salinity, and sometimes contains low levels of radiation. The wastewater goes to a settling pit, in which entrained oil disengages and floats to surface. The oil is skimmed off and sold. The skimmed water is trucked to disposal wells and injected thousands of feet underground for permanent storage. However, if an incident forces chemicals into an aquifer, responders may be unprepared to take actions until they analyze the sample. Citizens groups are concerned about groundwater contamination. The *Texas Tribune* article cites reports of fracturing causing and earthquakes and an instance of associated aquifer contamination. Certain aspects of fracking were exempted from the US Safe Drinking Water Act by the Energy Policy Act of 2005.

In the downstream, refineries and chemical plants generate contaminated process water, oily runoff, and sewage. Water is used by just about every process unit, especially those that require wash water, condensate, stripping steam, caustic, or neutralization acids. Contaminated process water may contain suspended solids, dissolved salts, water-soluble chemical, phenols, ammonia, sulfides, and other compounds. As much as possible, waste-water steams are purified and re-used. Present requirements ensure that the water going out of a processing plant is at least as clean as the water coming in.

Table 18.3 summarizes the sources and destinations of waste water in the refinery. Wastewater steams are purified and reused as much as possible. Present requirements

Table 18.3 Refinery waste
water treatment

Designation	Source
Oil-free water	Oil-free storm runoff
	Steam turbine condensate
	Air-conditioner cooling water
	Cooling water from light-oil units (C$_5$-minus)
	Cooling-tower blowdown
	Clean water from treatment plants
Oily cooling water	Oily storm runoff
	Cooling water from heavy-oil units (C$_5$-plus)
	Uncontrolled blowdown
Process water	Desalter water
	Excess sour water
	Water drawn from oil-storage tanks
	Accumulator draws
	Treating plant waste
	Barometric condeners
	Slop-oil breaks
	Ballast water
Sanitary WATER	Employee locker rooms
	Cafeteria
	Office buildings
	Control rooms
	Destination
Oil-free water	Oil/water separator
Oily cooling water	Oil/water separator
Process water	API Separator, activated sludge treatment
Sanitary water	Municipal water treatment plant

in the US and EU state that the water going out of a process plant must be as clean as the water coming in.

18.5.1   Primary Treatment

Primary treatment uses a settling pond to allow most hydrocarbons and suspended solids to separate from the wastewater. The solids drift to the bottom of the pond,

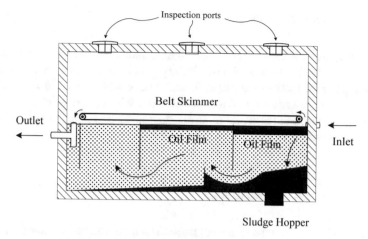

Fig. 18.3   API oil/water separator: simplified sketch

hydrocarbons are skimmed off the top, and oily sludge is removed. Difficult oil-in-water emulsions are heated to expedite separation.

Acidic wastewater is neutralized with ammonia, lime, or sodium carbonate. Alkaline wastewater is treated with sulfuric acid, hydrochloric acid, carbon dioxide-rich flue gas, or sulfur.

Figure 18.3 a simplified sketch of an API oil-water separator. The large capacity of these separators slows the flow of wastewater, allowing oil to float to the surface and sludge to settle out. They are equipped with a series of baffles and a rotating endless-belt skimmer, which recovers floating oil. Accumulated sludge is removed through sludge hoppers at the bottom.

18.5.2   Secondary Treatment

A small amount of suspended solids remains in the water after primary treatment. These are removed by filtration, sedimentation or air flotation. Flocculation agents may be added to consolidate the solids, making them easier to remove by sedimentation or filtration. Activated sludge, which contains waste-acclimated bacteria, digests water-soluble organic compounds, in either aerated or anaerobic lagoons. Steam-stripping is used to remove sulfides and ammonia, and solvent extraction is used to remove phenols.

18.5.3 Tertiary Treatment

Tertiary treatment removes specific pollutants, including traces of benzene and other partially soluble hydrocarbons. Tertiary processes include reverse osmosis, ion exchange, chlorination, ozonation, or adsorption onto activated carbon. Compressed air or oxygen can be used to enhance oxidation. Spraying water into the air or bubbling air through the water removes remaining traces of volatile chemicals such as phenol and ammonia.

18.6 Cleaning Up Oil Spills

Large spills of oil from tankers are not uncommon, but they can cause tremendous damage to the environment. Small spills come from leaks in tanks or mishaps during the loading or unloading of trucks, ships, or rail cars.

Oil spills can come from natural seepage, leaky storage tanks, petroleum exploration and production activities, the on-purpose flushing of fuel tanks at sea, and accidents such as those described in Sect. 18.8.

The cleanup of oil spills includes containment, physical and mechanical removal, chemical and biological treatment, and natural forces. Land-based spills are easier to clean than spills onto open water, which are spread quickly by currents and winds.

18.6.1 Natural Forces

Several natural forces tend to remove oil spills. These include evaporation, spreading, emulsification, oxidation, and bacterial decomposition.

Evaporation. A large portion of an oil spill may simply evaporate before other methods can be used to recover or disperse the oil. Rates of evaporation depend on the ambient temperature and the nature of the oil.

Spreading. The fact that spilled oil spreads quickly across the surface of water is a "good news, bad news" story. The good news is that spreading increases rates of evaporation and air oxidation. The bad news is that the more dispersed the oil becomes, the harder it is to collect.

As with evaporation, rates of oil-spill dispersion depend upon ambient conditions and the nature of the oil. Not surprisingly, oil disperses best in fast-moving turbulent water.

Oxidation. Freshly spilled crude oil has a natural tendency to oxidize in air. Sunlight and turbulence stimulate the process. Oxidation products include organic acids, ketones and aldehydes, all of which tend to dissolve in water. As a spill ages, oxidation slows as "easy" molecules disappear from the mix. Compared to other natural forces, oxidation plays a minor role in removing oil spills.

Emulsification. When crude oil spills at sea, it emulsifies rapidly. Two kinds of emulsions are formed—oil-in-water and water-in-oil. Oil-in-water emulsions, in which water is the continuous phase, readily disperse, removing oil from the spill. However, this kind of emulsion requires the presence of surface-active agents (detergents).

The composition of water-in-oil emulsions varies from 30 to 80% water. These are extremely stable. After several days, they form "chocolate mousse" emulsions, which are annoyingly unresponsive to oxidation, adsorption, dispersion, combustion, and even sinking. The most effective method for mousse emulsions is physical removal. Mousse contains roughly 80% water, so after a 40–50% loss of light-ends through evaporation, a spill of 200,000 barrels oil can form 400,000–500,000 barrels of mousse.

18.6.2 Containment and Physical Removal

Booms. When oil is spilled on water, floating booms may be used for containment. A typical boom extends 4 inches (10 cm.) above the surface and 1 foot (30 cm.) below. Foam-filled booms are lightweight, flexible, and relatively inexpensive. Typically used for inland and sheltered waters, they are made from polyvinyl chloride (PVC) or polyurethane. Rectangular floats allow them to be wound onto a reel for storage.

Inflatable booms use less storage space and can be deployed from ships or boats in open water. Towed booms (Fig. 18.4) are good for preventing dispersion of oil by winds and currents.

Fig. 18.4 Boom barrier towed by ships

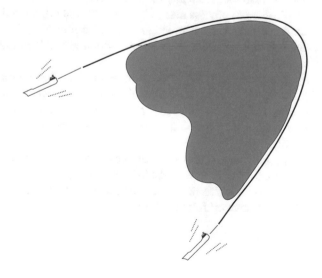

Fig. 18.5 Free vortex
skimmer for oil recovery

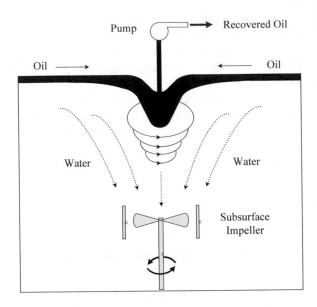

Beach booms are modified for use in shallow water or tidal areas. Water-filled tubes on the bottom of the boom elevate it above the beach when the water level is low and allow the boom to float when the water level rises.

Skimmers and pumps. After a spill is contained, skimmers and pumps can pick up the oil and move it into storage tanks. Weir skimmers are widely used because they are so simple. Modern designs are self-adjusting and circular so that oil can flow into the skimmer from any direction. They can be fitted with screw pumps, which enable them to process many different types of oil, including highly viscous grades. Cutting knives keep seaweed and trash from fouling the pumps. Screw pumps can develop high pressure differentials, which gives them higher capacities than other kinds of pumps.

Another kind of skimmer is an oleophilic ("oil loving") endless belt. The belt picks up oil as it passes across the top of the water. As of the belt returns to its starting point, it is squeezed through a wringer. Oil recovered by the wringer flows into containers, where it is stored until it can be moved to a land-based processing facility.

Another method uses a subsurface impeller to create a vortex, similar to the vortex that forms as water flows out of a bathtub (Fig. 18.5). This vortex funnels water down through the impeller and creates a bowl of oil in the middle of the vortex. A pump is used to remove oil from the bowl.

18.6.3 Adsorbents

Oil spills can be treated with absorbing substances or chemicals such as gelling agents, emulsifiers, and dispersants.

When applying adsorbents, it is important to spread them evenly across the oil and to give them enough time to work. When possible, innocuous substances should be used. Straw is cheap, and it can absorb between 8 and 30 times its own weight in oil. When it is saturated, the straw is loaded into boats with rakes or a conveyor system and transported to land. Oil can be recovered from the straw by passing it through a wringer.

Synthetic substances may also be used for adsorbing oil spills. Polymers such as polypropylene, polystyrene, and polyurethane have been successful. Polyurethane foam is especially good. It can be synthesized onsite easily, even aboard a ship. It adsorbs oil readily and doesn't release its load unless it is squeezed. Best of all, a batch of polyurethane can be used again and again.

18.6.4 Dispersion Agents

Dispersion chemicals act like detergents. One part of the molecule is oil-soluble while the other is water-soluble. In effect, the oil dissolves in water and diffuses quickly away from the spill. Dispersants reduce the tendency of oil to cling to partly immersed solids, such as walls, docks, buoys and boats.

18.6.5 Non-dispersive Methods

Non-dispersive methods for removing oil spills include gelling, sinking, and burning.

Gelling. Fatty acids and 50% sodium hydroxide can be added to a spill to trigger a soap-forming reaction. The resulting gel does not disperse. Instead, it remains in place to block the spread of non-gelled oil.

Oil sinking. Sinking an oil spill keeps it from reaching shorelines, where it can devastate marine life. Sinking is best used in the open sea. In shallow water near the coast, it can cause more problems than it solves. Amine-treated sand is the most common sinking agent. To initiate sinking, a sand/water slurry is sprayed onto the oil slick through nozzles. The required sand/oil ratio is about 1-to-1. The oil sticks to the treated sand, which sinks toward the bottom of the sea. According to many experts, the oil-coated sand remains in place for many years. According to others, it can damage fragile ecosystems on the ocean floor and/or lead to shellfish contamination.

During a full-scale test 15 miles from the coast of Holland, Shell Oil used amine-treated sand to sink a 100-ton slick of Kuwaiti crude in less than 45 min. Oil removal exceeded 95%.

Other materials have been used for oil sinking. These include talc, coal dust, cyclone-treated fly ash, sulfur-treated cement, and chalk. In general, a 1-to-1 ratio of sinker weight to oil weight is needed to sink a fresh spill. If weathering takes place, the density of the oil increases and less sinking agent is needed. It has been estimated that the cost for sinking is similar to the cost for dispersion. However, in open seas under high winds and waves, it may be difficult to spread the sinking agent.

Burning. Freshly spilled crude oil in a confined area may be combustible. However, after several hours, the spill may have thinned due to spreading and the most volatile components may have evaporated. If so, it may not be possible to ignite the remaining material. The addition of gasoline or kerosene can restore combustibility.

Burning is not used much anymore. It seldom removes much oil, and it can generate concentrated, unpredictable pockets of atmospheric CO and SO_2, both of which are poisonous.

18.6.6   Cleanup of Oil Contaminated Beaches

For cleaning oil from beaches, farm machinery and earthmoving equipment have been used to good effect. In many cases, a layer of straw is spread across the oil. After a few days, the oil-laden straw is raked onto a conveyor, screened to drop out sand, and sent to wringer. Recovered oil is trucked away to a refinery. The spent straw, which still contains some oil, can be blended with coal and burned in a power plant, or simply incinerated. The separated sand is washed and returned to the beach.

When beach pollution is severe, oily sand is removed with earthmoving equipment. When beach pollution is mild, sand removal may not be needed. Instead, detergents can be used. They must be applied cautiously to minimize harm to marine life. Wave action does a great job of mixing detergent into the sand. Usually, the detergent is applied about one hour before high tide. When the tide comes in, the washing begins. If high-tide washing is inappropriate, high-pressure hoses can be employed. Hoses are also effective for cleaning oil off of walls and rocks.

Froth flotation. In the froth-flotation process, oil-soaked sand from a polluted beach is poured into a vessel, where it is mixed with water and cleaned with a froth of air bubbles. Aided by chemical or physical pre-treatment, the froth strips oil away from the sand. Due to its low density and the action of the bubbles, the oil floats to the top of the vessel, where it is drawn off and sent to a separating chamber, where entrained water is removed. Tests show that sand containing 5000 ppm of oil can be cleaned down to 130 ppm, generating effluent water with 165 ppm of oil. Usually, the cleaned sand is returned directly to the beach.

Hot water cleanup. When milder methods fail to give the desired degree of cleanup, hot water can mobilize some of the oil still trapped within the polluted sand. This method is used as a last resort because of the damage it does to inter-tidal ecosystems.

18.7 Solid Waste

Solid wastes from coal and petroleum processing may include the following:

- Coal fines
- Cuttings from oil-well drilling
- Used drilling mud
- Spent catalyst ad catalyst fines
- Acid sludge from alkylation units
- Sludge from the bottom of storage tanks
- Miscellaneous oil-contaminated solids.

Waters that cannot be recycled are cleaned on site, sent to landfills, or transported to reclamation facilities.

18.7.1 Solid Waste Recovery and Disposal

Contaminated solids are produced during the drilling of oil wells, the transportation of crude oil, and in oil refineries. All oil-contaminated solids are considered hazardous and must be sent to hazardous-waste landfills. The transportation and disposal of hazardous waste costs an order of magnitude more than the transportation and disposal of sanitary waste. Thus, there is a huge economic incentive to remove oil from contaminated solids before they leave a site.

Considerable amounts of solid wastes can be generated during oil drilling. Oil contaminated cuttings must be cleaned before disposal. Table 18.4 shows the sources of solid wastes in a modern oil refinery. These data, provided by the American Petroleum Institute, are based on a "typical" 200,000 barrels-per-day high-conversion refinery. A plant this size produces about 50,000 tons per year of solid waste and about 250,000 tons per year of waste water. As discussed above, all waste water must be purified before it leaves the plant.

18.7.1.1 Super-Critical Fluid Extraction

A supercritical substance exists as a single fluid phase. Simultaneously, it can have liquid-like solvating powers and gas-like diffusivity and viscosity; its surface tension is zero. Ammonia, argon, propane, xenon, water, CO, and CO_2 are used for supercritical extraction. The phase diagram in Fig. 18.6 shows that the super-critical region for CO_2 lies above 88 °F (31 °C) and 1058 psig (7.4 MPa). The CO_2 triple point, where solid, liquid, and gas phases exist simultaneously, occurs at −69.9 °F (−56.6 °C) and 57.8 psig (0.5 MPa).

With non-toxic gases such as CO_2, the SFE process is simple. Untreated solids are placed in an extraction chamber and CO_2 is added. Pressurizing the CO_2 converts it into an effective solvent. By manipulating temperature and pressure, operators can extract the material of interest with high selectivity. After the extracted material dissolves in the CO_2, it goes to a collection vessel, where the pressure is reduced. At low pressure, CO_2 loses its solvating power and separates from the extract. The CO_2 is recovered, condensed and recycled. In SFE with CO_2, no liquid solvents are used. The CO_2 is recycled, so the process is not considered to be a contributor to global warming.

Table 18.4 Breakdown of refinery solid wastes

Solid waste	Percent of total (wt%)
Pond sediments	8.4
Sludge from biological treatment	18.9
Solid from DAF/IAF treaters	15.4
API separator sludge	13.2
Miscellaneous sludge	17.5
Off-grade coke	1.9
Spent catalysts	6.4
Slop oil emulsion solids	4.4
Spent solvents, chemicals	3.8
Contaminated soils and other solids	6.7
Heat exchanger cleaning sludge	0.1
Tank sludge	3.4

DAF dissolved air flotation; *IAF* induced air flotation

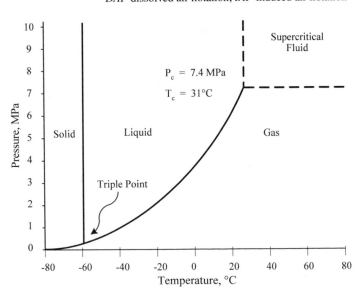

Fig. 18.6 Phase diagram for carbon dioxide

SFE is common in the food, pharmaceutical and cosmetic industries, where it extracts caffeine from coffee beans, bitter from hops, tar and nicotine from tobacco, and other natural compounds from spices, flowers, aromatic woods, and medicinal plants.

Since about 1990, SFE with CO_2 has been used in oil fields to remove oil from drill cuttings. CO_2 is especially good for this purpose because it can be used onsite. It is non-toxic (except when it suffocates), relatively unreactive, and it doesn't burn. On drill cuttings and other oil-contaminated soils and sands, it penetrates the mineral structure, removing both the free-oil phase and the oil trapped in the solid matrix. The process removes more than 99.9% of the oil, leaving no toxic residue. The required time is short—about 10 to 60 min per batch.

18.7.1.2 Sludge

According to RCRA, oil-containing sludge from storage tanks and refinery water treatment facilities is by definition hazardous, and should be sent to a hazardous land fill. In most cases, a lot of the sludge can be dissolved with detergents and/or or solvents (such as hot diesel oil) and blended into crude oil. Alternatively, dissolved sludge can go to delayed cokers, asphalt plants, carbon black plants, or cement kilns.

In one method, hot water and a chemical are circulated through the tank. On top of the water, a hydrocarbon such as diesel is added. The density difference between warm water and the hydrocarbons in the sludge causes the sludge to rise, allowing the chemical to strip out water and solids. The method recovers good-quality oil, which can be processed in the refinery, and leaves behinds a relatively clean layer of solids on the tank bottom.

Tanks associated with slurry oil from FCC units present an interesting challenge. It can be very expensive to take these tanks out of service, the sludge is loaded with finely divided FCC catalyst particles, and slurry-oil sludge is difficult to dissolve. Recently, a process called Petromax has been used to liquefy slurry-oil sludge, allowing it to be pumped out of tanks with ordinary equipment, even while the tanks are still in use [46].

Some service companies use robots, cutting wands, and other sophisticated devices to clean tanks completely without sending people inside. These methods are especially valuable when the tank contents are toxic. One such company is Petrochemical Services, Inc., which pioneered the use of robots in tank-cleaning operations.

Blending mobilized sludge into crude oil is limited by specifications on basic sediment and water (BS&W). This seems equivalent to moving waste from one place to another, but it really isn't. The dissolution of tank sludge separates useful oil from inorganic solids (sand, clay, salts, and metal oxides) and refractory organics (asphaltenes, long-chain waxy paraffins, kerogen, and coke).

18.7.1.3   Spent Catalysts

Many refinery catalysts are regenerated several times. The regeneration of FCC catalysts occurs in the (aptly named) regenerator. Catalysts from fixed-bed units can be regenerated in place, but usually they are sent off-site to a facility that specializes in catalyst regeneration.

Eventually, even the hardiest catalysts reach the end of their useful life. When this happens, they are sent to a metal reclaimer, which recovers saleable products such as alumina, silica, MoO_3, V_2O_5, nickel metal, and various forms of cobalt.

The reclamation company makes money on both ends of the plant—from the refiner who must dispose of the catalyst, and from the customer who buys reclaimed products.

18.8   Significant Accidents and Near-Misses

Pollution-causing incidents can be divided into four major categories:

1. **Usual-practice pollution**. In this case, waste is discharged routinely. The consequences are known, but someone has concluded that the cost of decreasing the discharge is too high. To different groups, "risk" and "cost" can mean entirely different things. Usually, nothing changes until some government body decides to force a change.
2. **Accidental pollution**. This category includes industrial accidents, several of which are described below. Post-disaster investigations often show that the accidents were preventable.
3. **Inappropriate response to pollution**. Many industrial accidents would have been far less harmful if the people involved had been prepared. Studies conclude that careful emergency planning coupled with thorough employee training (and re-training) prevents accidents and leads to faster, better responses to accidents.
4. **Malicious acts**. These include illegal "midnight dumping," cover-ups, sabotage, and acts of war.

The examples described below cover all four categories.

After significant accidents, damage is mitigated by the heroic efforts of doctors, nurses, firemen, cleanup crews, and volunteers. After the Chernobyl and Fukushima nuclear power plant meltdowns, the people toiling to control the reactors knew they going to die, but they kept going. When giving statistics on lives lost, reporters seldom enumerate the lives saved by these heroes.

18.8.1 Coal Mining: Explosions and Black Lung Disease

Coal mine explosions and maladies such as black-lung disease have plagued the industry since the start of the Industrial Revolution. In 1906, in a mine explosion in Courrières, France, 1099 miners died, including children. Courrieres was the worst mine accident in Europe. In April 1942, in a coal-mine explosion in Benxi, Liaoning, China, 1549 workers died in the worst coal mine accident ever.

Mine explosions are caused by methane and/or coal dust. "Damp" is a catch-all term for gases other than air in coal mines. In addition to firedamp (methane), there is blackdamp (carbon dioxide), stinkdamp (hydrogen sulphide), and afterdamp (carbon monoxide). Firedamp is explosive in air at concentrations between 4 and 16%.

In May 1812, Sir Humphry Davy developed the Davy lamp, in which a low flame was surrounded by iron gauze. Due to the gauze, the lamp did not ignite external methane, but methane could pass into the lamp and burn safely above the flame. At first, the Davy lamp improved safety by eliminating open-flame lighting. But it led to the mining of areas that had previously been closed for safety reasons, sometimes with disastrous results [47]. Black lung disease is actually a set of maladies associated with coal dust [48]. It was not well-understood until the 1950s. In the Federal Coal Mine Health and Safety Act of 1969, the US Congress set limits on coal dust exposure and created the Black Lung Disability Trust. Mining companies agreed to a clause which guaranteed compensation for workers who had spent 10 years doing mine work, coupled with X-ray or autopsy evidence of lung damage. Financed by a federal tax on coal, by 2009 the Trust had distributed over $44 billion in benefits to disable miners or their widows.

18.8.2 Lakeview: Blowout (March 1910–September 1911) [49]

The Lakeview gusher remains the largest accidental oil leak or spill on land in history. It released 1.2 million tons of crude in 544 days. Drilling started in January 1909. Initially, the only product from the well was natural gas. On March 14, 1910, pressurized oil blew through the well casing above the bit. At the time, blowout prevention technology didn't exist, so the well gushed unabated. The initial flow rate was 18,800 barrels per day. The peak flow rate was 90,000 barrels per day. The blowout created a river of crude, which crews tried to contain with sand bag dams and dikes. About half of the oil was saved. The remainder seeped into the ground or evaporated.

18.8.3   Texas City, Texas: Ammonium Nitrate Explosion (1947)

On April 16, 1947, the *SS Grandcamp* exploded in port at Texas City, Texas [50]. The ship was loaded the 7700 tons of ammonium nitrate. The initial blast, together with fires and explosions on other ships and nearby oil-storage facilities, killed 581 people and injured up to 8000. The force of the blast was equivalent to 3.2 kilotons of TNT. At about 8:00, a fire was spotted in the cargo hold of the ship. For roughly an hour, attempts to extinguish the fire failed as a red glow returned after each effort to douse the fire. The cause of the initial fire remains unknown. The event drew attention to (a) the dangers of storing massive amounts of hazardous materials and (b) storing hazardous materials in polluted areas.

18.8.4   London: Killer Fog (1952, '56, '57, '62)

In December 1952, thick fog rolled across many parts of the British Isles. In the Thames River Valley, the fog mixed with smoke, soot and sulfur dioxide from coal-burning homes and factories, turning the air over London into a dense yellow mass. The smog was so thick that buses could not run without guides walking ahead of them carrying lanterns. Fog stayed put for several days, during which the city's hospitals filled to over-flowing. According to the Parliamentary Office of Science and Technology [51], more than 3000 people died that month because of the polluted air. Similar, less-severe episodes occurred in 1956, 1957, and 1962. The 1956 event killed more than 1000 people.

London's deadly smog was caused by "usual-practice" pollution. Due to the widespread use of cheap, high-sulfur coal, the air in London had been bad for decades, but post-war growth made it worse than ever. In response to the incidents, Parliament passed Clean Air Acts in 1956 and 1962, prohibiting the use of high-sulfur fuels in critical areas.

18.8.5   Amoco Cadiz: Oil Spill (1978) [52]

On March 16, 1978, the supertanker *Amoco Cadiz* was three miles off the coast of Brittany, France when its steering mechanism failed. The ship ran aground on the Portsall Rocks.

For two weeks, severe weather restricted cleanup efforts. The wreck broke up completely before any remaining oil could be pumped out, so the entire cargo—more than 1.6 million barrels of Arabian and Iranian crude oil—spilled into the sea.

The resulting slick was 18 miles wide and 80 miles long. It polluted 200 miles of coastline, including the beaches of 76 Breton communities, harbors and habitats for marine life. On several beaches, oil penetrated the sand to a depth of 20 inches. Piers and slips in small harbors were covered with oil. Other polluted areas included the pink granite rock beaches of Tregastel and Perros-Guirrec, and the popular bathing beaches at Plougasnou. The oil persisted for only a few weeks along exposed rocky shores, but in the areas protected from wave action, the oil remained as an asphalt crust for several years.

At the time, the *Amoco Cadiz* incident caused more loss of marine life than any other oil spill. Cleanup activities on rocky shores, such as pressure-washing, also caused harm. Two weeks after the accident, millions of dead mollusks, sea urchins, and other bottom-dwelling organisms washed ashore. Nearly 20,000 dead birds were recovered. About 9000 tons of oysters died. Fish developed skin ulcerations and tumors.

Years later, echinoderms and small crustaceans had disappeared from many areas, but other species had recovered. Even today, evidence of oiled beach sediments can be seen in sheltered areas, and layers of sub-surface oil remain under many impacted beaches.

A 2.5 mile permanent boom protected the Bay of Morlaix. Although it required constant monitoring, the boom functioned well because the bay was protected from severe weather and the brunt of the oil slick. In other areas, booms were largely ineffective due to strong currents, and also because they were not designed to handle such enormous amounts of oil. Skimmers were used in harbors and other protected areas. Vacuum trucks removed oil from piers and boat slips where seaweed was especially thick. "Honey wagons"—vacuum tanks designed to handle liquefied manure—were used to collect emulsified oil along the coast.

Oil-laden seaweed was removed from the beaches with rakes and front-end loaders. Farm equipment was used to plow and harrow the sand, making it more susceptible to wave and bacterial action. Prior to harrowing, chemical fertilizers and oleophilic bacteria were applied to the sand.

At first, authorities decided against using dispersants in sensitive areas and along the coastal fringe. Meanwhile, the spill formed a highly stable water-in-oil emulsion ("chocolate mousse"). On the open sea, the French Navy applied both dilute and concentrated dispersants, but good dispersion was hard to achieve because in some places the mousse emulsion was several centimeters thick. If dispersants had been dropped from the air at the source of the spill—in days instead of weeks—the formation of mousse emulsion might have been prevented.

About 650 metric tons of chalk was applied in an effort to sink the oil. But after one month at sea, the oil was so viscous that the chalk just sat of top of it. Rubber powder made from ground-up tires was applied to absorb the oil. The French Navy used water hoses to spread most of the powder. Some was applied manually from small fishing boats. Because it stayed on top of the oil, the rubber powder had little effect; wave action wasn't strong enough to mix it into the oil, most of which was trapped inside the chocolate mousse emulsion.

During the third and fourth months of the cleanup, high-pressure hot water (fresh water at 2000 psi, 80–140 °C) was very effective in cleaning oil from rocky shores. A small amount of dispersant was applied to prevent oiling of clean rocks during the next high tide. The mouths of several rivers contained oyster beds and marshes that required manual cleaning. Soft mud river banks were cleaned with low-pressure water. To improve oil-collection efficiency, a sorbent was mixed with water and poured in front of the wash nozzles. The oil was collected downstream by a local invention called an "Egmolap." This device was good at collecting floating material in sheltered areas.

18.8.6 Bhopal, India: Methyl Isocyanate Leak (1984)

On December 2, 1984, a worker observed a build-up of pressure in a storage tank at the Union Carbide chemical plant near Bhopal, India. The tank contained about 15 tons of methylisocyanate (MIC), a chemical used to make pesticides. MIC is flammable, and at high concentrations it is deadly. At low concentrations, it causes lung damage and blindness.

The pressure increase probably was caused by water inside the tank. Water reacts with MIC to form methylamine and gaseous carbon monoxide. The reaction releases heat, which would have contributed to pressure rise in the tank by vaporizing some of the MIC. Normally, a refrigeration unit would have controlled the temperature, but that unit had been out of service for several months. Eventually, the pressure rise opened a safety valve. When this happened, the vented gas should have been routed to a caustic scrubber, which would have absorbed and hydrolyzed the MIC, rendering it harmless. Instead, the vented gas went to the flare. The flare should have converted the MIC into relatively harmless CO_2, H_2O and N_2. But the flare system failed, perhaps because it wasn't designed to handle such a large surge of gas. Consequently, tons of MIC poured into the air and spread across the countryside, covering 25 square miles (65 km^2). If operators had noticed the leak immediately, they might have been able to stop it before it did much damage. But the leak did not show up on their monitors because a critical panel had been removed from the control room.

How did water get into the tank? The MIC was stored under a blanket of dry nitrogen. Some experts suggest that the nitrogen was wet. Others guess that a water hose was inadvertently connected to the nitrogen line. A Union Carbide official suggested possible sabotage.

Even if the incident on December 2, 1984 was an accident, the MIC unit at Bhopal was a disaster waiting to happen. During a press conference, a Union Carbide executive acknowledged that the unit was in a state of sorry disrepair, and that its condition was so poor that it shouldn't have been running.

Up to 200,000 people were exposed to MIC in Bhopal and surrounding towns. More than 2500 died. Thousands more suffered permanent lung and/or eye damage. All told, there were 524,000 personal injury claims, 2800 lost-cattle claims, 4600 business claims, and 3400 wrongful death claims. Eventually, Union Carbide reached a US$470 million settlement with the Parliament of India.

18.8.7   Chernobyl, Ukraine: Nuclear Disaster (1986)

The literature contains many summaries of the causes and lingering impact the Chernobyl nuclear disaster. The International Nuclear Safety Advisory Group (INSAG) published a thorough report in 1986 and followed it with a revision in 1992 [53]. The INSAG documents remain the most generally accepted. The oft-vilified Anatoly Dyatlov, who was the Deputy Chief Engineer at the plant when the incident occurred, challenged the INSAG report. An English version of his perspective appeared in 1995 [54]. There are numerous summaries and timelines on the Internet, but they often disagree with each other and the facts in the INSAG document. A factual summary of health effects was published by the World Nuclear Association in 2009 [55]. Other articles on health effects are outlandish. For this chapter, choices had to be made. Please accept apologies for any errors.

A "SCRAM" is an emergency shutdown of a nuclear reactor. According to lore, it is an acronym for "safety control rod axe man." Supposedly, Enrico Fermi coined the term for the first nuclear reactor at the University of Chicago where graphite control rods were raised and lowered into the reactor core with a rope on a pulley. The axe man stood ready to cut the rope in an emergency, dropping all control rods all the way down, shutting down the reactor.

After a SCRAM at the Chernobyl nuclear power plant near Pripyat, Ukraine, the cooling water pumps stopped due to lost electrical power. Emergency power was supplied by diesel generators, but starting the diesels took almost a minute. During the delay, residual heat remained in the reactor until the cooling pumps were running again.

To Anatoly Dyatlov, this delay was unacceptable. He thought of a solution. Right after a SCRAM, he knew, the plant's power turbines kept spinning for several minutes due to their massive inertia. Dyatlov calculated that, as the turbines slowed down, they could produce enough power to run the water pumps until the diesels were up and running. He designed a complex network of switches to keep the current steady as the turbines lost momentum.

Dyatlov talked to the Chief Engineer of the plant, who gave him permission to run his experiment. The test began on April 25, 1986, as Unit 4 was shutting down for long-delayed maintenance. With Dyatlov's approval, the operators decided to run the test manually instead of using the unit's "unimaginative automatics."

At 2:00 PM, in violation of one of the most important safety regulations, the operators switched off the emergency cooling system. Just after midnight on April 26th, they switched off the reactor's power-density controls.

Under manual operation, the reactor became unstable. At 1:07 AM, the power output suddenly dropped to 0.03 gigawatts (GW), far less than the specified minimum of 0.70 GW. To generate more power, the operators raised the graphite control rods to the maximum height allowed by regulations. Dyatlov ordered them to raise the rods further. When the operators balked, Dyatlov insisted, and the operators complied.

At 1:23 AM, the power output seemed to be stable at 0.2 GW. The operators then violated THE most important safety regulation by disabling the emergency SCRAM, an automatic interlock designed to stop the reactor whenever the neutron flux exceeds a safe limit. (In modern nuclear power plants, it is physically impossible to disable this control.)

Now the experiment could begin. To see how long the electricity-generating turbine would spin without a supply of steam, they closed the valve that channeled steam from the reactor to the turbine. Steam that should have been taking heat out of the reactor was now trapped inside.

In less than 45 s, the reactor started to melt. Super-hot pellets of uranium-oxide fuel ruptured their zirconium-alloy containers, coming into direct contact with cooling water. The water flashed into steam, causing the first of two explosions that blew the top off the reactor. The second blast was caused by a H_2-CO-air explosion. The H_2 and CO were generated by reactions of zirconium and graphite with super-heated steam:

$$Zr + 2H_2O \rightarrow ZrO_2 + 2H_2$$
$$C + H_2O \rightarrow CO + H_2$$

About 15 tons of radioactive material from the reactor core rocketed into the atmosphere, where it spread across Europe. More than 36 h after the accident, plant personnel told local officials about the accident. About 14 h after that—50 h after the accident—radiation from the explosion was detected by technicians at the Forsmark nuclear power plant in Sweden. That measurement was the first notification to the world outside the Soviet Union that something had happened in Pripyat.

At the site, more than 30 fires broke out, including an intensely hot graphite fire that burned for 14 days. About 250 people fought the various fires. During 1800 helicopter sorties, pilots dropped 5000 tons of lead, clay, dolomite and boron onto the reactor. Near term, the explosion and high-level radiation killed 31 people, including operators, fire-fighters, and helicopter pilots. Thousands more died later.

During a trial that ended in August 1987, the Chernobyl plant director (Brukhanov) was convicted and sentenced to 10 years in a labor camp. The chief engineer (Fomin), and Dyatlov received shorter sentences. The two operators were acquitted, but both of them died soon afterwards from radiation poisoning. More than 60 other workers were fired or demoted.

18.8.7.1 Long-Term Consequences

In June 1987, 14 months after the disaster, some 27 villages within the restricted zone were still heavily contaminated, because the cleanup operation had stopped. Nearby cities and towns were reporting dramatic rises in thyroid diseases, anemia, and cancer. Hardest hit were children. Frequently, calves were born without heads, eyes, or limbs.

According to government figures, more than 4000 Ukrainians who took part in the clean-up had died, and 70,000 were disabled by radiation. About 3.4 million of

Ukraine's 50 million people, including some 1.26 million children, were affected by Chernobyl [56]. According to Gernadij Grushevoi, co-founder of the Foundation for the Children of Chernobyl, the long-term danger is even worse in Belorussia. In 1992, in testimony at the World Uranium Hearing [57], he said, "… 70% of the radioactive stuff thrown up by the explosion at Chernobyl landed on White Russian territory. There is not a centimeter of White Russia where radioactive cesium cannot be found."

18.8.7.2 The Rest of the Plant Kept Running for Nine Years

The three remaining nuclear reactors (Units 1, 2 and 3) continued to run. Despite ambient radiation levels 9 times higher than widely accepted limits, workers lived in newly built colonies inside the "dead zone."

In 1991, Unit 2 was damaged beyond repair by a fire in the turbine room. That left Units 1 and 3, which kept going for the next nine years. They kept going because the power was needed, and no funds were available to build a replacement.

On December 14, 2000, the plant was shut down. The shut-down was expedited by the European Commission, which approved a US$585 million loan to help Ukraine build two new reactors, and by the European Bank for Reconstruction and Development, which provided US$215 million.

18.8.8 Sandoz, Switzerland: Chemical Spills and Fire (1986) [58]

18.8.8.1 The Incident

On November 1, 1986, a fire broke out in a riverside warehouse at the Sandoz chemical plant in Schweizerhalle, Switzerland. While extinguishing the flames, firemen sprayed water over exploding drums of chemicals, washing as much as 30 tonnes of pesticides, chemical dyes, and fungicides into the Rhine River. Up to 100 miles (160 km) downstream from Schweizerhalle, the Rhine was sterilized. All told, more than 500,000 eels and fish were killed. More than 50 million people in France, Germany, and The Netherlands endured drinking water alerts.

Afterwards, while checking the Rhine for chemicals as it rolled through Germany, officials discovered high levels of a herbicide (Atrazine) that wasn't on the list provided by Sandoz. Eventually, Ciba-Geigy admitted that it had spilled Atrazine into the river just a day before the Sandoz fire. As monitoring continued, more chemicals were discovered, alerting German authorities to the fact that many different companies were secretly and knowingly discharging dangerous chemicals into the river. BASF admitted to spilling more than a tonne of herbicide, Hoechst discovered a chlorobenzene leak, and Lonza confessed to losing 2000 gallons (4500 L) of chemicals.

18.8.8.2   Recovery

In response to the Sandoz disaster, companies all along the Rhine joined the Rhine Action Program for Ecological Rehabilitation, agreeing to cut the discharge of hazardous pollutants in half by 1995. Although many experts thought the target could never be reached, samples showed that from 1985 to 1992, mercury in the river at the German town of Bimmen-Lobith, near the Dutch border, fell from 6.0 to 3.2 tonnes, cadmium from 9.0 to 5.9 tonnes, zinc from 3600 to 1900 tonnes, and polychlorinated biphenyls (PCBs) from 390 to 90 kilograms.

In December 1990, for the first time in 3 years, a large Atlantic salmon was fished from the Sieg, a tributary of the Rhine in West-Central Germany. The catch proved what officials had hoped: If you clean up your mess and clear the way, someday life will return. The event gave impetus to the Salmon 2000 project, and further success followed. In 1994, researchers found recently hatched salmon in the Sieg, and in 1996 a salmon was hooked near Baden-Baden. In 1998, encouraged by the success of the Rhine Action Program, targets were set to designate a large protected ecosystem from streams in the Jura Mountains, the Alps, the Rhine mountains, the Rhineland-Palatinate, the Black Forest, and the Vosges to the mouth of the Rhine in The Netherlands. Meanwhile, not all of the Rhine's pollution problems have been solved. One of the most serious is a huge basin in the Netherlands, into which toxin-laden mud dredged from the Port of Rotterdam has been dumped since the 1970s. Contamination levels are falling, but several toxins are very stubborn.

All along the Rhine, the main source of remaining pollution comes from farm fertilizers, which seep into the river every time it rains.

18.8.9   Exxon Valdez: *Oil Spill (1989) [59]*

On March 23, 1989, the 987-foot supertanker *Exxon Valdez* left port carrying more than 1.2 million barrels of Alaskan North Slope crude. The ship was headed south toward refineries in Benicia and Long Beach, California. At 10:53 PM, it cleared the Valdez Narrows and headed for Prince William Sound in the Gulf of Alaska. To avoid some small icebergs, Captain Joseph Hazelwood asked for and received permission to move to the northbound shipping lane. At 11:50 PM, just before retiring to his cabin, the captain gave control of the ship to the third mate, Gregory Cousins, instructing him to steer the vessel back into the southbound lane after it passed Busby Island.

Cousins did tell the helmsman to steer to the right, but the vessel didn't turn sharply enough. At 12:04 AM, it ran aground on Bligh Reef. It still isn't known whether Cousins gave the order too late, whether the helmsman didn't follow instructions properly, or if something went wrong with the steering system.

Captain Hazelwood returned to the bridge, where he struggled to hold the tanker against the rocks. This slowed the rate of oil leakage. He contacted the Coast Guard and the Alyeska Pipeline Service Company. The latter dispatched containment and skimming equipment. According to the official emergency plan, this equipment was supposed to arrive at a spill within 5 h. In fact, it arrived in 13 h—eight hours late.

The *Exxon Baton Rouge* was sent to off-load the un-spilled cargo and to stabilize the *Valdez* by pumping sea water into its ballast tanks. The oil transfer took several days. By the time it was finished, more than 250,000 barrels of oil had spilled into the Sound. Eventually, 33,000 birds and 1000 otters died because of the spill.

Eleven thousand workers treated 1200 miles (1900 km) of shoreline around Prince William Sound and the Gulf of Alaska, using 82 aircraft, 1400 vessels, and 80 miles (128 km) of oil-containing booms.

In response to the disaster, the U.S. Congress passed the Oil Pollution Act of 1990. The Act streamlined and strengthened the ability of the U.S. Environmental Protection Agency (EPA) to prevent and react to catastrophic oil spills. A trust fund, financed by a tax on oil, was established to pay for cleaning up spills when the responsible party cannot afford to do so. The Act requires oil storage facilities and vessels to submit plans to the Federal government, telling how they intend to respond to large oil discharges. EPA published regulations for above-ground storage facilities, and the U.S. Coast Guard published regulations for oil tankers. The Act also requires the development of area contingency plans to prepare for oil spills on a regional scale.

Captain Hazelwood was tried and convicted of illegally discharging oil, fined US$50,000, and sentenced to 1000 h of community service. Exxon spent US$2.2 billion to clean up the spill, continuing the effort until 1992, when both the State of Alaska and the U.S. Coast Guard declared the cleanup complete. The company also paid about US$1 billion for settlements and compensation.

On July 10, 2004, USA Today reported: "Not one drop of crude oil spilled into Prince William Sound from oil tankers in 2003—the first spill-free year since the ships started carrying crude from the trans-Alaska pipeline terminal in 1977."

18.8.10 Gulf War: Intentional Oil Spill, Oil Well Fires (1991)

18.8.10.1 Mina Abdulla Spill

On January 25–27, 1991, during the occupation of Kuwait, Iraqis pumped 4–6 million barrels of oil from stations at Mina Al-Ahmadi into the Arabian Gulf. They did so to preempt an expected beach invasion by allied troops. The size of the spill was 16–25 times greater than the *Exxon Valdez* spill. On January 27, allied bombers stopped the spill by destroying the pumping stations.

Ad Daffi Bay and Abu Ali Island experienced the greatest pollution. The spill damaged sensitive mangrove swamps and shrimp grounds. Marine birds, such as cormorants, grebes, and auks, were killed when their plumage was coated with oil. The beaches around the shoreline were covered with oil and tar balls.

Despite the ongoing war, the clean-up of the oil spill proceeded rapidly. Kuwaiti crude is rich in light ends, and water in the Arabian Gulf water is relatively warm. For these reasons, about half of the spilled oil evaporated, leaving behind a thick emulsion which eventually solidified and sank to the bottom of the sea. Another 1.5 million barrels were recovered by skimming. Operators of sea-water cooled factories and desalination plants were concerned that the oil might foul their intake systems. To prevent this, protective booms that extended three feet (1 m) below the surface were installed around intakes in Bahrain, Iran, Qatar, Saudi Arabia, and the United Arab Emirates.

18.8.10.2 Oil Well Sabotage

On February 23–27, 1991, retreating Iraqi soldiers damaged three large refineries and blew up 732 Kuwaiti oil wells, starting fires on 650 of them. Up to 6 million barrels per day were lost between February 23 and November 8, 1991. Crews from 34 countries assembled to fight the oil-well fires. Initially, experts said the fires would rage for several years. But due to the development of innovative fire-fighting technology, the job took less than 8 months.

The oil-well fires burned more than 600 million barrels, enough to supply the United States for more than a month. The fire-fighting effort cost US$1.5 billion. Rebuilding Kuwait's refineries cost another US$5 billion. In all, Kuwait spent between US$30 and US$50 billion to recover from the Iraqi invasion.

18.8.11 Tosco Avon Hydrocracker: Deadly Pipe Rupture (1997) [60]

On January 21, 1997 a temperature excursion led to the rupture of a reactor outlet pipe at the Tosco Avon Refinery in Martinez, California. The consequent release of hydrogen and hydrocarbons ignited on contact with air, causing an explosion and fire. One worker was killed and 46 others were injured. There were no reported injuries to the public.

18.8.11.1 Context

Fixed-bed catalytic hydrocracking converts vacuum gas oil, and other petroleum fractions with similar boiling ranges, into diesel, jet fuel, gasoline and other products [61]. Commercial fixed-bed units employ two kinds of catalyst—hydrotreating and hydrocracking – which usually are loaded into separate reactors.

Fig. 18.7 Reaction triangles for combustion and hydrocracking, showing factors required for the chemical reactions. Removing any one factor stops the reactions

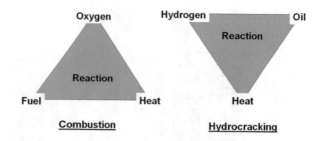

Hydrotreating catalysts remove sulfur, nitrogen and other contaminants, converting the sulfur and nitrogen into H_2S and NH_3. It is especially important to remove the organic nitrogen, because it neutralizes the acid sites of hydrocracking catalysts, suppressing their activity. Ammonia also inhibits cracking, but to a significantly lesser extent. Nitrogen slip (N slip) is the amount of organic nitrogen remaining in the feed after it undergoes hydrodenitrogenation (HDN). For high-activity cracking catalysts, typical N-slip targets range from 4 to 40 ppmw. Above about 200 ppmw, cracking catalysts might have nil activity.

Catalyst particles have roughly the same size as rice. They are tiny compared to the drum-shaped reactor beds into which they are loaded. The diameters of large reactors can reach 6 meters; reactor heights can exceed 30 m. Hydrocracking beds are shorter than hydrotreating beds. Most reactors have multiple beds, with quench zones in between. In the quench zones, hot fluids from the bed above mix with relatively cold hydrogen-rich quench gas. The cooling is required, because hydrotreating and hydrocracking are highly exothermic. If not controlled, temperatures can run away, becoming high enough to melt through inches of steel.

Figure 18.7 compares hydrocracking with combustion. Combustion requires oxygen, fuel, and heat. Hydrocracking requires hydrogen, oil, and heat. If one of the three is removed, the reaction stops. A wood fire can be smothered with a blanket or carbon dioxide, or cooled with water or carbon dioxide. The fastest and most effective way to stop a hydrocracker temperature excursion is to remove hydrogen and other reaction fluids by depressuring the unit through emergency depressuring (EDP) valves. Typically, there are two EDP valves. One depressures the unit at an initial rate of 7 bar (100 psi) per minute. The other depressures the unit three times faster.

Either EPD valve can be activated manually. All hydrocracking process licensors specify automatic depressuring at low rate when a reactor temperature (or any two reactor temperatures) reaches some limit, typically 427–440 °CC (800–825 °CF). High-rate depressuring is activated automatically at some higher temperature. If temperatures exceed the metallurgical limit, typically 465 °CC (870 °CF), the steel degrades. Certain events, such as a power failure or loss of the recycle gas compressor, automatically initiate high-rate depressing. Alert operators act before limits are reached, by adjusting quench flow and furnace firing. Early action forestalls emergency action.

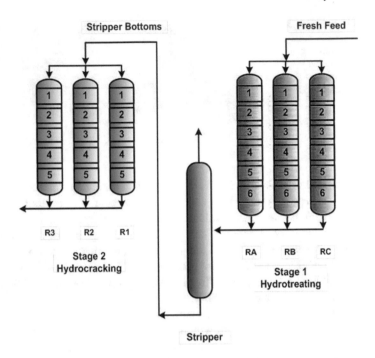

Fig. 18.8 Arrangement of reactors in the Tosco Avon hydrocracker (numbers in the reactor are bed numbers)

18.8.11.2   The Incident

Figure 18.8 shows the reactor layout for the Tosco Avon hydrocracker. The unit has 3 parallel Stage 1 hydrotreating reactors with 6 beds each, and three parallel Stage 2 cracking reactors with 5 beds each. The effluents from the Stage 1 reactors are combined and sent to a stripper. The stripper removes light gases, including methane, ethane, propane, and hydrogen sulfide. The stripper liquids go to three parallel Stage 2 hydrocracking reactors.

On January 19, two days before the fatal event, operators managed to defeat a temperature excursion in Reactor 1 (R1), without depressuring, by aggressively manipulating quench flow. The outlet pipe temperature indicator (TI) exceeded 482 °C (900 °F) versus the metallurgical limit of 465 °C, and the center TI in Reactor 1 Bed 4 (R1B4) reached 537 °C (998 °F) versus a specified maximum of 440 °C. During the excursion, an operator went outside to get TI readings from a field panel underneath the outlet pipe. The panel TIs could measure higher temperatures than the control-room readings.

Table 18.5 shows a timeline for the accident on January 21. The problem began several hours earlier when one of the Stage 1 hydrotreating reactors was brought down to repair a leak. The total unit feed rate was kept the same. This decreased the residence time of oil in the remaining two Stage 1 reactors by 50%, decreasing

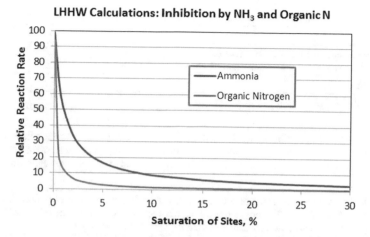

Fig. 18.9 Inhibition of hydrocracking by ammonia and organic nitrogen, where $K_{(organic\,N)} = 7 \times K_{(NH_3)}$. The trends are a function of percentage of inhibited catalyst sites at steady state, which is related to, but not the same as, the concentrations of ammonia or nitrogen in the feed

denitrogenation in considerably. By 10:00 AM, cracking had stopped. Almost no light products were being formed, and the exotherms in the Stage 2 hydrocracking reactors were very low. The lab reported an N-slip of 352 ppmw versus a target of 15 ppmw; higher N slip from Stage 1 meant less hydrocracking in Stage 2. At 11:30 PM, operators responded appropriately by increasing Stage 1 temperatures to decrease N-slip.

The second most dangerous decision of the day was taken at 1:10 PM, when Stage 2 temperatures were raised to "burn off" the nitrogen from the Stage 2 catalyst.

Operating guidelines should have included the following prohibitions: During normal operation, never raise temperatures in hydrotreating and hydrocracking catalysts at the same time. Never raise hydrotreating catalyst temperatures unless the cracking catalyst temperatures are at steady state, and vice versa. Figure 18.9 shows why. When concentrations of inhibitors are high, reaction rates are not especially sensitive to modest changes. As with applying the brakes of an automobile, when the vehicle is stationary, stepping harder on the brake pedal changes nothing. But at low inhibitor concentration, small changes have a huge effect.

At 5:38 PM, the *reported* N-slip was 47 ppmw. But the *true* N-slip at that time was considerably lower. On average, the N-slip had fallen 300 ppmw in 7½ hours, an average of 40 ppmw per hour. If the lag time between taking a sample and getting results was typical—say, about 1 h—the true nitrogen slip at 5:38 PM could have been <10 ppmw and close to zero at 7:34 PM.

Recall: Due to the "burn off" decision, cracking reactor temperatures were higher than they had been 16 h before, when the N-slip was 15 ppmw. If the temperature had been as high back then, it would have detonated an excursion even more serious than the one which actually occurred.

Table 18.5 Abbreviated timeline with comments for the Tosco Avon incident

4:50 AM	
One of three Stage 1 hydrotreating reactors was brought down to repair a leak. The total unit feed rate was kept the same, which reduced, the extent of HDN	
10:00 AM	
Nitrogen slip had risen from <15 to 352 ppmw	
11:30 AM	
Due to the high nitrogen slip and resulting inhibition of Stage 2 cracking activity, there was no temperature rise in one cracking reactor and very little rise in the other reactors. No light products were being formed. Stage 1 temperatures were raised to decrease nitrogen slip	
1:10 PM	
Stage 2 hydrocracking reactor temperatures were raised to "burn off" inhibitory nitrogen. This was a horrendous mistake. Treating and cracking temperatures must never be raised at the same time. Figure 18.4 shows how very sensitive activity is to organic nitrogen	
5:38 PM	
Nitrogen slip from Stage 1 result was reported to be 47 ppmw. The actual nitrogen slip when the result was received was considerably lower than 47 ppmw. Since 10:00 AM, the nitrogen slip had fallen 300 ppmw in 7½ hours, an average of 40 ppmw per hour. If the lag time between taking a sample and getting results was 1 h as usual, the true nitrogen slip was <10 ppmw at 5:38 PM and probably close to zero at 7:34 PM	
7:34 PM	
In Reactor 3, Bed 4 (R3B4), one TI jumped 108 °C in 40 s, reaching 439, 10 °C higher than allowed by procedure but not higher than in previous events	
7:34:20 PM	
Quench to R3B5 opened 100%. Data logger temperatures cycled between −18, 649, and 343 °C. In well-run units, the associated uncertainty would have compelled someone to bring the unit down immediately (if it had not been depressured previously due to high temperature)	
Sometime before 7:37 PM	
An operator went outside to verify temperatures at the field panel	
7:37–7:39 PM	
All R3B5 outlet TIs were >415 °C. One reached 679 °C	
After 7:40 PM	
An operator phoned an instrument technician to work on the fluctuating data logger	
7:41:20 PM—seven minutes after the first temperature jump	
Explosion	
The operator at the field panel died	
A board operator activated the EDS	

Time (pm)	Bed 4 Outlet Temp (°F) Pt-133C2	Bed 5 Inlet Temp (°F) Pt-140C3	Bed 5 Outlet Temp (°F) Pt-134C1	Bed 5 Outlet Temp (°F) Pt-134C2	Bed 5 Outlet Temp (°F) Pt-134C4	Bed 5 Outlet Temp (°F) Pt-134C5	Rx 3 Inlet Temp (°F) Pt-125C	Rx 3 Outlet Temp (°F) Pt-141C
7:33:00	628.3	636.9	648.6	646.3	656.8	645.9	632.2	641.3
7:33:20	628.3	636.9	648.6	646.3	656.8	645.9	632.2	641.3
7:33:40	636.5	658	648.6	646.3	656.8	645.9	632.2	641.3
7:34:00	823.2	720.7	648.6	646.3	656.8	645.9	632.2	641.3
7:34:20	732	859.5	648.6	646.3	656.8	645.9	632.2	641.3
7:34:40	732	859.5	648.6	646.3	656.8	645.9	632.2	641.3
7:35:00	637.3	792.4	624.2	627.7	633.4	645.9	632.2	641.3
7:35:20	637.3	715.7	624.2	627.7	633.4	615.4	632.2	641.3
7:35:40	637.3	664.4	624.2	640.3	633.4	615.4	632.2	641.3
7:36:00	637.3	633.4	650.3	647.9	623.6	615.4	632.2	649.9
7:36:20	637.3	633.4	663.9	667.7	623.6	615.4	632.2	649.9
7:36:40	637.3	660.7	672.9	667.7	623.6	625.6	632.2	658.9
7:37:00	637.3	660.7	681.1	676.8	656	645.8	632.2	658.9
7:37:20	637.3	660.7	681.1	676.8	672	673.5	632.2	760.7
7:37:40	637.3	660.7	697.3	707.5	690.7	673.5	640.2	760.7
7:38:00	637.3	660.7	717.2	876	690.7	673.5	640.2	684.6
7:38:20	637.3	660.7	1255.7	0	783	705.8	648.8	701.8
7:38:40	637.3	660.7	0		0	744	660	788.8
7:39:00	637.3	648	0		0	889	693	983.1
7:39:20	637.3	648	0		0	0	754.7	1219
7:39:40	637.3	655.6	0		0	0	826.5	0
7:40:00	637.3	655.6	0		0	0	889.1	0
7:40:20	637.3	655.6	0		0		960.7	0
7:40:40	637.3	645.7	0		1397.1	879.9	1233.5	0
7:41:00	637.3	645.7	0		1398.4	694.9	0	0
7:41:20	0.0 ????	0.0 ????	0.0 ????	0.0 ????	0.0 ????	0.0 ????	0.0 ??	0.0 ????
7:41:40	0.0 ????	0.0 ????	0.0 ????	0.0 ????	0.0 ????	0.0 ????	0.0 ??	0.0 ????

Fig. 18.10 Reactor 3 temperatures during the TOSCO Avon incident. Temperature indicator Pt-133C2 jumped by 104 °C (186 °F) in 20 s, reaching 439 °C (823 °F). The excursion cascaded through the quench section into the next catalyst bed, causing a jump of 223 °F (124 °C) in 40 s at Pt-140C3, reaching 460 °C (860 °F)

Figure 18.10 shows Reactor 3 temperatures during the Tosco temperature excursions. At 7:34 PM, a temperature excursion struck R3B4 and cascaded into R3B5, peaking at 439 °C (823 °F). According to operating guidelines from the technology licensor, the unit should have been depressured when any bed TI reading reached 427 °C (800 °F). But instead of depressuring, operators made their most dangerous decision: To keep the unit running. To do what they had done two days before and several times before that. They tried to control the excursion by cutting heater firing and manipulating quench gas. As before, an operator went outside to get readings from the field panel beneath the reactor outlet pipe.

Operators said they failed to act on January 21, because they didn't know whether or not an excursion was occurring. They had to reconcile temperature readings from three different sets of displays. Two sets were in the control room, and the other, as mentioned, was in the outside field panel. One control-room system was the data logger, which included most temperature readings. After the initial excursion, the data-logger readings started to fluctuate wildly, leading to further confusion.

In other refineries, uncertainty is not an excuse for unsafe behavior. Operators are told: When in doubt, assume the worse; bring the unit down.

18.8.11.3   Pipe Failure Analysis

Analysis of the ruptured pipe showed that the failure did not occur at weld, elbow or reducer. The pipe simply got too hot, reaching 1700 °F (927 °F). At the point of failure, the circumference of the 10-inch (245 mm) diameter pipe expanded by 5 inches (120 mm) before it failed. The wall thickness at the rupture was 0.3–0.4 (7–10 mm), about 50% lower than the thickness of the rest of the pipe.

18.8.11.4   Root Causes and Recommendations

Root causes for the Tosco Avon accident include the following:

- **Management directives**. On July 23, 1992, the unit was depressured at low rate to stop an excursion. A grass fire started at the flare. A worker told the author (PRR), in a private conversation, that after this event, 1992, upper management strongly discouraged use of depressuring. A fire might spread into the nearby grassy hills and the John Muir Historic Site, creating a public relations nightmare. Fixing the flare would have required investment, which in those days at Tosco was not an option.
- **Confusion**. Operators used three different instrumentation systems to obtain temperature data. Not all the data were immediately accessible, and the data often disagreed. Monitoring points capable of reading the highest temperatures were underneath the reactors and not connected to the control room. (In the same private conversation with the author of this chapter, a worker said that operations had lobbied hard to connect field-panel measurement to the control room, but management declined the requests to save money.)
- **Inadequate supervision and process training**. Supervisory management and operator training were inadequate. Operators and engineers did not understand that default values of zero on the data logger might mean extremely high temperatures. They did not understand that a decrease in makeup hydrogen flow indicated an extreme temperature excursion. They did not understand the temperature sensitivity of cracking catalyst activity at low N-slip.
- **Misunderstanding of sampling dynamics**. Operators and engineers did not comprehend the dynamics of sampling, i.e., the lag time between taking a sample, getting results, and what might have happened to the process in the meanwhile.

- **Poor procedure documentation**. Written procedures were outdated, incomplete, and contained in different documents in different locations.

The investigation team, led by EPA, developed the following recommendations for consideration by owners of all hydroprocessing facilities:

- Management must ensure that operating decisions are not based primarily on cost and production.
- Facility management must set safe, achievable operating limits and not tolerate deviations from these limits.
- Management must provide an operating environment in which operators can follow emergency shutdown procedures without fear of retribution.
- Process instrumentation and controls should be designed to consider human factors consistent with good industry practice. Hydroprocessing reactor temperature controls should be consolidated with all necessary data available in the control room. Backup temperature indicators should be used so that the reactors can be operated safely in case of instrument malfunction.
- Each alarm system should distinguish between critical alarms and other operating alarms.
- Adequate operator supervision is needed, especially to address critical or abnormal situations.
- Facilities should maintain equipment integrity and discontinue operation if integrity is compromised.
- Management must ensure that operators receive regular training on the unit process operations and chemistry. For hydrocrackers, this should include training on reaction kinetics and the causes and control of temperature excursions.
- Operators need to be trained on the limitations of process instruments and how to handle instrument malfunctions.
- Facilities need to ensure that operators receive regular training on the use of the emergency shutdown systems and the need to activate these systems.

18.8.12 BP Texas City Isomerization Unit: Explosion (2005)

On March 23, 2005, an explosion in an isomerization unit at the BP Texas City, Texas refinery [62], along with related explosions and fires, killed 15 people and injured another 180. A shelter-in-place order was issued, requiring 43,000 people to remain indoors. Houses were damaged as far away as three-quarters of a mile. Resulting financial losses exceeded $1.5 billion.

18.8.12.1 Context

The isomerization process converts straight-chain n-pentane and n-hexane into branched isomers with higher octane. The resulting isopentane and isohexane are blended into gasoline. Raffinate from the process is a mixture of straight-chain hydrocarbons. A raffinate splitter separates the light raffinate from the heavy raffinate. During normal operation, an automatic level controller maintains liquid level in the splitter by sending heavy raffinate to a storage tank when the level is too high. The level controller depends on feedback from a level transmitter. As a backup, level can be measured manually with a sight glass. The tower overhead is cooled and drawn off or sent back to the splitter for temperature control. If the tower pressure gets too high, three pressure relief valves can open, one at a time or in combination, sending vapors and liquids to an emergency blow-down drum. In the unit at BP Texas City, the drum should have been connected to a flare, but instead, it vented to the atmosphere.

18.8.12.2   The Incident

The accident occurred during the restart of the isomerization unit, after maintenance on the raffinate splitter. Before the restart, there were several unresolved safety-critical issues, including a faulty pressure control valve, a defective high-level alarm in the splitter, a defective (opaque) sight glass used for level measurement, and an un-calibrated level transmitter. During the morning of March 23, operators noticed that the heavy raffinate storage tanks were nearly full. Consequently, the night-shift supervisor agreed that the startup should not continue until some liquid was withdrawn from the tanks. But the day-shift supervisor arrived late and did not receive a full status report, including information about the raffinate tank level. Under his direction, the startup resumed at 9:30 AM. To reduce level in the raffinate splitter, the level control valve was opened, sending liquid to the already-nearly-full heavy storage tank. At about 10:00 AM, the level control valve was shut. As the startup proceeded, more raffinate entered the tower. Without a way to withdraw product, the tower started filling. The un-calibrated level transmitter indicated that the level was below 100%. The opaque sight glass was useless, so it wasn't possible to confirm the level visually.

As the startup proceeded, burners in the furnace were started to heat the inflowing raffinate. The procedure specified that the temperature of return flow to the tower should be raised to 135 °C (275 °F) at 10 °C (18 °F) per hour, but this time the procedure was altered. The return flow temperature reached 153 °C (307 °F) with an increase rate of 23 °C (41 °F) per hour. The defective transmitter was indicating a high (but still-safe) level of 93%. But due to the closed control valve, the true level was higher. Heat was increasing, causing expansion of the raffinate. Pressure started to build, but the operations team thought the increase was due primarily to overheating, which was a known start-up issue. The pressure was relieved and the furnaces were turned down.

At 12:42 PM, the level transmitter showed 78%. But the true level was higher, and due to heat from heat exchange, the liquid was expanding. Liquid overflowed into the pressure-relief system. At 1:13 PM, all three pressure relief valves opened, sending liquid to the blow-down drum. An operator opened the level control valve and shut down the furnace completely. But he did not stop the inflow of raffinate. In the blow-down drum, a "geyser" of hot hydrocarbon liquid vented directly into the air, then flowed down the outside of the drum, where it hovered above the ground and drifted away from the unit. It soon reached an idling pick-up truck about 9 m (30 feet) away. Investigators concluded that the truck must have ignited the hydrocarbons. The 15 people who died were in or around a nearby row of contractor trailers. I (PRR) was scheduled to be participating in a hydrocracker startup in the refinery on day the explosion. The startup was delayed. Otherwise, I might have been in one of those trailers.

18.8.12.3   Root Causes

According to the investigation report, the Texas City disaster was caused by organizational and safety deficiencies at all levels of the BP Corporation. Warning signs of a possible disaster were present for several years, and reported to officials by employees. But company officials did nothing. The extent of the safety culture deficiencies was further revealed when the refinery experienced two additional serious incidents just a few months after the March 2005 disaster. In one, a pipe failure caused a reported $30 million in damage. The other resulted in a $2 million property loss.

The report cited the following local deficiencies at the Texas City refinery:

- A work environment that encouraged operations personnel to deviate from procedure
- Lack of a BP policy or emphasis on effective communication for shift change and hazardous operations (such as during unit startup)
- Malfunctioning instrumentation that did not alert operators to the actual conditions of the unit
- A poorly designed computerized control system that hindered the ability of operations personnel to determine if the tower was overfilling
- Ineffective supervisory oversight and technical assistance during unit startup
- Insufficient staffing to handle board operator workload during the high-risk time of unit startup
- Chronic worker fatigue, due to the lack of a human fatigue-prevention policy
- Inadequate operator training for abnormal and startup conditions
- Failure to establish effective safe operating limits.

The industry learned from the BP Texas City disaster, but not everyone learned, and some who might have learned forgot: similar behavior occurred during the following event.

18.8.13 Deep Water Horizon *Drilling Rig: Blowout (2010)* [63]

On April 20, 2010, on the *Deepwater Horizon* drilling rig killed 11 people and gushed 560–585 thousand tons of crude oil into the Gulf of Mexico. On July 2, 2015, a settlement was announced between BP, United States, and five US states. BP agreed to pay up to US$18.7 billion in penalties [64].

18.8.13.1 Context

Deepwater Horizon floated in 4992 feet of water just beyond the continental shelf in the Mississippi Canyon. Below the seabed was a large reservoir of oil and gas. A long well—it was 2.5 miles long—had been drilled through the subsurface rock to the reservoir.

The well had just been cemented. After a well is drilled, it is completed by casing the well bore with steel pipe and cementing the casing into place. Casing prevents the well from collapsing. Cementing fills the otherwise empty annular space between well bore and the casing. If the annular space is totally blocked, oil and gas can flow out of the formation only through the casing. At the top of the casing, valves, blowout preventers, etc., control the flow of oil and provide a route for future activities, such as perforation and well stimulation.

Figure 18.11 illustrates how centralizers improve cementing. They keep the casing from resting against the surrounding rock, possibly blocking the flow of cement and jeopardizing the well-completion process. According to industry acquaintances, poor cement jobs are the leading cause of post-completion blowouts. Among other things, poor cementing can result from using low-quality cement and not using enough centralizers. A Halliburton model called for 21 centralizers. But there were only 6 on the rig. In short order, 15 more were flown to the rig, but they were conventional rather than custom-made. A BP engineer informed management that the new centralizers were conventional, and that it would take 10 h to install them. In the end, BP installed only six centralizers.

To test the integrity of cementing, two important tests are conducted: a positive-pressure test and a negative-pressure test. During the positive-pressure test, the pressure is increased in the steel casing and seal assembly to see if they are intact. The negative-pressure test determines the integrity of the cement at the bottom of the hole. Pressure is reduced to 0 psi and well evacuation is stopped. If pressure remains low, the test is deemed successful. If flow occurs or the pressure increases, reservoir fluids are entering the well, probably through gaps in the casing and/or the cement. A failed negative pressure test requires remedial work to re-establish well integrity. If the remedial work is not done, flow through well will be impossible to control.

Fig. 18.11 Role of centralizers in oil well completion. The centralizer keeps the casing away from the rock while allowing cement to flow through. The drawing is, to say the least, not to scale

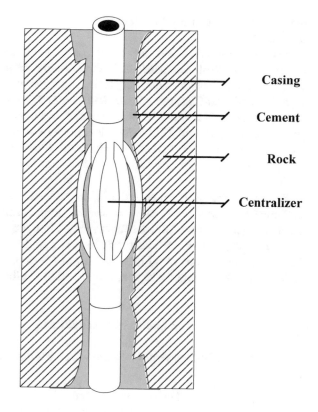

Casing

Cement

Rock

Centralizer

18.8.13.2   The Incident

At about noon on the day of the blow-out, a helicopter arrived. Among its passengers were four VIPs from Houston—two from Transocean, the owner of the drilling rig, and two from BP. The executives were there to conduct a 24-hour management visibility tour. Their presence contravened two lessons that should have been learned from the BP Texas City disaster. During a highly critical activity, (1) non-essential personnel must be removed from the vicinity, and (2) supervisory oversight must be provided by at all times by technically qualified personnel. Few activities are more critical than well completion, and few people are less essential during well completion than corporate executives. These particular VIPS were planning to stay on the rig overnight. They had to be fed, watered, and entertained. Their presence contributed to the flouting of Item (2) by monopolizing the time of highly capable experts, who otherwise might have been paying more attention to the drilling crew.

At 5:00 PM, the rig crew began the negative-pressure test. The pressure repeatedly built back up after gases were evacuated. Between 6:00 PM and 7:00 PM, the crew discussed how to proceed. A senior BP engineer insisted on another negative pressure test. An assistant tool pusher said pressure could be measured through another line—the "kill line"—where the pressure did indeed stay low. For whatever reason, the crew decided that no flow through the kill line equalled a successful negative pressure test.

The crew proceeded, preparing to set a cement plug deep in the well. They began to replace dense drill mud, which had counteracted reservoir pressure during drilling, with sea water. Sometime later, people on the bridge of the rig felt a high-frequency vibration and heard a hissing noise. The senior BP engineer got a call from the assistant tool pusher, who told him that mud was coming out of the well and that gas was coming up too. The senior BP engineer grabbed his hard hat and stepped outside, where saw that mud and seawater were "blowing everywhere." The assistant tool pusher called the senior tool pusher and said: "We have a situation ... the well is blown out."

Soon thereafter, at roughly 9:45 PM, an explosion shook the rig, starting a fire. A second explosion followed, then several more. Meanwhile, people were being evacuated into life rafts. Others simply jumped into the water.

Eleven people died. Over the next several months, the blown-out well gushed 560–585 thousand tons of oil into the Gulf of Mexico, causing economic and ecological harm throughout the region, from Mexico to Florida.

18.8.13.3 Root Causes and Recommendations

The Presidential Commission concluded with the following (author-abridged) observations:

- The explosion could have been prevented.
- The immediate causes were a series of mistakes made by BP, Halliburton, and Transocean. The mistakes revealed systematic failures in risk management. Tests and procedures were changed without consultation with technical experts.
- Deepwater energy exploration and production involve risks for which neither industry nor government were adequately prepared, but for which they can and must be prepared in the future.
- Regulatory oversight of leasing, energy exploration, and production require reforms, even beyond the significant reforms already initiated since the Deepwater Horizon disaster.
- Those in charge of oversight must have political autonomy, technical expertise, and authority to fully consider environmental protection concerns.
- Regulatory oversight alone will not be sufficient to ensure adequate safety, so the oil and gas industry will need to take its own, unilateral steps to increase safety throughout the industry, including self-policing mechanisms that supplement governmental enforcement.
- The technology, laws and regulations, and practices for containing, responding to, and cleaning up spills lag behind the real risks associated with deep water drilling into large, high-pressure reservoirs of oil and gas located far offshore and thousands of feet below the ocean's surface. Government must close the gap, and industry must support rather than resist that effort.

- Scientific understanding of environmental conditions in sensitive environments in deep Gulf waters, along the region's coastal habitats, and in areas proposed for more drilling, such as the Arctic, is inadequate. The same is true of the human and natural impacts of oil spills.

18.8.14   Lac-Mégantic Quebec, Canada: Oil Train Derailment (2013) [65]

At about 11:00 PM on July 5, 2013, a 74-car freight train carrying Bakken crude oil stopped in Nantes, Quebec. Nantes is about 7 miles west of Lac-Mégantic. The train was owned by the MM&A, the Montreal, Maine and Atlantic Railway, for which Nantes is a crew-change point.

The train was named MMA-2. With a length of 1433 m (4700 feet), it stretched for almost a mile. Pulled by five locomotives, it weighed more than 11,000 short tons. Each of the 72 tank cars carried about 30,000 gallons (714 barrels) of crude oil. The oil was produced in North Dakota and was being shipped to the Irving Oil Refinery in Saint John, New Brunswick.

While riding in his usual taxi from the train to his usual hotel, the engineer expressed concern about the lead locomotive, because it was smoking and leaking oil. Per MM&A policy, he had left the engine running, mainly to power the compressor that maintained pressure in the air brakes.

At about 11:45 pm, the troubled engine caught fire. Local residents called 911, and within five minutes, firefighters and police arrived. They saw dark smoke coming from the engine, where a broken piston had started a good sized blaze.

After dousing the fire, a Nantes Fire Department member contacted the rail traffic controller for MM&A in Farnham, Quebec, 125 miles away. The controller sent two maintenance workers to the scene. They arrived at 12:13 AM on July 6. It wasn't safe to restart the damaged engine, so they decided to leave it off and set hand brakes to augment the air brakes.

Without the compressor, the air brakes slowly released. The hand brakes alone couldn't restrain the 11,000-ton train. It began to roll toward Lac-Mégantic down a 1.2% grade. When it arrived at about 1:15 AM, the train was traveling at high speed. It derailed. The oil spilled out. The consequent fire and explosion killed 47 people.

The glut of oil being produced from the Bakken Shale in North Dakota is putting tremendous stress on truck and rail transportation. Old and poorly maintained equipment is being pressed into service. Until the equipment is replaced with new equipment or a pipeline—or oil production decreases—the stage is well-set for future accidents.

18.8.15   Tianjin, China: Storage Station Explosions (2015) [66]

On Wednesday, August 12, 2015, at around 22:50 local time, flames were reported in a warehouse in the Binhai district of Tianjin, China. Situated on Bohai Bay 150 km southeast of Central Beijing, the Binhai New Area is a Special Economic Zone (SEZ), similar to those in Shenzhen and Pudong in Shanghai. Several international companies, including Motorola and Airbus, have constructed facilities in the Binhai SEZ; in 2009, Airbus opened an assembly plant there for A320 airliners.

The warehouse was owned by Ruihai Logistics. It occupied 46,000 m² (500,000 square feet) and included multiple areas for storing containers of hazardous chemicals. A regulation specified that such warehouses must be located at least 1 km away from public facilities, but in this case the regulation was ignored. Local officials, including emergency responders, did not know for sure what was stored at the site. The situation was aggravated by major errors, including errors of omission, in customs documents.

According to post-mortem reports, the warehouse held more than 40 kinds of hazardous chemicals, including toluene diisocyanate, "vast quantities" of calcium carbide (CaC_2), 800 tons of ammonium nitrate, 700 tons of sodium cyanide, and 500 tons of potassium nitrate. CaC_2 reacts with water to form calcium hydroxide and acetylene, a highly flammable gas. Sodium cyanide reacts with water to form sodium hydroxide and hydrogen cyanide gas, which is both toxic and flammable. Ammonium nitrate is the chemical responsible for the massive explosion in Texas City, Texas in 1947.

First responders to the fire were unable to keep it from spreading. Unaware of what they were facing, they tried dousing the flames with water. The consequent explosion, possibly due to acetylene, was equivalent to 3 tons of TNT. Less than a minute later, the ammonium nitrate detonated, causing a blast equivalent to 20 tons of TNT. The blaze continued for several days, in part because of a pause in firefighting due to lingering uncertainty about the kinds and quantities of chemicals at the site. On August 15, the spreading flames triggered eight more explosions, which, fortunately, were relatively minor.

Photographs showed extensive destruction around the warehouse, with a massive crater at the center. Seven nearby buildings were destroyed, and others were rendered structurally unsafe. More than 17,000 nearby apartment units were damaged; 6000 people were evacuated.

On August 18, the first rain after the fire coated nearby streets with white chemical foam. People complained that contact with the rain caused burning sensations and rashes.

Also on August 18, the Central commission for Discipline Investigation (CCDI) initiated an investigation of Yang Dongliang. Yang was Director of the State Administration of Work Safety—China's highest-ranking work safety official. He had been Tianjin's vice mayor for 11 years. In 2012, he had issued an order that loosened rules for the handling of hazardous substances. On October 16, he was expelled from the Communist Party for corruption and other offences.

On August 20, tons of dead stickleback fish washed up on the land about 6 km from the explosion site. Officials downplayed the danger, claiming that the fish probably died due to oxygen depletion, not cyanide poisoning.

On August 25, another rain brought more complaints of skin burns, and white foam again appeared on the streets. The director of Tianjin's environmental monitoring center claimed that the foam was "a normal phenomenon when rain falls. Similar things have occurred before."

According to a report issued on September 12, the human toll at Tianjin was 173 killed, 797 hospitalized, and 8 still missing.

18.9 Personal Observations

During the past several years, in my technical support role, I (PRR) have witnessed shutdowns, near-misses, and loss of containment due to known (but un-amended) process design flaws, inadequate maintenance, inadequate training, lack of technical supervision, contravention of procedures, and worker fatigue. DIBS—Did It Before Syndrome—played a role in three of the four examples in this section, and in many of the examples described above. DIBS leads workers and managers to disregard safety rules and procedures, because they got away with doing so before.

18.9.1 Hydrocracker Startup: Loss of Containment

The refinery in which this incident occurred was owned by a company known for its dedication to safety. Behind the hydrocracker area in the control room was a large sign that said, in effect, "All employees are obliged to stop any operation he or she believes is unsafe." At midnight, it was time for the initial wetting of the catalysts with oil.

Initial wetting occurs once and lasts about an hour, but it releases heat due to enthalpy of adsorption. In multiple-bed hydrocrackers (see Fig. 18.9) wetting causes the Bed 1 outlet temperature to rise, within minutes, by as much as 50 °C. Wetting exotherms can be just as high in subsequent beds. Oil adsorbs heat, so the oil rate is kept high, usually between 50 and 70% of the design maximum. If the oil rate is lower, the temperature rise is higher.

Operators can prepare for the exotherms with post-quench. When oil reaches the inlet to Bed 1, they start sending cold gas into the quench deck between Bed 1 and Bed 2. It can take 5–8 min for oil to pass through the bed. When the oil arrives at the Bed 1 outlet, it mixes with the cold post-quench gas, which quickly drops the temperature. Without post-quench—that is, if operators didn't add cold gas until they saw the exotherm—the temperature wave would pass through the quench box

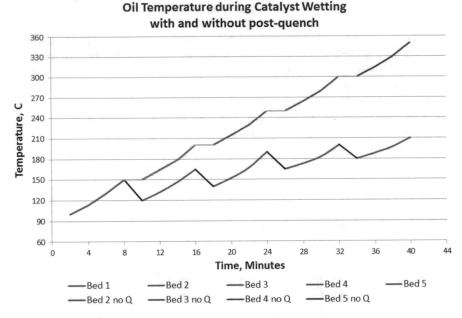

Fig. 18.12 Comparison of oil temperatures during initial catalyst wetting, with and without post-quench. The black and gray segments are for temperatures in the quench boxes. The "no Q" trends do not include post-quench

to the inlet of Bed 2 unabated. The heat of adsorption in Bed 2 would increase the temperature by another 50 °C. The total increase across in Beds 1 and 2 could be 100 °C. And so on.

Figure 18.12 compares wetting with and without post-quench. Without post-quench, wetting heat accumulates. Industry experience shows that temperatures can get high enough to initiate cracking and even temperature excursions. After the catalyst is wet, temperatures quickly fall to pre-wetting level.

Just after midnight in the subject refinery, the hydroprocessing Area Manager, unit engineer, were discussing how to proceed despite some known problems. As at BP Texas City, there were three known problems in a separation tower: the level transmitter wasn't working, the sight glass was foggy, and the offgas compressor wasn't working. There was no way to observe level or control pressure in the tower. Repairs work couldn't start before 8:00 AM, and then it would take an unknown time to fix the compressor. An in-house technician probably could repair the level controller in parallel with other work.

Two VIPs—the refinery Engineering Manager and Operations Superintendent—appeared in the control room and immediately started firing questions. They were concerned that wetting was being delayed. They asked operator to show panel displays irrelevant to the job at hand. Observing the operator's distress, I directed the VIPs to a nearby conference room. The room contained an observation-only panel display, with which they could browse without disturbing the startup team. When I returned to the control room, the unit operator and operator thanked me effusively.

The Area Manager decided to proceed despite the known problems. He concluded that with post-quench operation, the catalyst wouldn't get hot enough to induce hydrocracking. When wetting was over, oil could be circulated for a few hours to flush particulates out of the catalyst. By then, it would be 7:00 AM and the unit could be brought down for repairs.

Wetting started at about 3:00 AM. When liquid feed was introduced, a 4th problem became apparent: the feed pump valve was "sticky," taking up to three minutes to respond to a change in setpoint. Inability to control feed was crucial, due to the impact of feed on the size of exotherms. The operator called across the control room to the Area Manager. "This is not safe," he said. "I want to shut it down." According to company policy, he should have brought the unit down immediately, without asking permission. The

Area Manager said: "Steady as she goes. We can't go back. It's okay. We've done this before."

During wetting, the oil flow was lower than specified. With less oil to pick up heat, exotherms were larger than ever before. But thanks to post-quench operation, there was never any danger of a temperature excursion.

However, the downstream tower overflowed, sending hydrocarbons to the flare. The flaring lasted an hour or so, but it had to be reported to environmental authorities. The startup was delayed for days. Saving five hours cost the refinery more than $2 million.

The Area Manager had gotten away with operating the tower before without the overhead compressor. He had also gotten away with running without the level controller. He had never run the unit with both of those problems while also lacking reliable control of feed rate.

As for me (PRR): I was consulted by the operator during the event, but not by his boss. I did report the event to friend, a corporate subject matter expert. He was appalled. I never heard what action he took.

18.9.2 Sub-standard Equipment. Disabled Safety Interlock. Stagnation

A serious excursion was the latest of several in a new hydrocracker. The ultimate cause was a decision by the purchasing department to order substandard items for the associated hydrogen plant. One by one, the faulty items failed and were replaced with proper items. Meanwhile, during its first 6 months of operation, the unit never ran continuously for more than one month.

Each hydrogen plant failure upset the hydrocracker. Some failures sent liquid into the suction of the hydrocracker recycle gas compressor (RGC), causing the RGC to shut down. The licensor's EDS interlock gave operators 15 min to restart

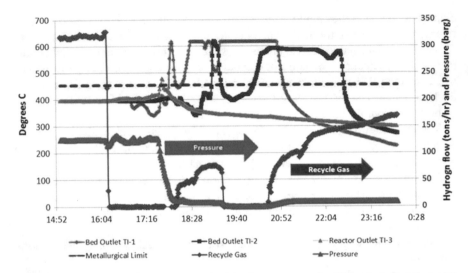

Fig. 18.13 Pressure, recycle gas flow, and selected temperatures for Example 2. Bed Outlet TI-1 flat-lines at 618 °C, because 618 °C is top-of-range. Actual temperatures were far higher, as evidenced by the fused catalyst in the bed when the reactor was unloaded

the compressor before automatically activating the low-rate depressuring valve. Each depressuring event shut the unit down for several hours if not days. To give themselves extra time to restart the RGC, the unit engineer and operators decided to disable the EDS/RGC interlock, assuming that depressuring triggered by high temperature would be sufficient.

This demonstrated ignorance of several fundamental concepts. The most egregious were:

- Residence time has a significant impact on reaction rates.
- The flow of fluids is the only significant way to remove heat from a hydroprocessing reactor.
- With zero or low flow, bed outlet TI readings are meaningless.

It is crucial to understand the difference between TI readings and temperature. A previous example shows that one never knows the highest or lowest temperature in a catalyst bed. But when reactants are flowing through at 50–100% of design capacity, one at least knows that bed outlet TIs correspond to the temperatures above them. When flow stops, no TI reading is relevant to any temperature more than a few inches away. Reactions do not stop when flow stops. Heat accumulates in catalyst beds, often without affecting the bed outlet TI readings. Without gas flow, hot oil pools due to gravity at low points. Eventually, the pooling oil will reach one or more bed outlet TIs and/or reactor wall TIs, providing the first tangible indication that something is wrong. Undoubtedly, that is what happened in this unit. Table 18.6 shows a timeline for this incident. A visual representation is presented in Fig. 18.13.

Table 18.6 Timeline for personal observations

16:14 h
Recycle gas compressor failed
Pressure = 127 barg
Maximum bed TI = 402 °C
Reactor outlet TI = 398 °C
17:38 h
Reactor Outlet TI-3 exceeded 440 °C
Automatic low rate depressuring
17:44 h
Reactor Outlet TI-3 continued up to 478 °C
18:00 h
Recycle gas compressor was restarted
18:27 h
A bottom reactor wall temperature reached 436 °C
19:17 h
Recycle gas compressor failed
18:20–20:54 h
Bed Outlet TI-1 jumped 214 °C in 7 min, stayed >618 °C for 154 min
20:28
Recycle gas compressor was restarted successfully.
18:56–22:30 h
Bed Outlet TI-2 reached >618 °C, fell, rose to >550 °C until 22:30

After the first compressor failure, it took 144 min for some overheated oil to affect TI reading at the reactor outlet. At 17:38 h, an upward spike in Reactor Outlet TI-3 triggered depressuring when it reached 440 °C. TI-3 continued to rise until it reached 487 °C. Depressuring seemed to make things worse. It decreased TI-3, but as the pressure fell, the exiting gas pulled formerly stagnant hot oil through the reactor. Between 17:48 and 17:55 h, Bed Outlet TI-1 jumped from 404 to 618 °C, an increase of 214 °C in 7 min.

The readings dropped after a restart of the RGC. But then, most likely due to the arrival of more overheated oil, TI-1 jumped back up to 618 °C, where it flat-lined for more than 2½ hours. Actual temperatures were higher than 618 °C, which is top-of-range. It was thought that the TI-1reading was faulty, but it wasn't; it returned to reading on-scale at 20:54. A second attempt to restart the RGC succeeded. TI-1 readings started to fall. But as it did, Bed Outlet TI-2 rose above 550 °C and stayed there for another two hours.

Afterwards, the unit ran at reduced rates for six months, with strict adherence to licensor-suggested temperature limits. The limits were tighter due to fears that the reactor outlet pipe had been compromised. During the subsequent turnaround, workers found that the reactor internals were damaged. They also found partially fused silica balls near the reactor wall. Pure silica melts at about 2000 °C. Impurities drop the fusion temperature considerably. But even so, it must have gotten incredibly hot in that reactor.

18.9.3 Botched Refinery Maintenance

Refineries schedule maintenance shutdowns for several units at once, often more than a year in advance. Each process unit depends on several others, serving as a destination for some streams or as a source of feedstock for other units. When a hydrocracker comes down, the need for hydrogen goes down, and major units normally fed by the hydrocracker have to be run differently. To do special work, equipment replacements must be ready to install, catalysts must be onsite, oil inventories must be adjusted, and maintenance contractors are hired to conduct repairs, provide maintenance, unload and load catalysts, etc. An unplanned shutdown is far more expensive than a planned shutdown, because other units aren't prepared, and maintenance personnel may not be instantly available.

The following instance was caused (mainly) by a decision to reduce the number of crews employed to service heat exchangers. In shell-and-tube exchangers, one gas or liquid stream is pumped through a bundle of parallel tubes. As it flows through a bundle, the "tube-side" fluid exchanges heat with "shell-side" fluid, which flows in the opposite direction, inside the shell but outside the tubes. A key activity is the cleaning or replacement of fouled tubes. Due to the shortage of people, exchanger maintenance was taking a lot longer than expected. It was clear that the pre-ordained schedule could not be met, given the original scope of work. A decision was made to clean just the exchangers most in need of cleaning. For an important set of exchangers at the hydrocracker, the feed/effluent exchanger (F/E), the crew cleaned only some tubes—the ones most badly fouled. Upon restart, flow was even in the F/E, where it preferentially followed the path of least resistance—most likely through the clean tubes. Within weeks, under-performance of the F/E required an unplanned hydrocracker shutdown and costly changes in the operation other units. The shutdown, service work, and subsequent restart required about two weeks at a cost of more than US$ 1 million per day—a steep price to pay for saving less than $50,000 per day from reducing tube-cleaning personnel.

The decision to clean only some tubes in the F/E displayed fundamental ignorance. Not knowing about the crew's decision revealed slip-shod management. No wonder. The refinery was chronically understaffed. During maintenance shutdowns, engineers had to work as many as 30 consecutive 14-hour days. Staffing decisions are made by refinery management, who are ultimately responsible for the consequences.

18.9.4 Fatigue-Related Near Miss

Despite the harsh comments in the BP Texas City report about worker fatigue and insufficient training, a small but significant number of U.S. refineries remain under-staffed, especially for shutdown and startup activities. Those refineries aren't alone. Many U.S. service companies also are under-staffed, especially during "turnaround season" in March through May. Refineries want to complete maintenance before June, so they can be up and running throughout the summer, when gasoline consumption is highest. Members of a vendor technical service team had been working long hours during the spring startup season, often more than 100 h per week. In addition to spending 12–14 h per day supervising startups, team members were required to handle questions from other customers, write proposals, and prepare for upcoming customer visits. Four hours of sleep per night was the rule, not the exception.

The group had always been busy, but the workload increased after the business unit was reorganized. The main reason was the switch from a leader who managed the team's workload by screening sales-support work, to a person who chased every apparent sales opportunity, no matter how unlikely.

One night, while driving from a refinery to his hotel, a team member fell asleep and drove off a country road. He wasn't injured. For that reason and others, he didn't report the incident until several months later, when another member of the team persuaded him to complain.

In response to the complaint, which included mention of possible consequences, a high-level manager said: "What are you, a lawyer? Mister HSE?" followed by "Why didn't you say something sooner?" The manager later admitted knowing about the over-work for several months and had considering adding staff. "We didn't," the high-level manager said, "because we might have had to lay people off if business slowed down." Within a year of the reorganization, 60% of the original team had left, primarily due to over-work.

The high-level manager's response to what could have been a fatal accident was completely inconsistent with company policy. Accidents and near-miss reports were supposed to be treated seriously, no matter how or when they were reported. The fact that the manager knew about the over-work for so long—and did nothing—is astounding.

18.10 Lessons

Lessons are learned such events. But not everyone learns. Many of those who do learn forget. Or worse, they decide that the lessons don't apply to them.

18.10.1   Dibs

DIBS—did-it-before syndrome—is a fundamental characteristic of human behavior. People ignore safety rules because they got away with ignoring them before, not just in the workplace, but in homes and public places, especially on the highway.

Statistics on road accidents provide a down-to-earth example, especially for North Americans. In the United States, the National Highway Traffic Safety Administration (NHTSA) conducts periodic surveys of speeding attitudes and behaviors. The most recent survey was conducted in 2011. The corresponding report was issued in 2013 [67]. Data were collected via telephone interviews with 6144 U.S. households, with allowances made for land-line versus cell-phone use. The report classified 30% of U.S. drivers as non-speeders, 40% as sometime speeders, and 30% are speeders. About 9% had been stopped for speeding, and about 6% received speeding citations. Here are a few survey results:

- 82% agreed that driving at or near the speed limit makes it easier to avoid dangerous situations
- 79% agreed that driving at or near the speed limit reduces their chance of having an accident
- 17% said that driving at or near the speed limit annoys them.

Other reports [68–70] provide the following statistics for the United States:

- Registered drivers = 212 million
- Registered vehicles = 269 million
- Crashes causing only property damage = 4.1 million
- Injury-causing crashes = 1.6 million
- Fatal crashes = 32,719
- Speeding-related fatalities = 9613
- Speeding related crashes in which the driver was alcohol-impaired = about 42%
- Speeding related crashes in which the people killed were wearing restraints = about 50%
- Average miles driven per year per vehicle = 13,476
- Average length of trip = 9.75 miles
- Average trips per year = 1360
- Speeding tickets per driver per year = 0.17
- Average speeding ticket cost = $152.

Using these data, one can calculate the following:

- Accidents per trip = 0.156%
- Injuries per trip = 0.061%
- Deaths per trip = 0.00037%.

Speeding gets us to where we want to be faster, and the risks are very low. The most likely negative consequence is a speeding ticket, for which the cost is tolerable. The average driver takes 460 trips without a fender-bender, 1640 trips without an injury, 80,000 trips without dying, and 270,000 trips without dying due to speeding. So we speed, even though we know speeding is dangerous.

In hydrocarbon industries, accidents can be far worse, threatening the health and prosperity of thousands of people. In well-run facilities, the personal financial cost of breaking a rule can far exceed the cost of a speeding ticket; it can get a worker fired.

But as described in several of the examples above, workers, managers and executives—knowingly and willfully—continue to break critical rules. Why? One can only guess. Is it simply because the rules are inconvenient? Do they believe the rules don't apply to them? Or is it DIBS—concluding that their way is better, because they've done it that way before with no negative consequences?

How can DIBS be diminished? Experience shows that the most effective way to decrease DIBS is to threaten individuals. If people can lose a job or go to jail for violating a rule, they take the rule more seriously, whether or not they agree with it.

18.10.2 MOC

Perhaps a DIBS practitioner is correct. Maybe his/her way really is okay, or even better.

Management of Change (MOC) [71] is a formal process by which procedures can be improved. With design models, a written document, and perhaps a formal presentation, the originator presents an idea to everyone involved. For minor, localized change in procedure, the MOC process can be quick, requiring approval from two or more colleagues and subsequent documentation. For a significant change in hardware or general operating procedures, an MOC committee usually includes peers, supervisors, and internal experts. It might also include one or more outside consultants. Implementation might require detailed design, budget approval, and development and review of operating procedures.

Ineffective MOC is one of the leading causes of serious incidents, according to the U.S. Chemical Safety and Hazard Investigation Board (CSB) [72]. CSB stated: "In industry, as elsewhere, change often brings progress. But it can also increase risks that, if not properly managed, create conditions that may lead to injuries, property damage or even death."

Changes subject to MOC procedures can include:

- Personnel changes, including reorganization, staffing, and training.
- Equipment changes, including revamps, maintenance schedules, communications devices, control hardware, and control software.
- Procedure changes, including procedures for normal operation, startup, shutdown, and MOC.
- Material changes, including significant changes in catalysts, chemicals, and feedstock.

18.10.3 Comparison of Examples

The Lakeview blowout occurred before the advent of blowout-prevention technology. Blowouts were known before, but it's difficult to say what might have been done to prevent this one, other than not drilling the well at all.

The Texas City ammonium nitrate explosion could be attributed to war; it is assumed that the chemical was going to be used to make explosives. The dangers of exposing ammonium nitrate to fire were well known. The danger was compounded by having so much in one place.

The London killer fog was mostly attributed to culture. In 1952, not very long after World War II, Londoners might have learned to accepted discomfort, including noisome smog, as just another thing to endure.

In 1978, the world simply was not prepared for oil spills as large as the one caused by the *Amoco Cadiz*. The main lesson learned there was that quick response is critical. By the time help finally arrived, the Cadiz was destroyed and its entire cargo was in the sea.

At Bhopal, plant maintenance was abysmal. The refrigeration system had been down for months. A faulty valve that should have sent the leaking gas to a scrubber sent it instead to the flare system, which failed. Operators didn't detect the leak, because a panel was missing from the control room. As stated by a Union Carbide executive, the plant should not have been running. The company should have put safety first, not profits.

Chernobyl was caused by contravention of procedures by people who didn't understand the process. Three major safety rules were violated when operators (a) shut off the emergency cooling water system, (b) raised the control rods too high, and (c) disabled the emergency SCRAM. If kept in operation any one of the idled systems might have prevented the accident.

Chemical companies all along the banks of the Rhine dumped wastes into the river to avoid having to spend money on proper disposal infrastructure, in a pro-business culture.

On the *Exxon Valdez*, Captain Hazelwood violated a U.S. Coast Guard guideline when he left the third mate in charge before the ship reached open water. Despite the 8-hour delay in expected response time, action by Alyeska and Exxon was relatively fast. Most of the oil stayed inside the ship until it was off-loaded. Consequently, instead of losing the entire cargo of 1.2 million barrels, "only" 250,000 barrels spilled into the sea.

In the Gulf War, retreating Iraqis took "scorched earth" to a new level.

The Tosco Avon disaster was caused by management and management-induced DIBS. Management didn't provide sufficient training for engineers and operators. Management refused to invest in proper instrumentation. Management refused to fix the problem that had caused a fire during the 1992 depressuring event. Management strongly encouraged operators to avoid depressuring. Operators became so accustomed to defeating excursions without depressuring that it never occurred to them to act correctly, even when temperatures far exceeded metallurgical limits. Apparently, nobody in the plant understood the relationship between N-slip and hydrocracking catalyst activity.

At BP Texas City, operators proceeded with the startup despite the lack of critical level measurement, reflecting a combination of inadequate training, inadequate technical supervision, inadequate communications between shifts, and probably DIBS.

Similar factors played a role in the *Deepwater Horizon* incident. Fatal mistakes included the use of an insufficient number of centralizers and the assumption that pressure in the kill line was indicative of pressure in the well. The drilling crew did not understand the significance several consecutive failed negative pressure tests. The visit of non-essential VIPs during the most critical phase of well completion distracted key technical personnel, who quickly would have known the significance of the negative-pressure test.

The train derailment in Lac-Mégantic was caused by faulty maintenance, the failure of the engineer to tell MM&A about the failing locomotive before it caught fire, and the failure of authorities to act to ensure that oil transportation infrastructure is safe.

All of the other examples occurred during startups. All were influenced by cost-cutting and/or unreasonable schedules. The most inexcusable are those that were caused by rogue managers who countermanded company policy in companies that spend millions on health and safety training.

Of these cases:

- Seven occurred during a startup or shutdown.
- Nine were caused in large part by contravention of procedures. Of these, three involved willful disabling of safety interlocks.
- Eight were caused by cost-cutting.
- Eight were aggravated by inadequate training.

18.10.4 Additional Safety-Related Issues

Other safety-related issues include the following:

Whistle-blowing. Over the years, the U.S. Congress has passed at least seven laws to protect environmental whistle-blowers from retribution from their employers. Even so, according to Seebauer [73], "Studies confirm the intuitive perception that whistleblowers tend to face hostility within their organizations and commonly leave their jobs, either voluntarily or otherwise. Lengthy and costly litigation often follows." Seebauer goes on to say: "Some employers are happy to hire workers who demonstrate such a strong commitment to high ethical standards. Nevertheless, there is no guarantee of employment, especially in a slow market. The whistleblower may lose seniority and retirement benefits, and must often move to another city. The needs of family members must be considered." When faced with these realities, insiders are reluctant to report environmental wrong-doing.

Smart companies encourage whistle-blowing. They send employees to courses that teach them about safety and environmental protection. They establish internal whistleblower hot lines and encourage employees to use them when they think their concerns haven't properly been addressed by the normal chain of command. They recognize that, in today's litigious society, whistleblowers can save a company billions of dollars in fines and cleanup costs.

Automation. Control systems aren't perfect. On the other hand, safety-related interlocks must *never* be disabled. They are there for good reasons. Operators, engineers, *and managers* should be trained to understand exactly how the interlocks work and why they are critical.

Personal protection equipment. One of the first things a person learns in the oil and gas business is the importance of personal protection equipment (PPE). In the old days, it wasn't uncommon to see bare-headed and even bare-handed workers in short-sleeved shirts building scaffolding in a refinery without PPE. That doesn't happen now. We are required to wear appropriate PPE at all times.

PPE includes fire-proof clothing, hard hats, safety glasses, steel-toed shoes, gloves, and toxic gas detectors. Harnesses are used to protect against falls. Fresh-air breathing equipment or self-contained breathing apparatus (SCBA) improves safety in confined spaces and in areas which might be contaminated by hazardous gases and particulates. In recent years, communication between outside and inside workers has been enhanced by allowing the use of inherently safe radios. Laboratory workers face unique hazards; hard hats are seldom needed in labs, but for some analytical procedures and chemical experiments, special gloves and full-face shields protect against hazardous chemicals and substances, solvents, and acids that are usually placed inside an air-ventilation hood.

Pocket-sized toxic gas detectors are important PPE. They sound an alarm when CO or H_2S reach dangerous levels. Some sites require all workers to wear H_2S monitors at all times when they are inside the gate. So-called four-gas monitors detect H_2S, CO, combustible hydrocarbons, and oxygen.

In the United States, the Occupational Safety and Health Administration (OSHA) sets standards for PPE [74, 75]. The first standards were adopted in 1971. A major revision was issued in 1994, after OSHA discovered gaps in the existing standards. Data indicated that injuries were occurring at the same rate, whether or not employees were wearing PPE. Some regulations were so restrictive that they discouraged innovation. Present regulations are based on the effectiveness of PPE. Recent improvements are addressing the main cause of worker non-compliance before the revision: some PPE was cumbersome, uncomfortable, and even impractical.

Confined spaces. Workers often enter tanks, reactors, and other confined spaces to perform inspection or maintenance. In such environments, oxygen levels can become too low even if air is circulating. Required PPE for confined space entry includes fresh-air breathing equipment and a harness that allows colleagues to lift the worker out if necessary. Modern video technology enables outside personnel to monitor people inside directly.

18.11   U.S. Environment Protection Agencies

This section focuses almost exclusively on U.S. environment protection agencies and laws. Please understand that to include similar detail for other countries would have been infeasible in the space allotted.

In 1970, United States formed the Environmental Protection Agency (US EPA) [76] and the Occupational Safety and Health Administration (OSHA). Together, these agencies are responsible for dramatic improvements in air and water quality and increases workplace safety throughout the United States.

EPA is an amalgamation of departments, bureaus, agencies, etc., taken from the Interior Department; the Department of Health, Education, and Welfare; the Agriculture Department; the Atomic Energy Commission; the Federal Radiation Council; and the Council on Environmental Quality. EPA's mission is "to enforce federal laws to control and abate pollution of air and water, solid waste, noise, radiation, and toxic substances. It is also to administer the Superfund for cleaning up abandoned waste sites, and award grants for local sewage treatment plants."

After its creation, EPA quickly took the following actions:

- Established 10 regional offices throughout the nation
- Established National Ambient Air Quality Standards, which specified maximum permissible levels for major pollutants
- Required each state to develop plans to meet air quality standards
- Established and enforced emission standards for hazardous pollutants such as asbestos, beryllium, cadmium, and mercury
- Required a 90% reduction in emissions of VOC and carbon monoxide by 1975
- Published emission standards for aircraft
- Funded research and demonstration plants
- Furnished grants to states, cities, and towns to help them combat air and water pollution.

EPA's law-enforcement efforts are supported by the National Enforcement Investigation Center in Denver, Colorado, which gives assistance to federal, state, and local law enforcement agencies. This unit has clamped down on the "midnight dumping" of toxic waste and the deliberate destruction or falsification of documents.

18.12   Other Environmental Agencies

Today, almost every U.S. state has an environmental agency. Arguably, the most famous of these is the California Air Resources Board (CARB), which pioneered regulations to mitigate smog in Los Angeles. In addition to administering state programs for improving air and water quality, the Texas Commission on Environmental Quality (TCEQ) participates in making plans to prevent and react to industrial terrorism.

Table 18.7 Environmental Agencies around the World

Country	Agency
Australia	Department of the Environment
Austria	Ministry for the Environment
Brazil	Ministry of Environment
Canada	Environment Canada
China	Ministry of Ecology and Environment
Europe	European Environment Agency
Finland	Ministry of the Environment
Indonesia	Kementerian Lingkungan Hidup
India	Ministry of Environment and Forests
Japan	Ministry of the Environment of Japan
Kuwait	Environment Public Authority
Malaysia	Department of Environment
Mexico	Secretariat of Environment and Natural Resources
Norway	Ministry of the Environment
Saudi Arabia	Presidency of Meteorology and Environment
Singapore	National Environment Agency
South Africa	Department of Environmental Affairs and Tourism
Taiwan (Republic of China)	Environmental Protection Administration
Thailand	Ministry of Natural Resources and Environment
United Kingdom	Environment Agency
United States	Environmental Protection Agency

Most countries have environmental agencies. As shown in Table 18.7, some are combined with public health departments, some are combined with energy agencies, and at least one is coupled with tourism. In addition to handling internal issues, most of these agencies administer their country's participation in international treaties, such as the Kyoto Protocol [77].

18.13 Occupational Safety and Health Administration

Pollution control and safety are two sides of the same coin. In the United States, the Occupational Safety and Health Administration (OSHA) is part of the U.S. Department of Labor. Its legislative mandate is to assure safe and healthful working conditions by:

- Enforcing the Occupational Safety and Health Act of 1970
- Helping states to assure safe and healthful working conditions
- Supporting research, information, education, and training in occupational safety and health.

OSHA can levy fines against unsafe people and companies. Not surprisingly, a large percentage of the safety infringements cited by OSHA have caused environmental damage or put the environment at risk.

- In 1993 OSHA fined the Manganas Painting Company for exposing workers and the environment to lead during sand-blasting operations. The proposed fines totalled US$4 million. The contractor appealed, but in February 2002, it pled guilty to knowingly and illegally dumping 55 tons of lead-containing sandblasting material, in violation of the Resource Conservation and Recovery Act (RCRA) [78].
- BP was fined a then-record US$21 million for infractions related to the 2005 Texas City explosion. After a 2009 followup, OSHA found that BP had failed to abate a number of safety-related items and committed several willful violations of safety regulations. In 2010, BP agreed to pay US$50.6 million to settle the failure-to-abate violations and in 2012 it agreed to settle 409 of 439 willful violations for US$13 million.

BP pled guilty to 11 felony counts of misconduct or neglect and paid US$4.5 billion for the *Deepwater Horizon* incident. The payment included more than US$1.25 billion to settle the criminal fines.

18.13.1 Material Safety Data Sheets (MSDS)

Under the Occupational Safety and Health Act, employers are responsible for ensuring a safe and healthy workplace. One key requirement is that all hazardous chemicals must be properly labeled. Workers must be taught how to handle the chemicals safely, and material safety data sheets must be available to any employee who wishes to see them.

Material Safety Data Sheets (MSDS) include the following information:

Material identification. The name of the product and the manufacturer's name, address, and emergency phone number must be provided.

Hazardous ingredients. The sheet must give the chemical name for all hazardous ingredients comprising more than 1% of the material. It must list cancer-causing materials if they comprise more that 0.1%. Listing only the trade name, only the Chemical Abstract Service (CAS) number, or only the generic name is not acceptable.

If applicable, exposure limits are listed in this section of the MSDS. The OSHA permissible exposure limit (PEL) is a legal, regulated standard. Other limits may also be listed. These include recommended exposure limits (REL) from the National Institute for Occupational Safety and Health (NIOSH) and threshold limit values (TLV) from the American Conference of Governmental Industrial Hygenists (ACGIH). Sometimes, short-term exposure and/or ceiling limits are shown. The ceiling limit should never be exceeded.

Physical properties. These include the appearance, color, odor, melting point, boiling point, viscosity, vapor pressure, vapor density, and evaporation rate. The vapor pressure indicates whether or not the chemical will vaporize when spilled. The vapor density indicates whether the vapor will rise or fall. Odor is important because a peculiar smell is the first indication that something has leaked.

Fire and explosion hazard data. This section provides the flash point of the material, the type of extinguisher that should be used if it catches fire, and any special precautions.

The flash point is the lowest temperature at which a liquid gives off enough vapor to form an explosive mixture with air. Liquids with flash points below 100 °F (37.8 °C) are called flammable, and liquids with flash points between 100 and 200 °F (37.8 and 93.3 °C) are called combustible. Flammable and combustible liquids require special handling and storage.

The four major types of fire extinguishers are Class A for paper and wood, Class B for flammable liquids or greases, Class C for electrical fires, and Class D for fires involving metals or metal alloys.

Health hazard data. This section defines the symptoms that result from normal exposure or overexposure to the material or one of its components. Toxicity information, such as the result of studies on animals, may also be provided. The information may also distinguish between the effects of acute (short term) and chronic (long-term) exposure. Emergency and first-aid procedures are included in this section.

Reactivity data. This section includes information on the stability of the material and special storage requirements.

Unstable chemicals can decompose spontaneously at certain temperatures and pressures. Some unstable chemicals decompose when they are shocked. Rapid decomposition produces heat, which may cause a fire and/or explosion. It also may generate toxic gas. Hazardous polymerization, which is the opposite of hazardous decomposition, also can produce enough heat to cause a fire or explosion.

Concentrated acids and reactive metals are hazards when mixed with water. They should be stored separately in special containers.

Spill, leak, and disposal procedures. This part of the MSDS gives general procedures, precautions and methods for cleaning up spills and disposing of the chemical.

Personal protection equipment requirements. This section lists the protective clothing and equipment needed for the safe handling of the material. Requirements can differ depending on how the chemical is used and how much of it is used.

Protective equipment and clothing can include eye protection (safety glasses or face shields), skin protection (clothing, gloves, shoe covers), self-contained breathing equipment, and/or forced-air ventilation (fume hoods).

18.14   Key Regulations

In this section, important environmental legislation is described. Most examples come from the United States, but there is an excuse: for several decades, the United States has pioneered environmental regulations, and many countries have followed suit. Recently, the European Commission took the lead in promoting climate-control treaties, including the Rio and Kyoto Protocols. These are discussed in Sect. 18.15.

In the United States, since 2017, environmental regulations in the United States are being rolled back and, in some cases, overturned [79]. These actions are designed to decrease the quality of the air we breathe and the food we eat, diminish the cleanliness of the rivers that flow past us, and increase the pace at which the climate is changing. Various emission standards were relaxed. Restrictions on mercury emissions from coal were suspended in the name of "ending the war on coal." Such policies are already to have substantial impact on those who experience them close up—and often are economically dependent on the industries the present administration is trying to help. Coal miners are the primary victims of such mercury emissions and pollution due to leaking ash ponds. It is difficult to comment on such actions without getting into politics. But we would be remiss if we failed to point out that the near-term consequences of such actions are already being observed, and in some cases the long-term consequences are in essence irreversible.

18.14.1   Clean Air Acts

The first clear-air acts in the English-speaking world were implemented by Parliament in 1956, in response to the "deadly fog" incidents around London.

Since 1963, the United States government has passed several clean-air acts, including:

- Clean Air Act of 1963
- Motor Vehicle Air Pollution Control Act of October 20, 1965
- Clean Air Act Amendments of October 15, 1967
- Air Quality Act of November 21, 1967
- Creation of EPA: Clean Air Act of 1970
- Clean Air Act Amendments of 1975, 1977, and 1990.

For convenience, the entire package often is called just the Clean Air Act (CAA). A good source for general understanding is the Plain English Guide to the Clean Air Act, available from the U.S. EPA web site [80]. The Guide shows how the CAA has led to significant improvements in human health and the environment. For example, since 1970:

- The six most commonly found air pollutants have decreased by more than 50%
- Air toxics from large industrial sources, such as chemical plants, petroleum refineries, and paper mills have been reduced by nearly 70%
- New cars are more than 90% cleaner and will be even cleaner in the future, and
- Production of most ozone-depleting chemicals has ceased

At the same time, the U.S. economy and associated energy consumption has grown significantly:

- The U.S. GDP has tripled
- Energy consumption has increased by 50%
- Vehicle use has increased by almost 200%.

Under the CAA, EPA's role is to set limits on certain air pollutants, including ambient levels anywhere in the country. EPA limits emissions of air pollutants from point sources like chemical plants, utilities, and steel mills. Individual states or tribes may have stronger air pollution laws, but they may not have weaker pollution limits than those set by EPA.

EPA assists state, tribal, and local agencies by providing research, expert studies, engineering designs, and funding to support clean air progress. EPA must approve state, tribal, and local air pollution reduction plans. EPA can issue sanctions against states which fail to comply; if necessary, it can take over enforcing the CAA in non-complying areas.

The CAA required EPA to set National Ambient Air Quality Standards (NAAQS) for pollutants considered harmful to public health and the environment. EPA set two types of standards—primary and secondary. Primary standards protect against adverse health effects. Secondary standards protect against damage to farm crops, vegetation, and buildings. Because different pollutants have different effects, the NAAQS are also different. Some pollutants have standards for both long-term and short-term averaging times. The short-term standards are designed to protect against acute, or short-term, health effects, while the long-term standards were established to protect against chronic health effects. NAAQS are shown in Table 18.8 [81].

Hazardous air pollutants (HAPs) are regulated, too. Table 18.9 lists 20 "core" compounds selected from of the 188 HAPs regulated under Sect. 112 of the Clean Air Act.

Compliance with air quality standards is monitored by the Office of Air Quality Planning and Standards (OAQPS) [81]. OAQPS evaluates the status of the atmosphere as compared to clean air standards and historical information, using measurements acquired from many hundreds of monitoring stations across the United States.

Table 18.8 National Ambient Air Quality Standards (NAAQS) pollutants

Pollutant	Primary/Secondary	Averaging time	Level	Form
Carbon monoxide	Primary	8 h	9 ppm	Not exceeded more than once/year
Carbon monoxide	Primary	1 h	35 ppm	Not exceeded more than once/year
Lead	Both	Rolling 3 month	$0.15\ \mu g/m^3$	Not to be exceeded
Nitrogen dioxide	Primary	1 h	100 ppb	98th percentile of 1 h daily max concentrations, averaged over 3 years
Nitrogen dioxide	Both	Annual	53 ppb	Annual mean
Ozone	Both	8 h	0.075 ppm	Annual 4th-highest daily max 8-hr concentration, averaged over 3 years
Particulates, $PM_{2.5}$	Primary	Annual	$12\ \mu g/m^3$	Annual mean, averaged over 3 years
Particulates, $PM_{2.5}$	Secondary	Annual	$15\ \mu g/m^3$	Annual mean, averaged over 3 years
Particulates, $PM_{2.5}$	Both	24 h	$35\ \mu g/m^3$	Annual mean, averaged over 3 years
Particulates, PM_{10}	Both	24 h	$150\ \mu g/m^3$	Not to be exceeded more than once/year, on average, over 3 years
Sulfur dioxide	Primary	1 h	75 ppb	99th percentile of 1 h daily max concentrations, averaged over 3 years
Sulfur dioxide	Secondary	3 h	0.5 ppm	Not exceeded more than once/year

Table 18.9 Twenty of the hazardous air pollutants regulated by the Clean Air Act

Volatile organics (VOCs)	Inorganics (Metals)	Aldehydes
Benzene	Arsenic	Formaldehyde
1,3-butadiene	Beryllium	Acetaldehyde
Carbon tetrachloride	Cadmium	Acrolein
Chloroform	Chromium	
Propylene dichloride	Lead	
Dichloromethane	Manganese	
Perchloroethylene	Mercury	
Trichloroethylene	Nickel	
Vinyl chloride		

18.14.2 Reformulated Gasoline (RFG) in the United States [82]

In 1970, gasoline blending became more complex. The U.S. Clean Air Act required the phase-out of tetraethyl lead, so refiners had to find other ways to provide octane. In 1990, the CAA was amended. It empowered EPA to impose emissions limits on automobiles and to require reformulated gasoline (RFG). RFG was implemented in several phases. The Phase I program started in 1995 and mandated RFG for 10 large metropolitan areas. Several other cities and four entire States joined the program voluntarily. The current phase began in 2000. RFG is used in 17 States and the District of Columbia. In 2015, about 35% of the gasoline in the United States was reformulated.

Tier 1 reformulated gasoline regulations required a minimum amount of chemically bound oxygen, imposed upper limits on benzene and Reid Vapor Pressure (RVP), and ordered a 15% reduction in volatile organic compounds (VOC) and air toxics. VOC react with atmospheric NO_x to produce ground-level ozone. Air toxics include 1,3-butadiene, acetaldehyde, benzene, and formaldehyde.

The regulations for Tier 2, which took force in January 2000, were based on the EPA Complex Model, which estimates exhaust emissions for a region based on geography, time of year, mix of vehicle types, and—most important to refiners—fuel properties. As of 2006, the limit on sulfur in the gasoline produced by most refineries in the U.S. was 30 ppmw.

Initially, the oxygen for RFG could be supplied as ethanol or C_5–C_7 ethers. The ethers have excellent blending octanes and low vapor pressures. But due to leaks from filling station storage tanks, methyl-t-butyl ether (MTBE) was detected in ground water samples in New York City, Lake Tahoe, and Santa Monica, California. In 1999, the Governor of California issued an executive order requiring the phase-out of MTBE as a gasoline component. That same year, the California Air Resources Board (CARB) adopted California Phase 3 RFG standards, which took effect in stages starting in 2002. The standards include a ban on MTBE and a tighter cap on sulfur content—15 ppmw maximum. In the United States, Tier 3 vehicle emission and fuel standards lowered the allowed sulfur content of gasoline to 10 ppmw, beginning in 2017.

18.14.3 Conservation, Efficiency, and Renewable Fuel Mandates

The **Energy Policy Act of 2005** [83] was signed into law by President George W. Bush on August 8, 2005. The Act addresses energy production and use in the United States. It provides tax incentives and loan guarantees for specific types of energy production Topics include:

- Energy efficiency
- Renewable energy
- Oil and gas
- Coal
- Tribal energy
- Nuclear matters and security
- Vehicles and motor fuels, including ethanol
- Hydrogen
- Electricity
- Energy tax incentives
- Hydropower and geothermal energy
- Climate change technology.

Critics say that the Act is a broad collection of subsidies for United States energy companies; in particular, the nuclear and oil industries [84]. It authorized the following tax reductions:

- $4.3 billion for nuclear power
- $2.8 billion for fossil fuel production
- $2.7 billion to extend the renewable electricity production credit
- $1.6 billion in tax incentives for investments in "clean coal" facilities
- $1.3 billion for energy conservation and efficiency
- $1.3 billion for alternative fuels (bioethanol, biomethane, liquified natural gas, propane)
- $500 million Clean Renewable Energy Bonds for government agencies for renewable energy projects.

The bill exempted hydraulic fracturing from certain provisions of the Safe Drinking Water Act. Consequently, drilling companies don't have to disclose the chemicals used in fracking.

The most questionable provision of the Act increases the amount of biofuel (specifically corn ethanol) that must be blended with gasoline. Proponents say that displacing gasoline with ethanol achieves significant CO_2 production and decreases dependence on petroleum imports. Detractors point out that while corn can give 1.1–1.4 times more energy than is required for its growth, transportation, and processing [85], sugar cane in Brazil [86] gives an energy gain of 8.0. Large amounts of arable land are used to grow corn for ethanol. The ethanol mandate is increasing food prices and putting pressure on ground water. Furthermore, instead of specifying concentrations of ethanol in the gasoline blend, the law specifies the total amount of ethanol that must be consumed, even if gasoline sales decrease.

The **Energy Independence and Security Act of 2007** required improved vehicle fuel economy. It also increased production requirements for biofuels, specifically biomass-derived diesel. To be biomass-based diesel, the fuel must reduce emissions by 50% versus petroleum-derived diesel. It increased the total target volume for ethanol, and specified that a large portion should come from non-corn sources. The Act motivates energy savings with improved standards for appliances and lighting, and in industrial and commercial buildings.

18.14.4   Gasoline (Petrol) in the EU

Transportation fuel specifications in the European Union are described by EN 228, "Specification for unleaded petrol (gasoline) for motor vehicles." Key specifications include the following:

- Octane: minimum RON and MON are 95 and 85, respectively
- Volatility: minimum percent vaporized at 100 and 150 °C are 46 and 75, respectively,
- Hydrocarbon types: the maxima for olefins/aromatics/benzene are 18/35/1 vol.%, respectively.
- Oxygen: The maximum oxygen content is 3.7% m/m. Oxygenates can include different amounts of methanol, ethanol (with stabilizing agents), isopropyl alcohol, butyl alcohols, ethers such at methyl-t-butylether (MTBE), and other mono alcohols.
- Sulfur: 10 ppmw.

EU gasoline is just as clean if not cleaner than EPA gasoline, but it is easier to manufacture and regulate. No models are used, and the allowable sources of oxygen are based on science.

Table 18.10   Clean fuels: specifications for Euro V diesel

Euro V diesel—EN590

Property	Unites	Low limit	Upper limit	Test-method
Cetane index		46	–	ISO 4264
Cetane number		51	–	ISO 5165
Density at 15 °C	kg/m^3	820	845	ISO 3675, 12185
Polycyclic aromatic hydrocarbons	% (m/m)	–	11	ISO 12916
Sulphur content	mg/kg	–	10	ISO 20846, 20847, 20884
Flash Point	°C	55	–	ISO 20846, 20884
Carbon residue (on 90–100% fraction)	% (m/m)	–	0.3	ISO 2719
Ash content	% (m/m)	–	0.01	ISO 10370
Water content	mg/kg	–	200	ISO 6245
Total contamination	mg/kg	–	24	ISO 12937
Copper strip corrosion (3 h at 50 °C)	Rating	Class 1	Class 1	ISO 12662
Oxidation stability	g/m^3	–	25	ISO 2160
Lubricity. Wear scar diameter at 60°C	μm	–	460	ISO 12205
Viscosity at 40°C	mm^2/s	2	4.5	ISO 12156-1
Distillation recovered at 250 °C. 350 °C	% (V/V)	85	<65	ISO 3104
95% (V/V) recovered at	°C	–	360	ISO 3405
Fatty acid methyl ester content	% (V/V)	–	7	ISO 14078

18.14.5   Ultra-Low-Sulfur Diesel

At present, the most advanced specification for diesel fuel is Euro V, which is described in Table 18.10. Severe hydroprocessing is required to making Euro V from difficult feedstocks.

18.14.6   Rivers and Harbor Act, Refuse Act

One of the first environmental laws in the U.S. was the Rivers and Harbor Act of 1899. It was passed to control obstructions to navigation. The Act required congressional approval for the building of bridges, dams, dikes, causeways, wharfs, piers, or jetties, either in or over a navigable waterway. The Act also made it illegal to discharge debris into navigable water without a permit. In 1966, a court held that the River and Harbor Act made it illegal to discharge industrial waste without a permit, not just directly into navigable waters, but also into associated tributaries and lakes, i.e., just about every puddle of open water in the United States. This led to the Refuse Act Permit Program of 1970, under which specific kinds of pollution are allowed under permits issued by the Army Corps of Engineers. Every application for a permit is reviewed by EPA. If EPA concludes that the discharge described in the application will harm the environment, the Army Corp of Engineers denies the permit.

The penalties for violating permits are severe, including stiff fines and jail time. A corporation that employs the guilty person can be fined up to $1,000,000. False reports also can be punished by fines or imprisonment. In this context, "guilty person" refers to the corporate officer responsible for the facility from which the illegal discharge originates. In other words, a negligent act by a sloppy operator can send the Big Boss to jail.

18.14.7 Federal Water Pollution Control Act

The original Federal Water Pollution Control Act (FWPCA) was approved on July 9, 1956. The present Act includes the following:

- Pollution Control Act Amendments of July 20, 1961
- Water Quality Act of October 2, 1965
- Clean Water Restoration Act of November 3, 1966
- Water Quality Improvement Act of April 3, 1970
- Federal Water Pollution Control Act of 1972
- Clean Water Acts of 1977, 1981, and 1987.

The FWPCA of 1972 gave EPA greater authority to fight water pollution. While implementing the Act, EPA cooperates with the U.S. Coast Guard and the Secretary of the Interior. Individual states have primary responsibility for enforcing water quality standards, but if the states fail to meet expectations, EPA can take civil or criminal action under the Refuse Act.

The FWPCA prohibits the discharge of harmful amounts of oil into navigable waters. If oil is spilled, the owner or operator is liable for cleanup costs. Initially, the bill authorized $24.6 billion for water pollution control over three years. The goal of the law was to eliminate the pollution of U.S. waterways by municipal and industrial sources by 1985.

18.14.8   Clean Water Acts, Water Quality Act

The main objective of the Clean Water Acts (CWA) of 1977, 1981 and 1987 is to maintain the "chemical, physical, and biological integrity of the nation's waters." It seeks to have "water quality which provides for the protection and propagation of fish, shellfish and wildlife and provides for recreation in and on the water." Under these Acts, each state is required to set its own water quality standards. All publicly owned municipal sewage treatment facilities are required to use secondary treatment for wastewater. To help states and cities build new or improved water treatment plants, Congress provides construction grants, which are administered by EPA. EPA allocates funds to states, which in turn distribute money to local communities.

Community programs are monitored by EPA. They must meet treatment requirements to obtain permits under the National Pollutant Discharge Elimination System (NPDES).

The Water Quality Act of 1987 requires discharge permits for all point sources of pollution. More than 95% of all major facilities now comply with 5-year NPDES permits, which specify the types and amounts of pollutants that legally can be discharged. When permits are renewed, they can be modified to reflect more stringent regulations. Violators are subject to enforcement actions by EPA, including criminal prosecution.

The authority of the EPA was strengthened under the 1987 Water Quality Act. The allowable sizes of fines were increased, and violators found guilty of negligence could be sent to prison.

18.14.9   Marine Protection, Research, and Sanctuaries Act

The Marine Protection, Research, and Sanctuaries Act of 1972 gave authority to the EPA to protect oceans from indiscriminate dumping. The Agency designates sites at which dumping is allowed and issues dumping permits. Fines can be imposed for illegal dumping.

18.14.10   Safe Drinking Water Act [87]

Since the 1970s, the assurance of safe drinking water has been a top priority for EPA, along with individual states and over 53,000 community water systems (CWSs). The CWSs supply drinking water to more than 280 million Americans—about 90% of the population.

The Safe Drinking Water Act of 1974 was amended in 1977 and again in 1986. It empowered EPA to set national standards for drinking water from surface and underground sources. It also authorized EPA to give financial assistance to states, which are in charge of enforcing the standards. Aquifers are protected from wastes disposed in deep injection wells, from runoff from hazardous waste dumps, and from leaking underground storage tanks. In 1987, EPA also established maximum contaminant levels for volatile organics (VOC) and 51 manmade chemicals. Standards for other chemicals were added as their toxicity was determined. At present, health and safety standards have been established for 91 microbial, chemical, and radiological contaminants.

18.14.11   Resource Conservation and Recovery Act (RCRA)

In the United States, solid wastes are regulated by the following:

- Solid Waste Disposal Act of 1965
- Resources Recovery Act of 1970
- Resource Conservation and Recovery Act of 1976 (RCRA).

In 1965, the Solid Waste Disposal Act provided financial grants to develop and demonstrate new technologies in solid waste disposal. The Resources Recovery Act of 1970 emphasized recycling and by-product recovery.

By 1976, problems caused by the accumulation of large quantities of hazardous waste prompted Congress to pass the Resource Conservation and Recovery Act (RCRA). This legislation gave EPA the responsibility of developing a "cradle to grave" approach to hazardous waste, which culminated in the establishment of Hazardous Waste and Consolidated Permit Regulations (1980).

Under RCRA, hazardous waste is tracked from its source to every destination, including final disposal. Tracking is based on transportation manifests, other required records, and the issuance of permits.

EPA classifies wastes according to four measurable characteristics, for which there are standardized tests. The characteristics are:

- Ignitibility
- Corrosivity
- Reactivity
- Extraction procedure toxicity (EP).

In 1980, the Waste Oil Recycling Act (WORA) empowered EPA to encourage the development of state and local programs for recycling waste motor oil. That same year, WORA was amended to strengthen its enforcement provisions.

A *generator* of hazardous waste is responsible for the following:

1. Determining if a waste is hazardous
2. Obtaining a facility permit if the waste is stored on the generator's property for more than 90 days
3. Obtaining an EP identification number
4. Using appropriate containers and labeling them properly for shipment

5. Preparing manifests for tracking hazardous waste
6. Assuring through the manifest system that the waste arrives at the designated facility
7. Submitting to EPA an annual summary of activities.

Prior to shipping hazardous waste, the generator must prepare a manifest which includes the following:

1. Name and address of the generator
2. Name of all transporters
3. Name and address of the facility designated to receive the waste
4. EPA identification number of all who will handle the waste
5. Department of Transportation (DOT) description of the waste
6. Amount of waste and number of containers
7. Signature certifying that the waste has been properly labeled and packaged in accordance with EPA and DOT regulations.

To send waste away from its site, the generator must use EPA-approved transporters. It must also keep records of all hazardous waste shipments and immediately report those that fail to reach the facility shown on the manifest.

A *transporter* of hazardous waste must:

1. Obtain an EPA identification number
2. Comply with the manifest system for tracking hazardous waste
3. Deliver the entire quantity of hazardous waste to the facility specified by the manifest
4. Retain copies of manifests for three years
5. Comply with Department of Transportation rules for reporting discharges and spills
6. Cleanup any spills that occur during transportation. All spills must be reported to both EPA and the DOT.

Any person who owns or operates a hazardous waste facility must receive a permit from the EPA. Standards for facilities that treat, store, or dispose of hazardous waste are designed to:

1. Promote proper treatment, storage, and methods of disposal
2. Provide states with minimum standards acceptable to EPA
3. Provide technical support to states that lack hazardous waste management programs.

The RCRA amendments of 1984 and 1986 extended the act to cover underground storage tanks, especially those used for gasoline and other petroleum liquids. At the time, about 15% of the nation's 1.4 million gasoline storage tanks were leaking. Most of these leaks have since been repaired.

18.14.12 Superfund, CERCLA

Years ago, people were less aware of the dangers of dumping chemical wastes. On many properties, dumping was intensive and/or continuous. This created thousands of hazardous sites, many of which were uncontrolled and/or abandoned. On December 11, 1980, in response to public concern, Congress passed the Comprehensive Environmental Response Compensation and Liability Act (CERCLA) [88], which authorized EPA to locate, investigate, and remediate the worst of these hazardous sites.

CERCLA established the Superfund, which provides emergency cleanup funds for chemical spills and hazardous waste dumps. The Superfund allows the government to take immediate action to cleanup spills or dumps where the responsible party cannot be identified easily. The Superfund draws about 90% of its money from taxes on oil and selected chemicals. The remainder comes from general tax revenues.

Except in an emergency, state agencies are consulted before the federal government takes action. When it does so, it uses one of three approaches:

1. If the owner of the hazardous site cannot readily be identified, the federal government may proceed with the cleanup.
2. If the owner can be identified but refuses to clean the site, or if the owner's efforts are not up to par, the federal government can take charge of the cleanup. The owner must pay the cost, whatever it happens to be.
3. When the owner can be identified and decides to do the work, the federal government monitors the project and gives official approval when the work is completed according to standards.

CERCLA covers a wide range of sites. In addition to land-based dumps, it applies to spills into waterways, groundwater, and even the atmosphere. Initial funding for the Act was US$1.6 billion over 5 years. In 1986, the Superfund Amendment and Reauthorization Act (SARA) extended the program by five years and increased the fund to US$8.5 billion. It also tightened cleanup standards and enhanced EPA's enforcement powers. The Emergency Planning and Community Right-to-Know Act (EPCRA), also known as SARA Title III, encourages and supports emergency planning efforts at state and local levels. It also gives information to the public and local governments on potential chemical hazards in their communities. This legislation helped reduce pollution and improve safety all across the land.

In 1984, Hazardous and Solid Waste Amendments (HSWA) were passed because citizens were concerned about the potential contamination of ground water by hazardous waste disposal sites.

In 1978, the Federal Insecticide, Fungicide, and Rodenticide Act (FIFRA) of 1947 was amended to give EPA authority to control pollution from DDT, mercury, aldrin, toxaphene, parathion, and related chemicals. About 1 billion pounds of pesticides, fungicides, and rodenticides are used every year in the United States. While they contribute enormously to the success of agriculture, they can be harmful to animals, birds and humans if not used properly.

18.14.13   *Toxic Substances Control Act (TSCA)*

The Toxic Substances Control Act (TSCA) of 1976 gave EPA the authority to regulate the development, distribution, and marketing of chemical products. Manufacturers, importers, and processors must notify EPA within 90 days before introducing a new chemical to the market. Certain tests (e.g., fish-kill tests) must be conducted to determine toxicity. Approved chemicals must bear warning labels.

Many chemicals are restricted or banned under TSCA.

- The manufacture, processing and distribution of completely halogenated chlorofluorocarbons (CFCs) is banned, except for a small number of essential applications.
- Chromium (VI) may not be used as a corrosion inhibitor in comfort cooling towers (CCTs) associated with air conditioning and refrigeration systems.
- Nitrosating agents may not be mixed with metalworking fluids that contain specific substances.
- The import, manufacture, processing, or distribution of PCBs is banned unless EPA agrees that the PCBs will be "totally enclosed."

18.14.14   *Asbestos School Hazard Abatement Act*

The Asbestos School Hazard Abatement Act (SHAA) of 1984 was passed to encourage the removal of asbestos from schools. In 1986, the Asbestos Hazard Emergency Response Act was passed to correct in deficiencies in the previous Act. The final rule, issued in 1987, required local education agencies to:

1. Inspect school buildings for asbestos-containing materials
2. Submit asbestos management plans to state governors
3. Reduce or completely eliminate all asbestos hazards.

18.14.15   *Control of Dumping at Sea*

On November 13, 1972, the Convention on the Dumping of Wastes at Sea was agreed in London by representatives of 91 countries, including all of the world's principle maritime nations. The list of substances that may not be dumped includes biological and chemical warfare agents, certain kinds of oil, certain pesticides, durable plastics, poisonous metals and their compounds, and high-level radioactive waste. Enforcement is left to individual countries.

18.15 Climate Control: Rio, Kyoto, and Paris [89–91]

18.15.1 Rio Earth Summit

In 1992, during the "Earth Summit" in Rio de Janeiro, 154 nations plus the European community signed the United Nations Framework Convention on Climate Change (UNFCCC). At the time, the Earth Summit was the largest-ever gathering of Heads of State. Effective on March 21, 1994, the UNFCCC called on industrial nations to voluntarily reduce greenhouse gas emissions to 1990 levels by the year 2000. As of May 2015, there were 196 Parties (195 countries and 1 regional economic integration organization) to the Framework.

In many respects, the Rio Declaration resembled the Declaration on Human Environment issued by the Stockholm Conference in 1972. The 27 non-binding principles of the Rio Declaration included the "polluter pays" concept and the "precautionary principle." The latter recommends that, before a construction project begins, an impact study should be conducted to identify and forestall potential harm to the environment.

The declaration asserted that present-day economic development should not undermine the resource base of future generations. It also affirmed that industrial nations pollute more than developing countries. (For example, on a per capita basis, the United States emits 25 times more CO_2 than India.) On the other hand, industrial nations have advanced technology and greater financial resources, which enable them to contribute more to environmental protection.

18.15.2 Kyoto Protocol

The Kyoto Protocol extended the UNFCC. Adoption committed countries to accept that climate change was real, and that anthropogenic CO_2 is the primary contributor. The agreement was concluded in December 1997 and took force in February 2005. As of 2015, it had been accepted by 92 countires. Significantly, Canada withdrew its support in 2012 and the United States never adopted it. The Protocol obliges developed countries to reduce greenhouse gas emissions while giving allowances in transitional (developing) coiuntries. It discusses inter-government emissions trading.

18.15.3 Paris Accord

The Paris Accord was signed in November 2016 by 194 members of the UNFCC. As of December 2016, it had been ratified by 127 countries. The main difference between the Paris Accord and previous agreements are its bottom-up approach and its flexibility. Countries set their own targets voluntarily. Targets are not legally binding.

No distinct differences are made between developed and developing economies. The goal of the Paris Accord is to limit the increase in global average temperature to 1.5 °C above pre-industrial levels. The NASA GISS model mentioned above predicts that if the temperature increase reaches 2 °C, more than 150 million people could be displaced due to famine and rising sea level. President Barrack Obama signed the accord, but the person who followed him reneged.

18.15.4   Impact of Climate-Related Technology on Jobs

Green technology does more than keep the planet clean. It also generates jobs. The green workforce in the US rose to 8.1 million in 2015, and solar energy jobs overtook those in the oil and gas extraction sector. Similarly, more people work in renewable energy than in oil and gas in China [92].

18.16   Summary

Safety, reliability, and protecting the environment are inextricably linked. In modern industry, they are viewed as prerequisites to profit. Company executives frequently emphasize their strong commitment to worker health and safety. Their commitment derives to a certain extent from strict legislation.

Health and safety rules address personal protection equipment, toxic substances, equipment maintenance, worker training, and compliance monitoring. Environment regulations fall into four main categories: air pollution (particulate matter, volatile hydrocarbons, and harmful gases), waste water, spills and solid wastes. Harm from chronic exposure accumulates with time.

Harm from major accidents causes tremendous short-term destruction, which often is followed by damage that lingers for years. Many of the worst peace-time accidents occur in coal, oil, and chemical industries. Such accidents can teach hard lessons. But as the examples show, not everyone learns the lessons, and those who do sometimes forget or choose to ignore them.

Things have gotten better. Before the 1970s, black lung disease, dead forests downwind from coal-fired power plants, and pools of poison around abandoned strip mines were just part of the price we paid for cheap power and minerals. Smoky flares in refineries were the rule, not the exception. They "smelled like money." More often than not, news of a river catching fire caused laughter instead of outrage.

Today we believe that our fundamental rights include clean air, clean water, and a safe and healthy workplace. We are products of the same social movement that created EPA, OSHA, and similar agencies around the world. Governments are providing the Big Stick—steep fines and possible jail time for corporate executives—but the Carrot comes from a basic change in our fundamental values.

References

1. http://corporate.exxonmobil.com/en/company/worldwide-operations/safety-and-health/workplace-safety. Retrieved 1 July 2015
2. http://www.shell.com/global/environment-society/safety.html. Retrieved 1 July 2015
3. http://www.kbr.com/About/Code-of-Business-Conduct/Translations/English/COBC_English.pdf. Retrieved 1 July 2015
4. http://www.chevron.com/about/operationalexcellence/tenetsofoperation/. Retrieved 1 July 2015
5. https://en.wikipedia.org/wiki/History_of_labour_law. Retrieved 3 Feb 2015
6. https://en.wikipedia.org/wiki/Environmental_movement. Retrieved 3 Feb 2015
7. Mantoux P (2000) The industrial revolution in eighteenth century: an outline of the beginnings of the modern factory system in England. Harper & Row. http://www.hoodfamily.info/coal/law1775act.html. Retrieved 6 Nov 2010
8. https://en.wikipedia.org/wiki/Labor_history_of_the_United_States. Retrieved 3 Feb 2015
9. http://www.msha.gov/MSHAINFO/MSHAINF2.HTM. Retrieved 6 Mar 2015
10. https://en.wikipedia.org/wiki/1900_United_States_Census. Retrieved 3 Feb 2015
11. https://en.wikipedia.org/wiki/Child_labor_laws_in_the_United_States. Retrieved 3 Feb 2015
12. https://en.wikipedia.org/wiki/National_Child_Labor_Committee. Retrieved 3 Feb 2015
13. http://www.dol.gov/dol/aboutdol/history/flsa1938.htm. Retrieved 3 Feb 2015
14. Margulis L, Sagan D (2000) What is life? The eternal enigma. University of California Press, Berkeley
15. Commoner B (1971) The closing circle: nature, man, and technology. Knopf, New York
16. Nobel Prize Organization. "Linus Pauling—Biographical." www.nobelprize.org/nobel_prizes/peace/laureates/1962/pauling-bio.html. Retrieved 26 Dec 2014
17. Carson R (1962) Silent Spring. Houghton Mifflin, New York
18. http://www.panna.org/issues/persistent-poisons/the-ddt-story. Retrieved 5 July 2015
19. http://www2.epa.gov/aboutepa/epa-history. Retrieved 3 Feb 2015
20. https://en.wikipedia.org/wiki/Environmentalism#Early_environmental_legislation. Retrieved 3 Feb 2015
21. https://en.wikipedia.org/wiki/History_of_fire_safety_legislation_in_the_United_Kingdom. Retrieved 3 Feb 2015
22. https://en.wikipedia.org/wiki/Alkali_Act_1863. Retrieved 13 June 2015
23. https://en.wikipedia.org/wiki/Public_Health_Act_1875. Retrieved 3 Feb 2015
24. https://en.wikipedia.org/wiki/List_of_national_parks_of_the_United_States. Retrieved 3 Feb 2015
25. http://www.nps.gov/hosp/learn/historyculture/index.htm. Retrieved 3 Feb 2015
26. http://www.prod.exxonmobil.com/refiningtechnologies/
27. Couch KA, Seibert KD, Van Opdorp P (2003) Controlling FCC yields and emissions: UOP technology for a changing environment, AM-04-45, NPRA Annual Meeting, San Antonio, TX, 23–25 Mar 2003
28. AQMD Adopts Regulation to Reduce Particulate Emissions from Oil Refineries. AQMD Advisor, 7 Nov 2003
29. Cleaner Production Initiatives—BP Kwinana Refinery (2004) Department of the Environment and Heritage, Canberra, Australia
30. https://www.osha.gov/SLTC/hydrogensulfide/hazards.html. Retrieved 26 Jan 2016
31. U.S. Environmental Protection Agency (2004) Air trends: stratospheric ozone. National Service Center for Environmental Publications, Cincinnati, OH
32. U.S. Environmental Protection Agency (2015) Ozone layer protection. http://www.epa.gov/ozone/intpol. 3 Feb 2015
33. http://www.epa.gov/ozone/title6/downloads/Section_608_FactSheet2010.pdf. Retrieved 3 Feb 2015
34. http://www.epa.gov/ozone/2007stratozoneprogressreport.html. Retrieved 9 July 2015

35. Policy Implications of Greenhouse Warming (1992) Mitigation, adaptation, and the science base. National Academy Press, Washington, DC
36. McIntyre S, McKitrick R (2003) Corrections to the Mann et. al. (1998) proxy data base and Northern Hemispheric average temperature series. Energy Environ 14(6):751
37. Energy Information Administration, International Energy Statistics. https://www.eia.gov/beta/international/
38. Lower SK. Carbonate equilibria in natural waters. www.chem1.com/acad/pdf/3carb.pdf
39. Andersson AJ, Yeakel KL, Bates NR, de Putron SJ (2014) Partial offsets in ocean acidification from changing coral reef biogeochemistry. Nat Climate Change 4:56–61
40. GISTEMP Team (2015) GISS Surface Temperature Analysis (GISTEMP), NASA Goddard Institute for Space Studies. http://data.giss.nasa.gov/gistemp/. Retrieved 26 June 2015
41. http://www.bloomberg.com/graphics/2015-whats-warming-the-world/. Retrieved 26 June 2015
42. https://en.wikipedia.org/wiki/Steam_assisted_gravity_drainage. Retrieved 6 Dec 2014
43. Gates ID, Larter SR (2014) Energy efficiency and emissions intensity of SAGD. Fuel 115:706–713
44. American Chemical Society (2016) The science and technology of hydraulic fracturing. https://www.acs.org/content/acs/en/policy/publicpolicies/sustainability/hydraulic-fracturing-statement.html. Accessed 3 Nov 2018
45. Galbraith K, Henry T (2013) As fracking proliferates in Texas, so do disposal wells, Texas Tribune, 29 Mar 2013. http://www.texastribune.org/2013/03/29/disposal-wells-fracking-waste-stir-water-concerns. Retrieved 26 Jan 2016
46. World Energy Interviews Barry Rosengrant (2004) Chief Executive Officer of Petromax Technologies. World Energy 7(2)
47. https://en.wikipedia.org/wiki/Davy_lamp. Retrieved 5 Feb 2015
48. https://en.wikipedia.org/wiki/Black_lung_disease. Retrieved 10 Feb 2015
49. https://en.wikipedia.org/wiki/Lakeview_Gusher. Retrieved 5 Feb 2015
50. https://en.wikipedia.org/wiki/Texas_City_Disaster. Retrieved 5 Feb 2015
51. Air Quality in the UK. Postnote, Nov 2002 (188)
52. Proceedings of the symposium: "Twenty Years after the Amoco Cadiz," Brest, France, 15–17 Oct 1998
53. International Nuclear Safety Advisory Group (1992) The Chernobyl accident: updating of INSAG-1, Safety Series No. 75-NSAG-7. International Atomic Energy Commission, Vienna
54. Dyatlov A (1995) Why INSAG has still got it wrong. Nuclear Engineering International, Sept 1995
55. http://www.world-nuclear.org/info/Safety-and-Security/Safety-of-Plants/Appendices/Chernobyl-Accident—Appendix-2–Health-Impacts/. Retrieved 9 July 2015
56. Chernobyl Powers Down Permanently. CNN.com, 15 Dec 2000. http://www.cnn.com/2000/WORLD/europe/12/15/chernobyl.shutdown/. Verified 26 Jan 2016
57. Grushevoi G (1992) World Uranium Hearing, Salzburg, pp 259–260
58. Weber U (2000) The miracle of the rhine. UNESCO Courier, June 2000
59. Tankers Have Spill-Free Year in Alaska. USA Today, 11 July 2004
60. EPA Chemical Accident Investigation Report: Tosco Avon Refinery, Martinez, California. EPA 550-R-98-009, Nov 1998
61. Robinson PR, Dolbear GE (1996) Hydrotreating and hydrocracking Fundamentals. In: Chapter 7 in practical advances in petroleum processing. Springer, New York
62. U.S. Chemical Safety and Hazard Investigation Board (2005) Refinery explosion and fire, BP Texas City Refinery (15 Killed, 180 Injured), Report No. 2005-04-I-TX, 23 Mar 2005
63. National Commission on the BP Deepwater Horizon Oil Spill and Offshore Drilling, Report to the President: Deepwater: The Gulf Oil Disaster the Future of Offshore Drilling, Jan 2011
64. Nandakumar A, Bousso R, Finn K (2015) BP settles 2010 Gulf oil spill claims for $18.7 billion. Reuters: DailyFinance, 2 July 2015. http://www.dailyfinance.com/2015/07/02/bp-pay-damages-water-pollution-gulf-oil-spill/
65. https://en.wikipedia.org/wiki/Lac-Megantic_derailment. Retrieved 5 Feb 2015

66. "2015 Tianjin explosions". https://en.wikipedia.org/wiki/2015_Tianjin_explosions. Retrieved 19 Jan 2016
67. Schroeder P, Kostyniuk L, Mack M (2013) 2011 National survey of speeding attitudes and behaviors. Report No. DOT HS 811 865. National Highway Traffic Safety Administration, U.S. Department of Transportation, Washington, DC
68. National Highway Traffic Safety Administration. Traffic safety facts 2013: a compilation of motor vehicle crash data from the fatality analysis reporting system and the general estimates system. U.S. Department of Transportation, Washington, DC
69. Santos A, McGuckin N, Nakamoto HY, Gray D, Lisshttp S (2011) Summary of travel trends: 2009 National household travel survey. Report No. FHWA-PL-11-022, Federal Highway Administration, U.S. Department of Transportation, Washington DC, June 2011
70. Federal Highway Administration (2015) Our nations' highways. http://www.fhwa.dot.gov/ohim/onh00/bar8.htm. Retrieved 15 July 2015
71. http://ohsas18001expert.com/2007/07/18/what-is-management-of-change. Retrieved 29 Jan 2016
72. http://www.csb.gov/csb-safety-bulletin-says-managing-change-is-essential-to-safe-chemical-process-operations/. Retrieved 26 Jan 2016
73. Seebauer EG (2004) Whistleblowing: is it always obligatory? Chem Eng Prog 6:23
74. United States Department of Labor, Occupational Safety and Health Administration, Personal Protective Equipment for General Industry: Title 29 of the Code of Federal Regulations (CFR), Part 1910 Subpart I
75. https://www.osha.gov/OshDoc/data_General_Facts/ppe-factsheet.pdf. Retrieved 26 Jan 2016
76. http://www2.epa.gov/aboutepa/epa-history. Retrieved 3 Feb 2015
77. Depledge J, Lamb R (2003) Caring for climate: a guide to the climate change convention and the Kyoto protocol. Courir-Druck, Bonn
78. Bridge painting company and its president plead guilty to federal crimes relating to the dumping of lead waste. Business and Legal Reports, 15 Feb 2002
79. http://www.epa.gov/air/caa/peg/index.html. Retrieved 4 July 2015
80. National Air Quality Standards (NAAQS). http://www.epa.gov/air/criteria.html. Retrieved 4 July 2015
81. http://www.epa.gov/airquality/montring.html. Retrieved 4 July 2015
82. http://www.epa.gov/otaq/fuels/gasolinefuels/rfg/index.htm. Retrieved 4 July 2015
83. http://www2.epa.gov/laws-regulations/summary-energy-policy-act. Retrieved 4 July 2015
84. Grunwald M, Eilperin J (2005) Energy bill raises fears about pollution, fraud critics point to perks for industry. Washington Post, 30 July 2005
85. Shapouri H, Duffield JA, Wang M (2005) The energy balance of corn ethanol: an update. U.S. Department of Agriculture, Office the Chief Economist, Office of Energy Policy and New Uses. Agricultural Economic Report No. 813
86. Ethanol Fuel in Brazil. http://en.wikipedia.org/wiki/Ethanol_fuel_in_Brazil#cite_note-Wilson-2. Retrieved 11 Oct 2011
87. "2003–2008 EPA Strategic Plan (2003) Goal 2: clean and safe water". U.S. Environmental Protection Agency, Washington, DC
88. United States Environmental Protection Agency (2015) CERCLA Overview. http://www.epa.gov/superfund/policy/cercla.htm. Retrieved 19 May 2015
89. Kyoto Protocol to the United Nations Framework Convention on Climate Change (1998) UNFCC, Bonn, Germany
90. Pearce F (2003) New scientist, 10 Dec 2003
91. http://unfccc.int/2860.php. Retrieved 19 May 2015
92. Hirtenstein A (2019) Clean energy jobs surpass oil drilling for the first time in the US. https://www.bloomberg.com/news/articles/2016-05-25/clean-energy-jobs-surpass-oil-drilling-for-first-time-in-u-s. Accessed 9 Jan 2019

Index